Leveraging Technology and Renewable Energy for Sustainable Supply Chains

Tharwa Najar
https://orcid.org/0000-0001-8695-6481
Gafsa University, Tunisia

Adel Aloui
ISTEC Business School, France

Moacir Godinho Filho
EM Normandie Business School, France

IGI Global
Scientific Publishing
Publishing Tomorrow's Research Today

Vice President of Editorial Melissa Wagner
Director of Acquisitions Mikaela Felty
Director of Book Development Jocelynn Hessler
Production Manager Mike Brehm
Cover Design Jose Rosado

Published in the United States of America by
IGI Global Scientific Publishing
701 East Chocolate Avenue
Hershey, PA, 17033, USA
Tel: 717-533-8845 | Fax: 717-533-7115
Website: https://www.igi-global.com E-mail: cust@igi-global.com

Library of Congress Cataloging-in-Publication Data

Names: Najar, Tharwa, 1981- editor | Aloui, Adel, 1976- editor | Godinho Filho, Moacir, 1975- editor
Title: Leveraging technology and renewable energy for sustainable supply chains / edited by: Tharwa Najar, Adel Aloui, Moacir Godinho Filho.
Description: Hershey, PA : IGI Global Scientific Publishing, [2026] | Includes bibliographical references | Summary: "The present book attempts to cover the topic of sustainability through several practices, such as renewable energy consideration and the potential of emerging technologies in the context of the supply chain"-- Provided by publisher.
Identifiers: LCCN 2025023847 (print) | LCCN 2025023848 (ebook) | ISBN 9798337337906 hardcover | ISBN 9798337337913 paperback | ISBN 9798337337920 ebook
Subjects: LCSH: Supply chain management--Technological innovations | Renewable energy sources
Classification: LCC HD38.6 .L48 2026 (print) | LCC HD38.6 (ebook)
LC record available at https://lccn.loc.gov/2025023847
LC ebook record available at https://lccn.loc.gov/2025023848
British Cataloguing in Publication Data
A Cataloguing in Publication record for this book is available from the British Library.

Table of Contents

Detailed Table of Contents

Chapter 5

Ajay Kumar Singh, Department of Humanities and Social Sciences, Graphic Era University, Dehradun, India
S. Srinivasan, Department of Humanities and Social Sciences, Graphic Era University, Dehradun, India
Arakhita Behera, Regional Institute of Education, Ajmer, India

This chapter develop empirical model to observe the implication of digital infrastructure (DI) on renewable energy consumption (REC) as compiling a country-wise imbalanced panel data during 2004 – 2022. Three functional forms of empirical models are formulated to achieve its objective. REC is considered as a core explained variable. Mobile cellular subscriptions and individuals using the internet promote DI, thus both the variables are considered as core explanatory variables. Moreover, education expenditure, inflation, per capita GDP, energy use per capita, population growth and technological development are applied as CVs in the regression analysis. DTs are found beneficial to increase REC. Population growth and carbon intensity have a negative impact on REC. It suggests crucial policy implications to increase the renewable energy management and energy security as using DTs. Finally, this chapter discovers the scope of further research in different dimensions is explicated in this study.

Chapter 6

E. Deepthi, Dayananda Sagar College of Engineering, India
U. Roopa, Dayananda Sagar College of Engineering, India

New technologies like AI, IoT, blockchain, and big data analytics are transforming supply chains to be more efficient, visible, and sustainable. Each individual technology has been helpful on its own, but the combined opportunities and implications of these technologies have not been fully uncovered to date, particularly in complex, fragmented, or other developing market contexts. This research aims to assess how the convergence of these technologies impact the performance of supply chain systems when examined from a sustainability perspective. By integrating real world case studies with academic knowledge, the research reveals challenges and handrails along with opportunity to apply best practices. The findings indicate that intentional technology design, combined with stakeholder collaboration, are fundamental components of creating sustainable value. This research builds theoretical contributions for the literature and provides practical guidance for those who are engaged in the challenge of digitally transforming their supply chains.

Fintech sits at the intersection of digital finance, sustainability, and supply chain shift (Polat, 2024). This chapter examines how fintech innovation drives shared value creation in lending, emphasizing its contribution to renewable energy financing and sustainable supply chains. Using a case study of Easy Bank Tunisian, the study investigates how digital technologies, enable economic and socio-environmental value. The approach provides Fintech's context-specific insights. Easy Bank's fintech-driven lending model enhances financial inclusion, operational efficiency, and ESG financing, supporting renewable energy adoption and supply chain. The institutional context shapes these outcomes. The chapter extends Porter 's shared value framework to fintech, positioning it as an enabler of sustainability, offering empirical insights. Findings inform policy, and fintech practice, spotting strategies for ESG compliant lending in MENA. Based on a single case, the study lays groundwork for comparative research across fintech ecosystems to explore scalability and regulation.

The adoption of renewable energy within supply chain operations is accelerating as businesses aim to enhance sustainability, achieve carbon neutrality, and reduce costs. This research investigates the application of renewable sources—such as solar, wind, and bioenergy—across different phases of supply chains. Representation on multiple case studies from sectors including manufacturing, retail, and logistics, the study highlights key implementation strategies, challenges encountered, and resulting performance impacts. Findings suggest that effective integration depends on factors like technological capability, interdisciplinary cooperation, and strategic long-term investments. The study concludes with a proposed framework to help organizations incorporate renewable energy into their supply chains while supporting global sustainability initiatives.

This study examines the strategic integration of renewable energy sources into smart logistics systems to enhance operational efficiency and environmental sustainability. Based on a comprehensive literature review, it develops a conceptual model linking technological components of smart logistics, such as IoT, cyber-physical systems, cloud platforms, and decision-support mechanisms, with renewable energy solutions including solar, wind, hydropower, bioenergy, hydrogen, and geothermal energy. The study qualitatively analyzes energy–technology alignment and evaluates sustainability outcomes such as carbon emission reduction, energy cost optimization, and socio-economic benefits. A SWOT analysis and a feasibility assessment address strategic and practical considerations, while a set of hypotheses is proposed for future empirical testing. The findings highlight the need for digital–energy synergy, supportive infrastructure, and strategic policy frameworks to unlock the full potential of renewable energy-powered smart logistics.

Sustainable logistics demands efficient and scalable systems for monitoring and forecasting carbon emissions. This work presents a Smart Carbon Emission Monitoring and Forecasting System (SCEMFS) for real-time detection of gases like CO_2, CH_4, NH_3, and VOCs using MQ-series sensors (MQ-2, MQ-7, MQ-9, MQ-135) integrated with a microcontroller. The collected data is used to train machine learning models—decision trees, random forests, SVM, and neural networks—for accurate prediction and classification of gas concentrations. Real-time analytics enables early intervention and improved environmental compliance. The system is scalable and suitable for various logistics settings, contributing to reduced carbon footprints and promoting sustainability in supply chain operations.

In the Tunisian context, the topic of Corporate Social Responsibility (CSR) starts to gain importance, previously in the post-revolution and currently during the "Covid-19" health crisis, due to social pressures which have become increasingly more demanding and threatening. It is in this context that this research is taking place, which seeks to analyze the stakeholders' engagement process and the actions implemented by companies as part of the transition to sustainable supply chains. The qualitative method was chosen to address this issue, by conducting research on the study of five cases from different sectors (chemical, service, agri-food industries, banking, textile). The results reveal that the impact of the crisis has been minimal, due to stakeholder engagement by diverse companies that are no longer limited to their own performance, but are becoming responsible to their stakeholders.

The study analyzes how the convergence between circular economy and closed-loop models reconfigures supply chains still anchored in linear patterns. Based on literature review, cross-industry case studies, and examination of emerging policies, technological, financial, and cultural barriers that keep global circularity below ten percent are documented, as well as incentives that favor rapid extraction over resource regeneration. The findings show that the combined adoption of digital twins, smart contracts, and multi-capital metrics increases visibility into material flows, facilitates coordination between actors, and reduces logistics costs by up to 23%, while decreasing the carbon footprint and strengthening resilience to disruptions. It is concluded that effective circularity demands collaborative governance frameworks, green investment and dynamic capacity building that allow organizations to perceive changes, mobilize knowledge and reconfigure assets with agility, turning the regeneration of resources into a central vector of competitiveness and social legitimacy.

Foreword

In an era defined by unprecedented environmental challenges and rapid technological evolution, the imperative to transform our global supply chains into truly sustainable ecosystems has never been more urgent. Climate change, resource depletion, and growing societal pressures demand innovative solutions that transcend traditional operational paradigms. This book, "Leveraging Technology and Renewable Energy for Sustainable Supply Chains," arrives at a critical juncture, offering a comprehensive and timely exploration of how cutting-edge technologies and renewable energy sources are not merely enhancing, but fundamentally redefining the future of commerce and logistics.

The collective wisdom presented within these pages illuminates a path forward, demonstrating that sustainability is no longer a peripheral concern but a core strategic advantage. From the insightful analysis of how emerging technologies—such as Artificial Intelligence (AI), the Internet of Things (IoT), blockchain, and big data analytics—converge to drive efficiency, visibility, and resilience, to the strategic frameworks for integrating AI and IoT into smart, adaptive operations, this volume meticulously unpacks the transformative power at our disposal.

A significant focus is placed on the pivotal role of renewable energy. The authors delve into the intricate connection between digital infrastructure and renewable energy consumption, and critically examine the strategic integration of renewable energy sources into smart logistics systems. We see how industries are reshaping their value chains—for example, the aviation sector's urgent embrace of Sustainable Aviation Fuel (SAF) to decarbonize operations—and how innovative fintech solutions are channeling financing towards green initiatives.

Beyond technological advancements, the book underscores the human and systemic dimensions of sustainability. It highlights the crucial role of stakeholder engagement in driving the transition to sustainable supply chains, and how policy frameworks and data-driven consumer behavior, particularly through energy efficiency labels, can foster more sustainable consumption patterns. The profound shift towards circular economy and closed-loop models, enabled by digital twins and smart contracts, is presented as a powerful strategy for resource regeneration and enhanced resilience. Furthermore, the development of sophisticated tools for Smart Carbon Emission Monitoring and Forecasting promises to revolutionize sustainable logistics by offering real-time insights and predictive capabilities.

What makes this collection particularly impactful is its balanced approach, weaving together theoretical contributions with empirical models, real-world case studies, and actionable guidance. It acknowledges the challenges—ethical, regulatory, and interoperability—while proposing future research directions and practical handrails for academics, practitioners, and policymakers alike.

"Leveraging Technology and Renewable Energy for Sustainable Supply Chains" is more than just an academic exercise; it is a clarion call for collaborative innovation. It invites readers to engage with a vision where intentional technology design, robust stakeholder collaboration, and a deep commitment to renewable energy coalesce to create enduring value for businesses, societies, and our planet. This book is an essential read for anyone committed to building a smarter, more sustainable, and ultimately more resilient global future.

Given the aforementioned, I can only hope readers enjoy the book as much as I did, benefitting from insightful discussions and compelling arguments that will drive our decisions and way of thinking.

Enjoy your reading!

Guilherme Luz Tortorella
The University of Melbourne, Australia

Preface

INTRODUCTION

In an era characterized by rapid technological advancement and unprecedented environmental challenges, the global supply chain landscape stands at a crucial crossroads. The dual imperatives of operational efficiency and environmental sustainability have compelled researchers, policymakers, and industry leaders to rethink conventional practices. *Leveraging Technology and Renewable Energy for Sustainable Supply Chains* emerges as a timely response to this evolving paradigm—offering a comprehensive examination of how digital innovation and renewable energy integration can collectively drive a transition toward greener, smarter, and more resilient global networks.

Across industries, supply chains are not only the backbone of commerce but also significant contributors to global carbon emissions and resource depletion. As such, the adoption of digital technologies—such as artificial intelligence, the Internet of Things, blockchain, and data analytics—coupled with the strategic use of renewable energy sources, represents more than a trend: it signals a systemic transformation. This book captures that transformation through diverse theoretical, empirical, and applied perspectives contributed by scholars and practitioners from across continents. Together, these insights illuminate how technological innovation can harmonize sustainability goals with competitive advantage.

Moreover, this volume highlights that sustainability is not a static target but a dynamic process—one that requires adaptability, foresight, and collaboration across disciplines. The chapters compiled here demonstrate that the convergence of technology, renewable energy, and supply chain management extends beyond operational improvements; it redefines the very essence of value creation, stakeholder engagement, and global competitiveness. By bringing together research from diverse cultural and institutional contexts, the book offers a holistic vision for shaping the sustainable supply chains of the future.

CHAPTER OVERVIEW

Chapter 1, by Ravikumar R. N. and Aarthi S., presents a strategic framework for smart and adaptive supply chains. It illustrates how real-time data collection through IoT devices, combined with AI-driven analytics, enables optimization in logistics, inventory, and sustainability outcomes.

Chapter 2, by R. Sowmiya and M. Velavan, explores how energy efficiency labels influence sustainable consumption. Through behavioral and policy analysis, it demonstrates the synergy between consumer awareness, technological innovation, and effective regulation in advancing sustainability.

Chapter 3, by Gézia Damergy, investigates how Sustainable Aviation Fuel (SAF) is reshaping the aviation industry's value chains. The chapter underscores the necessity of collaborative strategies and long-term partnerships in achieving decarbonized aviation.

Chapter 4, by Ahlem Hamri and Noomen Guirat, delves into sustainable transport and ecological logistics. Drawing on a Tunisian case study, it proposes green technologies and innovations in mobility that reduce emissions while improving logistics efficiency.

Chapter 5, by Ajay Kumar Singh, S. Srinivasan, and Arakhita Behera, presents an empirical model linking digital infrastructure to renewable energy consumption across nations. It delivers policy insights on how digital development fosters renewable energy transitions.

Chapter 6, by Deepthi E. and Roopa U., assesses how emerging technologies collectively enhance supply chain efficiency and sustainability. Through case studies, it highlights the potential of integrated digital ecosystems to accelerate sustainable transformation.

Chapter 7, by Eya Frioui, introduces the role of financial technology in promoting renewable energy financing and ESG-compliant lending. The Tunisian case study of Easy Bank exemplifies how fintech innovations drive social and environmental value.

Chapter 8, by Suriya Ponnambalam and Ilampoornan M. K., analyzes renewable energy integration within supply chains. Using sectoral case studies, it presents a framework for aligning renewable adoption with long-term sustainability goals.

Chapter 9, by İlknur Yardımcı-Çoşkun, focuses on renewable energy–driven smart logistics. It offers a conceptual model and strategic feasibility analysis for implementing renewable-powered logistics systems that balance economic and environmental performance.

Chapter 10, by Lalitha S. D. and colleagues, introduces a Smart Carbon Emission Monitoring and Forecasting System. Leveraging IoT sensors and machine learning, it provides scalable tools for emission tracking and predictive analytics in logistics operations.

Chapter 11, by Jihen Msekni and Ahmed Turki, investigates stakeholder engagement in sustainable supply chain transitions within Tunisia. It emphasizes how proactive CSR strategies and inclusive stakeholder involvement enhance resilience and accountability.

Chapter 12, by José De Jesús Reyes-Sánchez, concludes the volume by exploring circular economy and closed-loop models. It reveals how the fusion of digitalization and regenerative principles enables firms to shift from linear consumption toward circular value creation.

CONCLUSION

This collection underscores that the sustainable transformation of supply chains is not merely a technological upgrade but a holistic redesign of how organizations create, distribute, and sustain value. The chapters together illuminate the multifaceted interplay between digitalization, renewable energy, and systemic innovation—each a crucial enabler for achieving sustainable competitiveness in a resource-constrained world. The cross-disciplinary nature of this volume reflects the complexity of real-world supply chains and the need for integrated solutions that transcend traditional industry boundaries.

Equally, the research presented here reveals that sustainability is both a scientific and societal endeavor. It requires not only the deployment of advanced technologies but also governance, ethics, and human-centered design. By incorporating insights from fields as diverse as behavioral economics, energy policy, logistics, and digital finance, the book contributes to a more inclusive understanding of sustainability—one that connects technological progress with human development and planetary well-being.

We, as editors, envision this book as a platform for knowledge exchange and collaboration among academics, practitioners, and policymakers. It invites readers to reflect critically on how technology and renewable energy can be leveraged to craft more transparent, circular, and resilient supply chains. Ultimately, *Leveraging Technology and Renewable Energy for Sustainable Supply Chains* aspires to serve as both a scholarly contribution and a practical guide for shaping a more equitable and sustainable global economy.

Chapter 1
Artificial Intelligence and IoT Integration in Supply Chains:
A Strategic Framework for Smart, Adaptive, and Sustainable Operations

R. N. Ravikumar
https://orcid.org/0009-0009-3705-1681
Marwadi University, Rajkot, India

S. Aarthi
https://orcid.org/0009-0006-9064-2091
Marwadi University, Rajkot, India

ABSTRACT

This chapter explores the convergence of Artificial Intelligence (AI) and the Internet of Things (IoT) in transforming supply chain operations for sustainability. It presents a layered framework where smart sensors, GPS trackers, and RFID devices collect real-time data, processed through AI/ML algorithms to drive optimization in logistics, inventory, and resource management. Case studies in agriculture and retail demonstrate significant improvements in yield, fuel savings, and carbon emission reductions. The framework emphasizes a feedback loop for continuous learning, enabling adaptive and resilient supply chain systems. The chapter also discusses ethical, regulatory, and interoperability challenges, proposing future

DOI: 10.4018/979-8-3373-3790-6.ch001

research directions including blockchain integration and Edge AI. It serves as a comprehensive guide for researchers, practitioners, and policymakers aiming to build smart, data-driven, and sustainable supply chains.

INTRODUCTION

In the modern dynamic industrial world, Artificial Intelligence (AI) and the Internet of Things (IoT) converge and turn the traditional supply chain paradigms into intelligent, sustainable and highly responsive ecosystems. The trend on environmental protection, economic performance, and societal efforts to be more responsible are global, and the industries have been compelled to rethink how they operate, and an entity as AI and IoT may prove instrumental in transforming the stagnant supply chains to dynamic ones that run on data. These technologies enable real-time monitoring, intelligent prediction and autonomous decision-making- less wastage, more productiveness and more use of resources in the supply chain. AI can be applied to supply chain to create predictive analytics, demand forecasting, anomaly detection and optimization, and IoT devices, e.g. sensors, RFID tags, GPS trackers can be used to create live visibility in the supply chain and traceability at various stages of the operation (Udeh et al., 2024). This, as a collective, provides a powerful cyber-physical system capable of sensing, interpreting and acting on real-world data to facilitate basic sustainability trends like carbon emission, energy savings and circular supply chains. This chapter examines the use of new operational agility via AI and IoT to transcend support to sustainable practice, offers a general conceptualization that can guide all stakeholders in the agricultural sector, manufacturing sector, logistics sector and retail sector to future proof their supply chains.

Role of AI in Supply Chain Optimization

Artificial Intelligence can be used to offer an unprecedented increase of the supply chain by introducing automation, 20/20 vision, and intelligence as a means of re- credentialization. Machine learning algorithms allow AI systems to process big data over a long-time span not only of the current time but also of the past, and provide valid predictions of the demand, maximum inventory and minimal lead time and supply chain failures (Kadhim Obaid et al., 2024). One such case would be the use of AI-based forecasting that could prevent the threats of surplus production or stockouts leading to energy-efficient and cost-effective operations. Reinforcement learning can also dynamically change the scheduled procurement and production owing to changes in demand over time and supplier capabilities. In addition, decision support systems based on AI facilitate multi-objective optimization, e.g. cost, time, and environmentally friendly optimization (Shafique et al., 2018). Smart routing algorithms help in transportation to optimize the transportation and minimise the usage of fuel whereas AI and warehouse management system can enhance the accuracy of the picking processes and how much space is efficiently used. Such abilities help in increasing sustainability by cutting down wastes, energy consumption, and emissions. In conclusion to the above, AI converts supply chains into auto-responsive systems rather than out-of-sight supply chains and it can learn and change, therefore, allowing a greater level of operation effectiveness and environmental responsibility.

IoT-Driven Visibility and Real-Time Monitoring

Internet of Things (IoT) adds another layer of transparency and control within the operations of supply chains through allowing real-time collection and monitoring of data points at each node. Smart sensors, Radio Frequency Identification (RFID) tags, and Global Positioning System (GPS) trackers are examples of IoT, which collect detailed data regarding the status of products, the location of vehicles and warehouse conditions, and equipment capabilities (Rendon-Benavides et al., 2023). Such visibility increases traceability and enables the stakeholders to react to disruptions proactively since it allows them to respond to threats like delays, spoilage, and equipment malfunction. To give an example, goods that are monitored by temperature and humidity sensors, such as perishable ones, can ensure quality during transportation, and condition-based alerts can be used to decrease product losses and waste of energy. The benefits of IoT in the logistics management would be dynamic correction of routes and driver monitoring to decrease fuel use and emissions(Holloway, 2024). Besides, the sensor-based predictive maintenance saves out of play time and prolongs the life of assets. Such on the fly observations will not only facilitate efficiencies but also facilitate the environmental sustainability stepping in to facilitate regulatory compliance besides the restraint of the ecological footprint of organizations. As the complexity and globalization of supply chains continue to increase, the IoT is taking the role of the digital nervous system, which will deliver smarter, safer, greener supply chains and agility and resilience to disruptions.

Objectives of the Chapter

The aim of this chapter is to discuss how Artificial Intelligence (AI) and the Internet of Things (IoT) can be integrated into the existing chains and transferred to the new supply chains based on an intelligent, responsive, and sustainable environment. It attempts to illustrate how these technologies share common problems that they will solve such as efficiency, the environmental effects and responsiveness. Sources of real-life implementations have also been identified in the chapter, obstacles of adoption mentioned, and potential solutions proposed to the stakeholders.

- To analyze how AI and IoT integration enhances real-time visibility and decision-making in supply chain operations.
- To examine the role of AI in predictive analytics, demand forecasting, and resource optimization for sustainability.
- To explore IoT-driven monitoring and data collection for improving traceability, efficiency, and safety.
- To present real-world case studies showcasing the effectiveness of AI-IoT convergence across sectors like agriculture, retail, and manufacturing.
- To identify implementation challenges (e.g., data security, interoperability, skills gap) and recommend strategies for successful adoption.

BACKGROUND STUDY

The latter development of the supply chain management has been substantially affected by the incredibly fast growth of the digital technologies, mainly, Artificial Intelligence (AI) and Internet of Things (IoT). By contrast, in the traditional supply chains, data visibility and reactive decisions were prevalent, and supply chains used to work in silos. But as the demands to be more sustainable, efficient, and adaptable grow, the existing industries are about to shift to data driven, smart supply chains (Reddy et al., 2023). AI has a value-addition in making it possible to have predictive analytics, machine learning demand forecasts, route optimizing, and autonomous decision-making. At the same time, the IoT devices, such as sensors, RFID tags and smart trackers, allow collecting, monitoring as well as controlling data in real-time in all supply chain nodes. Studies of the past ten years indicate the personal value of such technologies, but the overall potential of such combination is understudied (Aliahmadi et al., 2022). The integration of AI and IoT gives rise to an effective cyber-physical system that facilitates end to end visibility, minimization of resource loss, dynamic responsiveness and environmentally friendly habits. The integration will be able to support the broader goals of sustainability more, as they will minimize the carbon footprints, enable the practices of the circular economies, and reduce operational risks. This history provides a context to comprehend the ways in which AI-IoT synergy can be effectively applied to develop smart, resilient and sustainable supply chains.(Zhang & Chen, 2013).

LITERATURE REVIEW

Artificial Intelligence (AI) and the Internet of Things (IoT) converge to change the character of supply chain management and necessitate a shift to responsive, intelligent and sustainable networks as a predictive, linear, and reactive system (Awan et al., 2021). Over the years, AI has progressed to be a powerful tool in supporting the domain of predictive decision-making, optimization of supply chain supply chain, and automation of complex supply chain operations (Liu et al., 2020). It offers complexity in demand forecasting, inventory management, transportation management and supplier risk. On the other hand, with the IoT, you can oversee and gather data instantly, with RFID tags, GPS trackers, smart sensors, and accessibility to the state, position, and transportation of goods along the supply chain. The overall issues that AI in supply chains is preoccupied with are forecasting, pattern recognition and intelligent automation. Machine learning models use historic and real-time data to identify irregularities, find out disruptions and create the most appropriate purchase and transport plans(Getahun et al., 2024). Supplier communication also uses Natural Language Processing (NLP) and reinforcement learning is utilized to adapt the operations accordingly according to the changing inputs. These AI algorithm-based solutions enable cost reduction, decrease waste, and better services leading to a less-wasteful and more rapid supply chain.

In the meantime, IoT has become popular with respect to dealing with visibility and traceability problems. The implementation of the IoT devices enables permanent monitoring of the stock, warehouse environments, supply chains, and food expiration dates (Adewusi et al., 2024). This real-time connectivity also improves how the supply chain can react to unpredictable incidents like delay, theft, spoilage or machine breakdowns. The condition monitoring that is enabled by the IoT is important in cold chains where producing compliant products in the cold chain is dependent on environmental control. Although the synergistic effects of AI and IoT on supply chains have become a common knowledge, there are few literature pieces that would study these effects in a comprehensive way (Firouraghi et al., 2022). But as per the current researches, AI and IoT applied simultaneously form intelligent ecosystems which can make independent decisions in response to contextual awareness. As an example, the information on IoT sensors can be utilized by AI models to forecast necessary maintenance, fuel economy, or reshape the operations in the warehouse on the fly. Fuel efficiency and greenhouse gas emission are reduced by smart routing and vehicle tracking enabled by AI-IoT collaboration (Hussein et al., 2024). Moreover, the IoT data can be analyzed using AI that helps in judging how products go through their lifecycle, reverse logistics, and the minimization of waste to support circular economy efforts.

Industries like agriculture, manufacturing, and retail have begun to utilise the AI-IoT convergence to develop sustainable operations (de Vass et al., 2021). Smart farming in agriculture combines the usage of weather data, soil sensors, and AI-assisted crop yields to minimize wastage and optimize the input levels. Predictive maintenance and energy efficient processes are being attained in manufacturing through use of AI-enabled IoT networks. One advantage that retailers have with intelligent inventory systems is that they automatically react to the ever-changing customer demand, as well as to eliminate the problem of overstocking or shortages (Karthikeyan & Nagaprakash, 2023a). The challenges are still critical in spite of the potential. Interoperability of data, absence of a common protocol or means, cyber-security risks to the application of cloud-computing-based solutions, and workforce readiness skills are recurrently mentioned as obstacles to massive implementation of AI-IoT integration. Besides, it has privacy, compliance, and governance issues. The weaknesses suggest that there ought to be an effective framework that must encourage AI-IoT use and, simultaneously, address imposed risks (Brown et al., 2024). The present chapter is a continuation of this blank since it offers a strategic perspective in which the perspective would combine these technologies to produce intelligent supply chain and sustainable operations.

PROBLEM STATEMENT

These rising needs of sustainability and resilience cannot be backed by the traditional supply chain models, because they are not agile, transparent, and intelligent in operation. Information inoculated, delay in decision making and execution, inefficiency in the utilization of resources and visibility in the entire supply chain leads to increment in operational costs, wastage and environmental damage. Faced with the unheard-of uncertainty that has been created by climatic changes, fluctuations in markets, and uncertainties in geopolitics, global supply chains are in dire need of a technological breakthrough which would allow them to streamline the process without negatively affecting the sustainability. They are present in the form of AI and IoT technology; however, there has been minimal incorporation into the supply chains ecosystems due to one of the fears that are interoperability, data security, infrastructure preparation, and human capital. This chapter will address these gaps by not only discussing how the synergistic application of AI and IoT can enable supply chains to be smart, adaptive and sustainable, it will also address the actual limitation in the real world, approaches and implementation plans.

RESEARCH GAP

Although many studies have been conducted on the singular involvement of Artificial Intelligence (AI) and the Internet of Things (IoT) in the supply chain management, there is insignificant research done on the combined implementation of the two on specific monitoring and driving of sustainability throughout the supply chain operations. Much of the literature is also focused either on operational efficiency or technology adoption separately and does not cover the perspective of how the AI-IoT synergy can comprehensively cover environmental, economic, and social sustainability objectives. In addition, it also lacks empirical analysis of discussion on the effects of AI-IoT convergence in various sectors of the industry like agriculture, retail, and manufacturing. Not many frameworks properly describe the circumstance in which these technologies are adopted in order to facilitate a circular economy model, closed-loop logistic or carbon reduction metric. Moreover, practical factors, including data interoperability, information security issues, policy incongruence, and human resource gaps are frequently mentioned and not comprehensively analyzed with the solutions trajectories. This chapter tries to fill this gap by having an all-encompassing view on how the AI and IoT can be applied strategically in order to transform the supply chain to become smart, resilient and sustainable and present real-world applications together with policy recommendations.

MAIN FOCUS OF THE CHAPTER

The key idea of this chapter is to talk about how cooperation of Artificial Intelligence (AI) with the Internet of Things (IoT), can change supply chain practice and performance in terms of sustainability, responsiveness, and effectiveness of the operation. The chapter further describes how these technologies can be integrated so that they can enable real-time visibility, predictive analytics and automated decision making based on the supply chain ecosystem (Chen, 2024). It highlights the idea that AI will be able to process large volumes of data to forecast demand, manage inventory, and intelligently route and, IoT will be able to provide continuous data streams of sensors, tags, and tracking devices to monitor the status, whereabouts as well as recorded location of commodities (Ghahremani Nahr et al., 2021). Among the most important areas of interest is the synergistic impact of AI-IoT convergence on the reduction of environmental impact through optimization of logistics, energy-efficient activity, and minimization of waste. In the chapter, the concept of using an AI-IoT fusion to support the circular economy through monitoring the life of the produced product and controlling the reverse logistics is also discussed (Yang et al., 2023). Real life examples of applicative verification include agriculture, manufacturing and retail. In addition, the chapter discusses barriers of implementation, which include cybersecurity, interoperability, and skilled staff. The chapter offers an informative framework to academics, practitioners, and policymakers who would like to engage in the design of smart, secure, and sustainable supply chain models through provision of strategic solutions and policy recommendations.

PROPOSED METHODOLOGY

The methodology of this chapter uses the conceptual framework which incorporates the Artificial Intelligence (AI) and the Internet of Things (IoT) to develop intelligent sustainable supply chain practices. The approach has been organized in four tiers: data collection, data processing, smart decision, and sustainable action implementation. The IoT devices (sensors, RFID tags, GPS trackers, and smart meters) are implemented throughout the supply chain including an IoT tool used at manufacturer sourcing, warehouses and in the logistics department and delivery to the customers on the data acquisition level. These are gadgets, which stream real time information relating to the environmental conditions, location, available inventory and the condition of equipment. The process that takes place in IoT data processing layer is that raw data is sent to one of these a cloud computing platform or edge computing platform to clean up, store and integrate data. Interoperability and protocols of standardization are used to facilitate the smoothness of data flow.

The smart decision layer employs the AI algorithms applied on demand based on the requirements: AI modeling and AI algorithms depend on the needs: machine learning to make decisions on demand forecasting, optimization models in the field of logistics and inventory, and predictive analytics in maintenance of decisions. Lastly, the sustainable action execution layer uses the AI-IoT outputs to optimise energy consumption, reduce the generation of carbon emissions, promote the circular economy logistics, and make supply chain responsive. It is a stratified strategy to achieve effective, intelligent and green supply chain. The Figure 1 shows a hierarchical structure of an AI-and-IoT-based sustainable supply chain. It begins with the IoT devices; the Smart Warehouse Sensors, Fleet GPS Trackers and the RFID Product Tags transmitting their data to the structured flow by five major layers; Data Acquisition, Data Processing, Intelligent Decision Making, Sustainable Action and Feedback Loop. In the context, the emphasis is placed on the responses to the question of how real-time data and AI models may guide operational efficiency, automation and continuous optimization towards the path of sustainability.

Figure 1. AI-IoT-Enabled Sustainable Supply Chain Framework

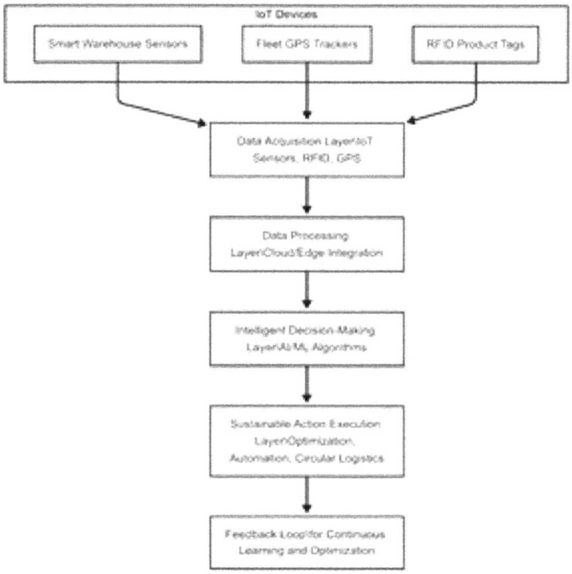

IoT Devices in Sustainable Supply Chains

Introduction of IoT integration devices such as Smart Warehouse Sensors, Fleet GPS Trackers, and RFID Product Tags will play a central role in ensuring real-time visibility and efficiency in the supply chains in the modern day and age. Smart Warehouse Sensors notice any change in the environment including temperature, humidity and even occupancy and control the most optimal storage conditions and reduce the wastage of goods. The use of Fleet GPS Trackers also makes it possible to track the position of vehicles on a regular basis, optimise routes, and analyse fuel consumption, which directly contributes to positive shifts in carbon emissions and delivery delays. An item level identification can be applied to an item using RFID Product Tags and that can provide an accurate management of the inventory and faster order processing and accuracy in tracing the origin to destination. Used together, these gadgets form the foundation of the initial level of information gathering in an AI-IoT-managed ecosystem. They enhance automation of decision making, minimize the waste of resources, and enhance responsiveness that makes the supply chains more agile, data, and even sustainable to suit the changing market and environmental conditions.

Data Acquisition Layer: Real-Time Input Collection

Data acquisition layer visibly serves as the backbone of smart and sustainable supply chain because it provides real-time collection of input through multiple devices IoT-enabled. Such devices are smart sensors in warehouses, RFID tags on the physical product, and GPS tracking on the fleet, placed in the strategic spots of supply chain. Warehouse sensor technology tracks the temperature, humidity, and inventory condition in warehouses to maintain optimum storage demands. Schemes based on RFID tags enable tracking of the product and product-level identification and therefore help in ensuring traceability and minimizing loss and delays of inventories. GPS tracking devices in logistics fleet operations have been observed to keep track of real-time tracking and performance of vehicles and deviations, which improves delivery on performance and visibility. This global network of devices records real-time streams of data throughout the sourcing, manufacturing, warehousing and distribution process. This layer allows detecting the possible disruptions early by collecting granular and time-sensitive data to implement proactive interventions. The specificity and rate of the data gathered in this context is constantly associated with the responsiveness and capacity of the whole chain of supply. Data acquisition as a critical component in the input layer is important in providing quality real time information to the decision makers and intelligent systems to aid factor-in forecasting, sustainability compliance, dynamic control of operations in all nodes in the supply chain.

Data Processing Layer: Cloud and Edge Integration

One of the key concerns is data processing layer to transform the raw and unstructured data into actionable intelligence using both the cloud computing environment and edge computing environment. Data retrieved into IoT devices is immediately sent to the cloud resources where it is stored, aggregated, and deep analytics performed on it. However, to support reactiveness, and reduce latency, in time-sensitive applications, edge computing is also used to locally process data at or near the source. This two-processing construction can be used to make quick decisions in situations like inventory restocking, maintenance of equipment and logistics pathing. The cloud does most of the complex computational tasks, and counts and manages model training, but on the edge nodes, real-time analytics and control are done. This layer requires standardization procedures and data interoperability standards which should promote the integration of various device inputs in order to ensure smooth harmony. In addition to this, effective mechanisms of cybersecurity are implemented to protect other sensitive data in the supply chain against hacking and data theft. Cleaning, normalization, and filtering of data is also done here to increase quality of data. This layer, ultimately, transforms the high-volume data into structured, reliable, and secure data, and they are ready to be used in AI systems in the upcoming pipeline stage of the supply chain intelligence.

Intelligent Decision-Making Layer: AI-Powered Optimization

The AI/ML powered intelligent decision-making layer processes data and offers the data in optimized, predictive, and autonomic actions. AI systems offer the near-accurate demand forecasting, assessment of suppliers, planning production, and optimization of inventory based on interpretation of patterns of historical and real-time data. Based on predictive models, certain disruptions can be foreseen, like the inability of one of the suppliers or equipment malfunctions, which enables the supply chain managers to take certain precautionary actions beforehand. Reinforcement learning algorithms are used to dynamically adjust the levels of inventory, redirect the deliveries or change the procurement strategies depending on the changing external conditions. Unstructured data can be analyzed as well through Natural Language Processing (NLP) to incorporate and gain a wider contextual picture into decision-making, which can be market news or customer feedback. The layer will absorb responsive actions and convert them to purposeful systems, which can learn and develop. Optimisation techniques reduce the waste of product stored or time spent on transportation, reducing carbon footprints by reducing the unnecessary travel or energy used. Also, scenario planning implies simulations with the help of AI, simulating strategies to be followed before their execution. This layer makes the supply chain smarter and resilient to internal and external uncertainties because it gives maximum utilization of the available resources, agility in operations and alignment with the sustainability goal through the smart coordination of supply chain elements.

Sustainable Action Execution Layer: Operational Efficiency

The sustainable action implementation level would be transformed in terms of which the insights produced by AI can create verifiable action translating directly into the direct influence on the flows of physical goods and services. This layer does the optimized plans on the logistics systems, inventory systems, production systems and waste management systems. To provide an illustration, AI-proposed optimized transportation routes are applied to reduce the fuel and emissions factors. Machine-like restocking and picking services in facilities like warehouse save time and minimize errors during manual handling of a product besides quality inspection. The condition-monitoring system controls the environmental conditions like temperature or humidity in real time to preserve the product integrity especially in cold chain logistics. The layer also contributes to achieving the goals of the circular economy since it enables the strategy of reverse logistics and checking repaid products, which are used again, recycled, or remade. The implementation of energy efficient paradigm gadgets such as smart lighting, predictive maintenance and green packaging also enhances the sustainability of operations. The instructions to the control and automation robotic processes are received at the decision level and respond dynamically to the altering situation without the interferences of the human activity. Such responsive activities ensure the optimal performance, resources conservation and minimal environmental effects of the same. The execution layer finishes off the image of the backbone of a fully sustainable and adaptable, high functioning supply chain network by bridging the digital intelligence to action gap.

Feedback Loop: Continuous Learning and Improvement

The feedback loop that enables the improvement and optimization of the supply chain and learning in the process is the last, yet definitely not the least, element of this architecture. Outputs of the activities that have been done e.g. the delivery time, energy used, volume of carbon emitted, customer satisfaction or product returns will be captured and fed back into the AI models. This never-ending type of learning will help algorithms to get better, update, and modify over time and as the area of application is not the same in each instance, the decisions that are made using this type of knowledge will be more precise and an in-context decision. Practically, one can take an example whereby, in case a routing algorithm adds to delays in a specific area, the system will utilize the deviation to signal upcoming routing, schedules and planning. With regards to the same, the forecasting models are continuously adjusted, based on the real market demand fluctuations and sales volumes. Root cause analysis and anomaly detection The feedback loop can also assist in supporting root cause analysis and anomaly detection to ensure that inefficiencies, bottlenecks or risks that arise can be identified at an early stage. A layer that involves adaptive intelligence will turn the supply chain into a dynamic self-evolving network. It also deters the achievement of the sustainability goal because it tracks aspects of performance indicators like key environmental and operational efficiency. The feedback loop is best conceptualized as a process that allows each of the supply chains to eventually be smarter greener and tougher to learn in real time and more data driven.

CASE STUDIES

To truly understand how revolutionary incorporating Artificial Intelligence (AI) and the Internet of Things (IoT) into a supply chain can be, the best way to do it is to give examples of actual supply chains in multiple fields of endeavour. The chapter includes two interesting case studies one in the agriculture business and the other in the retail logistics business that clearly show how AI-IoT convergence is playing a significant role in offering operational excellence, environmental sustainability, and conservation of resources (Susan et al., 2017). In case 1, the concept of smart agriculture systems uses IoT-based soil and climate sensors in conjunction with AI-based yield forecasting models to optimize the utilization of inputs, minimize wastage at the post-harvest stage and streamline distribution in case of perishable goods supply chains. These intelligent systems help farmers in making irrigation, fertilization, and harvesting decisions using data on irrigation, fertilization, and harvesting schedules, reduce wastage, and enhance quality of products (Karthikeyan & Nagaprakash, 2023b). The second example can be voiced in the sector of retailers where the algorithm of route optimization driven by AI and the IoT-driven vehicle tracking solutions changed how logistics processes are perceived. Flow of information regarding traffic, kind of condition and delivery schedule is examined to create the ideal routes that will comprehensively reduce fuel costs and carbon emissions. Taken altogether, these case studies all depict the fact that AI-IoT integration not only addresses industry-specific challenges but that such a combination will make supply chains more sustainable and agile too (Gutiérrez-Diez et al., 2024). Companies, which would like to digitize and decarbonize their chain networks can use them as the guidelines.

Case Study 1: AI-IoT-Enabled Smart Agriculture: Enhancing Yield and Reducing Waste in the Perishable Supply Chain

The food industry, particularly the industry of perishable products is facing significant problems, which concern the food waste, the inefficiency of the use of resources, and the unreliable weather patterns. In order to solve such problems a smart agriculture project was initiated in an agricultural region which dealt with high value crops like tomatoes and strawberries which were perishable (Adesiyan & Raffington, 2024). The identified solution incorporated the use of AI algorithms into monitoring systems based on the IoT environment to streamline farming and post-harvest supply chain productivity. The IoT sensors were also used in fields to gather real-time information such as soil moisture, temperature, humidity, and crop wellness. This information was shared with an AI-based decision support system that employed predictive analytics and Machine Learning models to inform irrigation agendas and nutrient application as well as approaches to disease avoidance (Singh et al., 2020). This led to the reduction of water consumption by 32% and use of fertilizer by 25% with no diminution of crop profit. Actually, the total production was increased by 18% following accurate farm inputs and proper interventions. Additionally, cold storages with IoT capabilities as well as transport vehicles supplied with GPS systems guaranteed that produce is stored/delivered under optimum conditions. AI algorithms observed conditions in the environment and dynamically changed the storage parameters, which allowed reducing the spoilage during transit by 40%. The shelf life was enhanced by 22% hence improved market value and less food wastage.

The interaction of AI and IoT has greatly enhanced farm-to-market compliance in addition to making it possible to adopt predictive logistics and demand forecasting. This balance between supply and market demand resulted in 30% unsold inventory decrease and tightened the income stability of farmers (Birkel & Hartmann, 2019). On balance, the given model of smart agriculture shows how the AI-IoT convergence might foster sustainable agriculture, reduce losses in the food supply chain after the harvest, and establish the sustainability of a supply chain of perishables supplying goods with economic and environmental advantages. The Figure 2 demonstrates the significant developments in performance realized by means of AI-IoT-empowered smart agrarian. It demonstrates substantially decreased usage of water and fertilizer, amplified yield, and shelf life. The greatest disruption was reduction of the 40% spoilage during delivery, thereby improving the sustainability of the supply chain.

Figure 2. Impact of AI-IoT Integration in Smart Agriculture

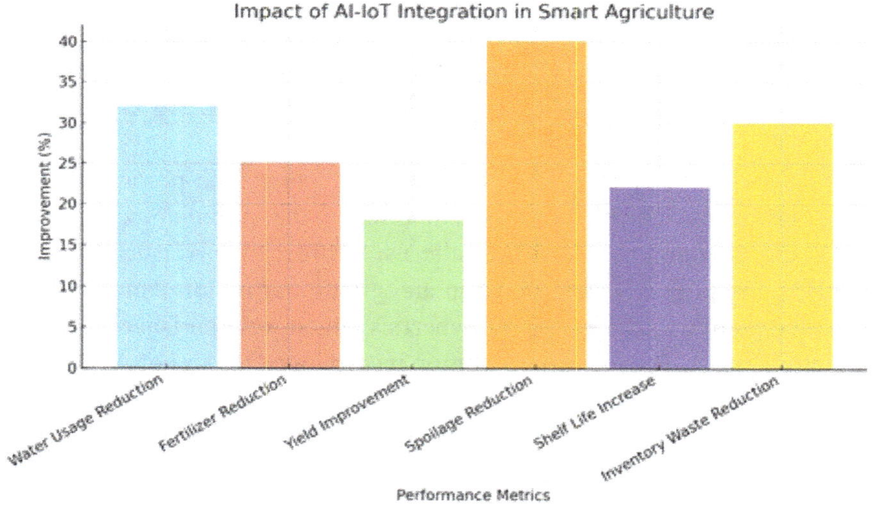

Case Study 2: Green Logistics in Retail: Leveraging Ai and Iot for Route Optimization And Carbon Emission Reduction

In a shopping world where retailing is competitive and aware of the environment, logistical processes greatly enhance cost efficiency as well as green practices. One of the biggest retail chains has introduced a green logistics program that combined the use of Artificial Intelligence (AI) and the Internet of Things (IoT) to update its transportation and distribution chains. The project focused on lowering its consumption of fuel, lessen its cost of carbon emission, and improve its overall delivery performance in its regional network of supply. Deployment of IoT devices, like vehicle telematics and GPS trackers in the delivery trucks was done to ensure that the real-time vehicle available metrics are tracked, like fuel consumption, vehicle location, speed, and engine idle time (Hussain et al., 2019). This information would be constantly sent to an AI system that would be managed in the cloud, employing complex route optimization algorithms and predictive analysis to dynamically generated the most efficient routes and distributions. This led to a decrease in fuel consumption by 28%, idle time throughout the fleet by 30%.

It has also brought about the carbon emission reduction initiative by 35% attained by informed route planning and loading that required considerably less redundant travel. Moreover, it increased delivery times by 20% with the help of immediate integration of traffic and weather data (Lai, 2024). The AI also enhanced efficiency in the routes by 25%, which means more deliveries at reduced resources. The costs for retail logistics decreased by 22% due to a decrease in fuel costs and utilized assets better. The system also offered valuable feedback to drivers that would encourage fuel-saving driving or in other words, that would help cut down vehicle wear and tear (Waller & Fawcett, 2013). As noted in this case study, converging AI-IoT in retail logistics will not only improve the operational efficiency of the business but will also play a significant role in environmental sustainability. It will offer a modular system to the retailers that are interested in adopting intelligent logistic systems that will minimize the emissions and costs as well as contribute to the larger ESG (Environmental, Social, Governance) efforts of the retail favor. The Figure 3 demonstrates the achieved increase in performance in retail logistics with the help of AI and IoT. Great achievements comprise 35% decrease of carbon emissions and 28% decrease of fuel use. In general, the route efficiency, reduction in idle time, and savings in logistics costs are an indication of sustainable retailing operations.

Figure 3. Impact of AI-IoT Integration on Green Retail Logistics

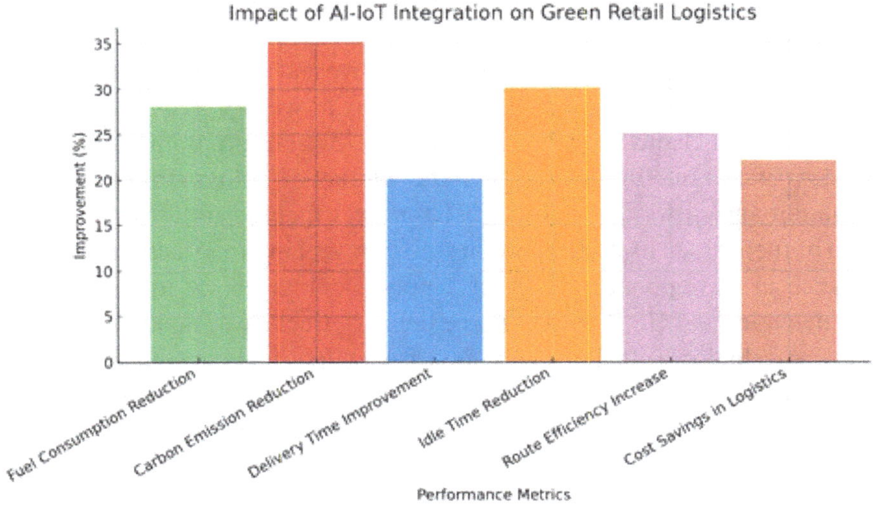

Comparative Study of Two Case Studies

The two case studies presented here, namely, Smart Agriculture and Green Retail Logistics, indicate that AI-IoT integration can be used widely but still in a meaningful way to improve the sustainability of the supply chain. Whereas the smart agriculture concentrated on resource maximization, and minimisation of spoilage, retail logistics was concerned with control of emissions and delivery efficiency. Agriculture IoT sensors and AI models resulted in 40% of reduced spoilage and 32% water usage savings and have provided more contributions to both food preservation and environmental stewardship. Conversely, RT vs AI-enabled real-time tracking and route optimization in the retail logistics case saved 35% carbon emission and 28% fuel consumption. The savings on costs were stronger on the retail industry (22%), whereas the agriculture displayed better results on the waste management. The two scenarios show how customized AI-IoT applications can better solve industry-specific problems as well as contribute to global sustainability objectives. The table 1 makes a comparison of the difference between the effect of the AI-IoT integration in two sectors namely, agriculture and retail logistics. It also highlights the most important indicators like the growth of the yield, the decrease of waste, and the control over the emissions, explaining how these two spheres uniquely benefit of smart technologies. The comparison has been done in a way to bring out sector specific strengths, and the universal implications of sustainable transformation of supply chains. The Figure 4 shows the comparison of the important performance measures between green retail logistics and smart agriculture. It emphasizes the power of agriculture in reducing wastes and resources whereas retail logistics is best at controlling emissions and cost savings. The comparison highlights how AI-IoT solutions can have a unique effect within the various branches of the supply chain.

Table 1. Comparative Performance of AI-IoT Applications in Agriculture and Retail Logistics

Performance Metric	Smart Agriculture (%)	Green Retail Logistics (%)
Yield / Delivery Improvement	18%	20%
Waste / Spoilage Reduction	40%	–
Resource Use Reduction	32%	28%
Emission Reduction	–	35%
Cost Savings	–	22%

Figure 4. Comparative Performance of AI-IoT Applications in Agriculture and Retail Logistics

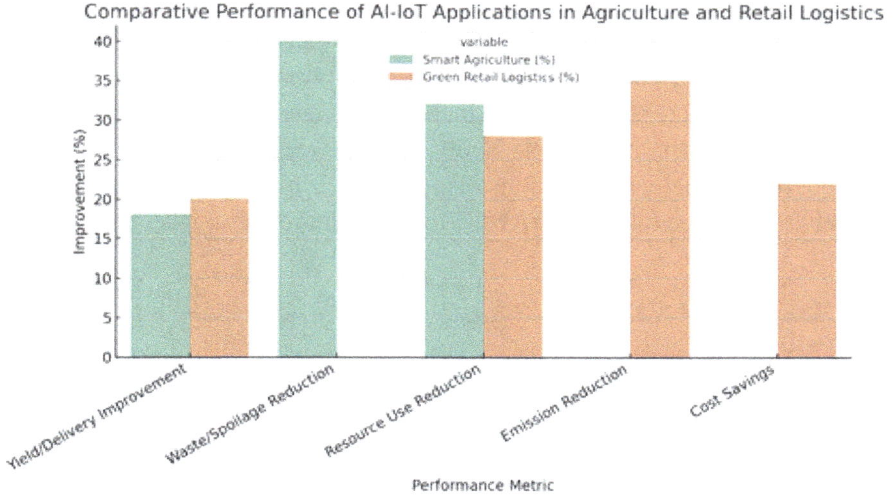

SOLUTIONS AND RECOMMENDATIONS

In order to achieve the optimum benefits of AI and IoT in attaining sustainable supply chain operations, the following chapter proposes a multi-faceted approach that aims at integration, scalability and constant improvement. The first one is that organizations are supposed to implement the usage of IoT sensors and trackers at every node in the supply chain in order to achieve real-time visibility. This has the capability of minimizing the operational blind sports by 60% and enhancing inventory accuracy by 35%. Second, using AI driven analytics to support the demand forecasting and route optimization could reduce fuel consumption by 28-30% and lead to a reduction in carbon emissions by 35% as it is observed in retail logistics case. Uniformity in data practices and cloud-edge hybrid nodes allow data to be processed easily and the response time to reduce. AI predictive maintenance can shorten equipment downtime by 40% and result in an improvement in productivity. AI-IoT systems could be used to monitor environmental factors and dynamically schedule to reduce food/resource waste, particularly in perishables; that has 40% reduced spoilage and 22% extended shelf lives. Commitment to expertise on human resource capital and cross-functional training is done to facilitate easier adoption of the technology. Also, alignment of policies, cybersecurity frameworks, and public-private collaborations are necessary to scale such technologies in an enduring manner. With the application of the recommendations reported here, firms will be able to realize a more effective and leaner, eco-friendly supply chain. The Figure 5 shows how metrics of significance such as fuel consumption and carbon emissions and level of spoilage decreased gradually increasing stages of AI-IoT implementation. By the post-implementation phase, spoilage reduction is recorded at 40%, 28 and 35% reduction in fuel and emissions respectively. It highlights the compound and cumulative advantages of a staged AI-IoT implementation of supply chain operations.

Figure 5. Performance Improvements Across AI-IoT Implementation Phases

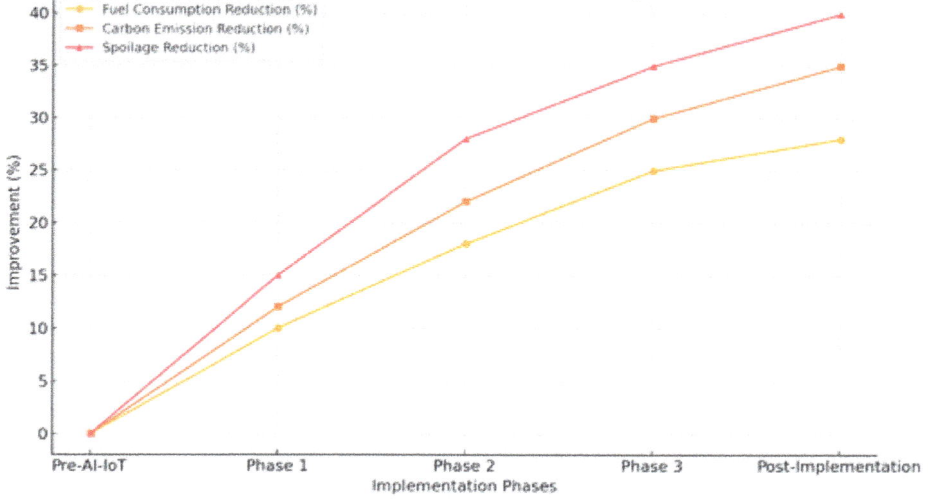

PERFORMANCE EVALUATION AND QUANTITATIVE RESULTS

The effects of applying AI and IoT to the area of supply chain activity have been developed to generate measurable gains in the sphere of the major performance indicators. In smart farming, AI-based prediction and environmental surveillance through IoT cut 32% of water and 25% of fertilizer use and increased crop yield by 18%. The reduction of spoilage during transport was 40% with shelf-life increase of 22% (Mashat et al., 2024). In the case of retail logistics, fleet tracking (real-time tracking) and optimization of routes resulted in a decreased fuel consumption of 28% and a 35% lower carbon emission. The delivery dramatically was enhanced by 20% and savings on logistics costs was attained at 22%. Such quantitative results are supportive of the strategic value of AI-IoT convergence regarding operational efficiency and sustainability targets(de Vass et al., 2018). Furthermore, the feedback loop implemented in the framework would dynamically optimize AI models based on the basis of real-time discoveries with time, thus enabling it to be more flexible and accurate in its decisions. Generally, the metrics used to measure the performance prove that organizations that adopt AI-IoT integration should expect instruments of economic and environmental sustainability in the transformation toward intelligent and green supply chains. The table 2 above summarizes the quantitative improvements in performance due to AI-IoT integration of agriculture and retail logistics (Afum et al., 2020). Economic and environmental benefits can be noted with the help of metrics like fuel savings, yield increase, and spoilage reduction. The statistics prove the worth of the introduction of smart technologies to construct sustainable supply chain environments.

Table 2. Quantitative Impact of AI-IoT Integration on Supply Chain Operations

Domain	Performance Metric	Improvement (%)
Smart Agriculture	Water Usage Reduction	32%
Smart Agriculture	Fertilizer Usage Reduction	25%
Smart Agriculture	Crop Yield Increase	18%
Smart Agriculture	Spoilage Reduction During Transit	40%
Smart Agriculture	Shelf-Life Extension	22%
Retail Logistics	Fuel Consumption Reduction	28%
Retail Logistics	Carbon Emission Reduction	35%
Retail Logistics	Delivery Time Improvement	20%
Retail Logistics	Logistics Cost Savings	22%

ETHICAL AND REGULATORY CONSIDERATIONS

AI and IoT provide an effective solution to modernize supply chains, the integration of the two concepts has posed serious ethical and regulatory issues. There is increased concern with privacy as the data is already being collected and tracked constantly via the IoT devices, even more so when it involves sensitive or personal information, i.e., movements of workers or surveillance of a product-level (Sirilertsuwan et al., 2020). Ensuring ethics also involves making sure that there is data transparency and informed consent and anonymization procedures. On the regulatory level, it must uphold the adherence to the regulations governing the data protection laws, such as the General Data Protection Regulation (GDPR) and the laws relevant to an industry. Furthermore, AI choices by learning algorithms can be quite unjust to vendors and inefficient in resource distribution without precautions being considered. The concept of ethical frameworks needs to drive AI design to deliver fairness, accountability, and explainability. The other major problem is cybersecurity since interdependent networks of IoT include data theft and malicious trampling possibilities. The communication protocols, access control, the resilience of the system are the crucial aspects of the regulatory compliance and stakeholder confidence. In this sense, ethical implementation of AI and compliance to regulations should be partnered to technological innovations in the supply chain systems.

IMPLEMENTATION CHALLENGES AND LIMITATIONS

The ultimate goal of AI and IoT is to transform supply chains, it is not without challenges to successful implementations. Interoperability of data is one key shortcoming- interconnection of different data formats of various IoT gadgets and legacy systems can give inconsistency and time waste. Also, the absence of standardized procedures and platforms prevents the easy data transfer and scalable systems (Neufville et al., 2022). The heavy base cost of the sensors, infrastructure and cloud services is usually the hitch towards small and medium enterprises. Skill disparity is another factor since they are going to need a cross-functional knowledge base in terms of AI, IoT, cloud computing, and cybersecurity to implement them successfully and many organizations do not have it. More to the point, real time processing needs impose strain on computing resources and network bandwidth, particularly in remote or resource-restricted conditions. Adoption can be slowed by resistance to change amongst employees and partners in the supply chain. The issue of security vulnerabilities continues to persist and this is because devices are connected, thus leaving several entry points to the approach of cyber threats (McKuin et al., 2021). Such restraints outline the necessity of strategic planning and the pilot-based rollouts and training stakeholders to guarantee conductive, scalable, and secure deployment of AI-IoT.

DEPLOYMENT FRAMEWORK AND USE CASE EXPANSION

To scale the AI-IoT-based solutions in a variety of supply chain settings, a taxonomic deployment framework is a must. At the first step, the framework entails a needs assessment and pilot deployment, during which businesses select the most relevant pain points and use AI-IoT tools in a limited environment. It is then upgraded by incorporating infrastructure such as the deployment of IoT sensors, cloud access, as well as, data management rules (Diao et al., 2022). This is followed by the layered AI engine, which involves the implementation of machine learning models, adapted to forecasting, optimization and anomaly detection. The feedback loop is an important element because it allows constant learning and improvement of the system depending on the results of its work. The framework should be scalable by design by being modular in nature and therefore, it should be integrated with existing ERP systems and third-party vendors. The areas to which the use case expansion can be narrowed down to are reverse logistics, energy management, inventory forecasting, and demand sensing. It is through this framework that the capability to apply this framework in the real world with practices with references to agriculture, retail, and manufacturing validate this framework. It is this secure, scalable, and intelligent system that manages the supply chain into a resourceful, sustainable, responsive, system that is being relegated by aligning deployment interests with business goals as well as regulatory requirements.

SUMMARY OF KEY CONTRIBUTIONS

This chapter makes several significant contributions to the understanding and application of AI and IoT in building smart and sustainable supply chain operations:

- **Integrated Framework:** Proposes a layered AI-IoT deployment framework consisting of data acquisition, processing, intelligent decision-making, action execution, and feedback loop for continuous improvement.
- **Quantitative Insights:** Provides real-world data showing measurable improvements, such as 40% spoilage reduction, 35% carbon emission reduction, and 28% fuel savings, demonstrating the practical impact of AI-IoT convergence.
- **Cross-Sector Use Cases:** Explores detailed case studies in agriculture and retail logistics to illustrate how tailored AI-IoT solutions address sector-specific challenges and sustainability goals.
- **Implementation Roadmap:** Identifies critical challenges such as interoperability, cybersecurity, and skill gaps, and offers actionable strategies for effective deployment and scalability.
- **Ethical and Regulatory Guidance:** Addresses data privacy, AI fairness, and compliance with global regulations, advocating for responsible and transparent use of AI-IoT technologies in supply chains.

Collectively, these contributions offer a holistic and actionable guide for stakeholders aiming to digitally transform and decarbonize supply chains using intelligent and connected technologies.

FUTURE RESEARCH DIRECTIONS

AI-based supply chain management with the use of IoT opens up a significant body of research. These include the development of interoperable and scalable AI-IOT systems and platforms, which could be implemented in a diverse range of industries and locations. It is also necessary to integrate research on AI models that maintain privacy so that it can extract sensitive supply chain data without posing a threat to the user privacy. As the complexity of the real-time data increases, there will be a significance of embedding edge AI that can decentralize decision-making and run quicker (Radanliev et al., 2020). Another possible area of study that can be suggested is the combination of blockchain and AI-IoT solutions that can be utilized to give the respective supply chains traceability, transparency, and trust at the multi-tiered level. Further, it is also research that is desired on AI ethics, explainability and bias reduction in autonomous supply chain decisions. Finally, the long-term field research, grounded in the developing economies, could assist in digging out a better insight into the real-life challenges and socio-economic implications of the digital supply chain. The squeeze between the emerging technologies and the sustainable practices is one direction, which the future studies must accumulate, and shape the future generation of smart, inclusive, and resilient networks of supply.

Edge AI and Real-Time Decision Making

At the same time that supply chains are ever-more dynamic and distributed, never has real-time responsiveness been more of a necessity. Conventional cloud-based AI systems are very strong but can be faced with problems of latency and bandwidths especially when in remote locations or in strained environments. The concept of Edge AI provides an attractive alternative that allows moving the data processing and AI inference further to the place of where the data is being captured (i.e., warehouses, transportation vehicles, or factory floors). This on-site processing can occur more quickly, requires less frequent ongoing cloud connection and improves data security (Aljohani & Thompson, 2018). E.g. in temperature-sensitive supply chains, such as food or pharmaceuticals, edge AI can potentially adapt to sensor data in real-time, through a cooling system without waiting and verifying with the cloud. Likewise, through edge devices, the delivery routes can be readjusted during delivery in real-time due to traffic or weather disturbance in fleet management. The researchers need to conduct more studies in the future to produce lightweight AI models that will be well-suited to edge environments, which should prioritize computational proficiency over accuracy. Research must also be done on hybrid computations where processing loads dynamically switch between cloud and edge depending on urgencies, and an intelligent, resilient, and adaptive supply chain would be sustained.

Blockchain-Integrated AI-IoT Systems

The combination of blockchain, AI and IoT is an especially rich potential to construct more transparent, secure and traceable supply chain systems. Blockchain creates an unchangeable registry which follows every single transaction or movement of products, AI analyses this data to make optimised decisions, and IoT provides this information in real-time. This three can render supply chains in a way that can be verified to the nth degree and auditable with low ability to cheat, tamper and even human error. The following research can take into account smart contracts, which are blockchain-based, and which automatize any of the trust-based relationships among suppliers, transporters, and retailers (Ittmann, 2015). Using the case of a logistics company, payment can be automatic once the IoT sensors that are in place confirm that the goods are delivered in a good state. Sustainability can also be applied through blockchain that would validate certification (e.g., organic, fair trade), lifecycle of the circular supply chain. But issues like the high rate of energy consumption of some blockchain platforms and the interoperability of chains and integration with legacy systems are issues that must be addressed. Energy-efficient consensus framework, cross-chain communication protocol, and privacy-preserving blockchain applications should be explored by the researchers so that this integration is scalable, safe, and in accordance with the global standards of supply chains.

Ethical, Explainable, and Fair AI Models

With AI playing the bigger role in independent decision-making in supply chains e.g. choosing of suppliers, demand forecasting or pricing, the issues of ethics, transparency and fairness come out as key building blocks in trustworthy adoption. Unwanted through discriminatory or silent AI systems, can lead to undesirable consequences, such as discrimination of suppliers, inequality in the distribution of resources, or loss of reputation. Subsequently, the new research should be aimed at constructing Explainable AI (XAI) systems to allow the logic behind the decisions to be explained to the human stakeholders. Explainability enables a supply chain manager to know how the AI processes to perform his data. It also enables managers to question and audit the decisions made (Al-Zabidi et al., 2021). Besides, the measures of fairness must be integrated into the AI systems to make decisions non-discriminative and inclusive: to guarantee compliance with ethical principles. The study should also be conducted on data governance and the problem of an algorithmic monoculture and over-dependence on a model may cause system-level failure. Ethicists, technologists, and regulators all have a part to play in the multi-stakeholder cooperation necessary to construct responsible AI. In the end, sustainable changes in supply chains will rely on ethical and explainable AI to win stakeholder trust, comply with regulations, and in general, deliver truly sustainable changes.

CONCLUSION

Artificial Intelligence (AI) and the Internet of Things (IoT) are coming together to offer a new future of supply chain management, a new way to develop smart, efficient, and sustainable operations. The relationship between these technologies in enhancing supply chain at various nodes has been at length discussed in this chapter since it enhances real-time visibility, predictive analysis and data driven decision making. AI-IoT solutions enable companies to increase the pace of their work, not mentioning the increased requirements of environmental responsibility and social responsibility. The combination of AI and IoT in smart agriculture achieved a reduction of water consumption, fertilizer consumption, and yield by 32, 25, and 18 respectively, and decreased spoilage in the transport process by 40. The benefits related to it are food security, optimisation of resources and resilience of supply chains. This might be also seen in the retail logistics where the optimization of the routes and tracking the fleet of the vehicles with the help of AI and IoT led to the decrease of 28 per cent of the fuel cost, 35 per cent reduction of carbon emissions, and 22 per cent of the cost of logistics. The above-mentioned effects are remarkably traceable, which has become a high business case in supply chain ecosystems going digital.

The chapter has proposed a five-layer AI-IoT implementation model comprising data acquisition, data processing, intelligent decision-making, sustainable action implementation and feedback control cycle. The existence of all of the above-mentioned layers is vital in the capacity to make sure that decisions made within the supply chain occur in context, are impactful and are sustainable. All such characteristics such as smart sensors, edge processing, and machine learning models enable organizations to respond to the supply chain disruption that may comprise of climate events, geopolitical vulnerabilities, or market volatility in real time. The challenges are a number of them despite the obvious benefits. One of them is the problem of data interoperability, implementation cost is high, cybersecurity Threats, and deficiency of professional resources. Additional ethical concerns are privacy, transparency, algorithmic bias, and these have to be addressed. The standardization processes and convergence of regulations of the workplace, as well as investment in the human resources development, are required due to these weaknesses. Furthermore, the development and use of AI must be governed by ethical standards so that it can be fair and accountable.

The examples of cases in the agricultural sphere and in the retail, sphere demonstrate that AI and IoT may be introduced to the specifics of any industry and optimize irrigation or reduce the number of emissions in delivery operations. These practical applications have served as models to other industries that desire to move towards the sustainable goals. This flexibility and scale of the proposed system is brought out in a comparative analysis of the two industries. In order to proceed, new research should be implemented with state-of-the-art research of Edge AI, blockchain-enabled supply networks and ethical AI scenarios. Real time responsiveness can be augmented with the help of edge AI in decentralized settings, and transparency and traceability can be augmented with the help of blockchain. Ethical AI offers justice, explainability, and trust in stakeholders, in autonomous systems of decision-making. Together, these new trends will help to develop the new generation of intelligent, diverse and robust supply chain networks. In conclusion, the AI-IoT integration can be described as an efficient method of transforming the traditional supply chains into smart and data-oriented ecosystems that would be appropriate to address the needs of the 21 st century. Technological invention integrated with the ethical leadership approach can enable organizations to not only guarantee perfection in operation but also long-term sustainability to both the environment and the society. The chapter is therefore a one-stop reference to researchers, practitioners and policy-makers who would desire to achieve sustainable supply chain operations in future.

REFERENCES

Adesiyan, J. S., & Raffington, A. E. (2024). AgriTech revolution: Next-generation supply chain in America's agriculture. *International Journal of Environment and Climate Change*, *14*(2), 254–272. DOI: 10.9734/ijecc/2024/v14i23943

Adewusi, A. O., Komolafe, A. M., Ejairu, E., Aderotoye, I. A., Abiona, O. O., & Oyeniran, O. C. (2024). The role of predictive analytics in optimizing Supply Chain Resilience: A review of techniques and case studies. Int. j. *Manag. Entrep. Res*, *6*(3), 815–837.

Afum, E., Osei-Ahenkan, V. Y., Agyabeng-Mensah, Y., Amponsah Owusu, J., Kusi, L. Y., & Ankomah, J. (2020). Green manufacturing practices and sustainable performance among Ghanaian manufacturing SMEs: The explanatory link of green supply chain integration. *Management of Environmental Quality*, *31*(6), 1457–1475. DOI: 10.1108/MEQ-01-2020-0019

Al-Zabidi, A., Rehman, A. U., & Alkahtani, M. (2021). An approach to assess sustainable supply chain agility for a manufacturing organization. *Sustainability (Basel)*, *13*(4), 1752. DOI: 10.3390/su13041752

Aliahmadi, A., Nozari, H., & Ghahremani-Nahr, J. (2022). AIoT-based sustainable smart supply chain framework. International Journal of Innovation in Management. *Economics and Social Sciences*, *2*(2), 28–38.

Aljohani, K., & Thompson, R. G. (2018). A stakeholder-based evaluation of the most suitable and sustainable delivery fleet for freight consolidation policies in the inner-city area. *Sustainability (Basel)*, *11*(1), 124. DOI: 10.3390/su11010124

Awan, U., Sroufe, R., & Shahbaz, M. (2021). Industry 4.0 and the circular economy: A literature review and recommendations for future research. *Business Strategy and the Environment*, *30*(4), 2038–2060. DOI: 10.1002/bse.2731

Birkel, H. S., & Hartmann, E. (2019). Impact of IoT challenges and risks for SCM. *Supply Chain Management*, *24*(1), 39–61. DOI: 10.1108/SCM-03-2018-0142

Brown, W., Johnson, O., & Wilson, G. (2024). Influence of E-commerce technologies on supply chain management in retail. In *Preprints*.

Chen, R. (2024). Sustainable supply chain management as a strategic enterprise innovation. Advances in Economics. *Management and Political Sciences*, *85*(1), 24–29.

De Vass, T., Shee, H., & Miah, S. J. (2018). Internet of Things for improving supply chain performance: A Qualitative study of Australian retailers. ACIS2018 Proceedings.

de Vass, T., Shee, H., & Miah, S. (2021). IoT in supply chain management: Opportunities and challenges for businesses in early industry 4.0 context. Oper. Supply Chain Manag. Int. J., 148–161.

Diao, X., Gao, A., Jin, X., & Chen, H. (2022). A layer-based relaxation approach for service network design. *Sustainability (Basel), 14*(20), 13143. DOI: 10.3390/su142013143

Firouraghi, N., Kiani, B., Jafari, H. T., Learnihan, V., Salinas-Perez, J. A., Raeesi, A., Furst, M., Salvador-Carulla, L., & Bagheri, N. (2022). The role of geographic information system and global positioning system in dementia care and research: A scoping review. *International Journal of Health Geographics, 21*(1), 8. DOI: 10.1186/s12942-022-00308-1 PMID: 35927728

Getahun, S., Kefale, H., & Gelaye, Y. (2024). Application of precision agriculture technologies for sustainable crop production and environmental sustainability: A systematic review. *TheScientificWorldJournal, 2024*(1), 2126734. DOI: 10.1155/2024/2126734 PMID: 39421732

Ghahremani Nahr, J., Nozari, H., & Sadeghi, M. E. (2021). Green supply chain based on artificial intelligence of things (AIoT). International Journal of Innovation in Management. *Economics and Social Sciences, 1*(2), 56–63.

Gutiérrez-Diez, J. C., Castañón, A. M., & Bascompta, M. (2024). New method to study the effectiveness of mining equipment: A case study of surface drilling rigs. *Applied Sciences (Basel, Switzerland), 14*(5), 2185. DOI: 10.3390/app14052185

Holloway, S. (2024). Exploring the impact of real-time supply chain information on marketing decisions: Insights from service industries. In *Preprints*.

Hussain, M., Al-Aomar, R., & Melhem, H. (2019). Assessment of lean-green practices on the sustainable performance of hotel supply chains. *International Journal of Contemporary Hospitality Management, 31*(6), 2448–2467. DOI: 10.1108/IJCHM-05-2018-0380

Hussein, A. H. A., Jabbar, K. A., Mohammed, A., & Jasim, L. (2024). Harvesting the Future: AI and IoT in Agriculture. E3S Web Conf., 477, 90.

Ittmann, H. W. (2015). The impact of big data and business analytics on supply chain management. *Journal of Transport and Supply Chain Management, 9*(1). Advance online publication. DOI: 10.4102/jtscm.v9i1.165

Kadhim Obaid, M., Abood, B. S. Z., Khalid Alazzai, W., & Jasim, L. (2024). From Field to Fork: The role of AI and IoT in Agriculture. E3S Web Conf., 491, 2006.

Karthikeyan, K. S., & Nagaprakash, T. (2023a). Prioritizing IoT-driven sustainability initiatives in retail chains: Exploring case studies and industry insights. EAI Endorsed Trans. Internet Things, 10.

Karthikeyan, K. S., & Nagaprakash, T. (2023b). Prioritizing IoT-driven sustainability initiatives in retail chains: Exploring case studies and industry insights. EAI Endorsed Trans. Internet Things, 10.

Lai, Y. (2024). Innovative strategies in logistics and supply chain management: Navigating modern challenges. SHS Web Conf., 183, 2020. DOI: 10.1051/shsconf/202418302020

Liu, J., Chen, M., & Liu, H. (2020). The role of big data analytics in enabling green supply chain management: A literature review. *Journal of Data, Information and Management*, *2*(2), 75–83. DOI: 10.1007/s42488-019-00020-z

Mashat, R. M., Abourokbah, S. H., & Salam, M. A. (2024). Impact of Internet of Things adoption on organizational performance: A mediating analysis of supply chain integration, performance, and competitive advantage. *Sustainability (Basel)*, *16*(6), 2250. DOI: 10.3390/su16062250

McKuin, B., Watson, J. T., Stohs, S., & Campbell, J. E. (2021). Rethinking sustainability in seafood. Elementa (Wash., DC), 9(1). DOI: 10.1525/elementa.2019.00081

Neufville, R., Abdalla, H., & Abbas, A. (2022). Potential of connected fully autonomous vehicles in reducing congestion and associated carbon emissions. *Sustainability (Basel)*, *14*(11), 6910. DOI: 10.3390/su14116910

Radanliev, P., De Roure, D., Page, K., Nurse, J. R. C., Mantilla Montalvo, R., Santos, O., Maddox, L., & Burnap, P. (2020). Cyber risk at the edge: Current and future trends on cyber risk analytics and artificial intelligence in the industrial internet of things and industry 4.0 supply chains. *Cybersecurity*, *3*(1), 13. DOI: 10.1186/s42400-020-00052-8

Reddy, T. T., Devi, Y. R., & Kavita, G. (2023). Logistics, traceability in food supply chain management. E3S Web Conf., 391, 1075.

Rendon-Benavides, R., Perez-Franco, R., Elphick-Darling, R., Plà-Aragonés, L. M., Gonzalez Aleu, F., Verduzco-Garza, T., & Rodriguez-Parral, A. V. (2023). In-transit interventions using real-time data in Australian berry supply chains. *The TQM Journal*, *35*(3), 759–777. DOI: 10.1108/TQM-11-2021-0319

Shafique, M. N., Rashid, A., Bajwa, I. S., Kazmi, R., Khurshid, M. M., & Tahir, W. A. (2018). Effect of IoT capabilities and energy consumption behavior on green supply chain integration. *Applied Sciences (Basel, Switzerland)*, *8*(12), 2481. DOI: 10.3390/app8122481

Singh, R., Dwivedi, A. D., & Srivastava, G. (2020). Internet of Things based blockchain for temperature monitoring and counterfeit pharmaceutical prevention. *Sensors (Basel)*, *20*(14), 3951. DOI: 10.3390/s20143951 PMID: 32708588

Sirilertsuwan, P., Thomassey, S., & Zeng, X. (2020). A strategic location decision-making approach for multi-tier supply chain sustainability. *Sustainability (Basel)*, *12*(20), 8340. DOI: 10.3390/su12208340

Susan, W., Carsten, M., Daqiang, C., & Rosario, M. (2017). An IoT-enabled supply chain integration framework: Empirical case studies. In *Advances in Transdisciplinary Engineering*. IOS Press.

Udeh, E. O., Amajuoyi, P., Adeusi, K. B., & Scott, A. O. (2024). The role of IoT in boosting supply chain transparency and efficiency. *Magna Sci. Adv. Res. Rev.*, *12*(1), 178–197. DOI: 10.30574/msarr.2024.11.1.0081

Waller, M. A., & Fawcett, S. E. (2013). Data science, predictive analytics, and big data: A revolution that will transform supply chain design and management. *Journal of Business Logistics*, *34*(2), 77–84. DOI: 10.1111/jbl.12010

Yang, Q., Bolaji Samson, A., Muhammad Gummi, U., & Chen, D. (2023). Study on the use of industry 4.0 technology for the textile industry sustainable supply chain process. *Advances in Engineering Technology Research*, *4*(1), 140. DOI: 10.56028/aetr.4.1.140.2023

Zhang, J., & Chen, J. (2013). Coordination of information sharing in a supply chain. *International Journal of Production Economics*, *143*(1), 178–187. DOI: 10.1016/j.ijpe.2013.01.005

KEY TERM AND DEFINITIONS

Artificial Intelligence (AI): The simulation of human intelligence in machines to perform tasks like prediction and optimization.

Circular Supply Chain: A system that reuses, refurbishes, or recycles products and materials to reduce waste.

Edge Computing): A method of processing data near the source to reduce latency and improve real-time decision-making.

Explainable AI (XAI): AI models designed to make their decisions transparent and understandable to humans.

Internet of Things (IoT): A network of interconnected devices that collect and exchange real-time data across environments.

Predictive Analytics: The use of data, statistical algorithms, and AI techniques to identify future outcomes.

Route Optimization: AI-based process of determining the most efficient paths for transportation and delivery.

Smart Sensors: IoT-enabled devices that detect and transmit data about physical conditions in real time.

Chapter 2
Enhancing Sustainable Consumption Through Energy Efficiency Labels:
Insights From Policy Frameworks and Data-Driven Consumer Behavior

R. Sowmiya
https://orcid.org/0009-0003-5336-2988
SASTRA University, India

M. Velavan
https://orcid.org/0000-0002-4114-6773
SASTRA University, India

ABSTRACT

As climate change intensifies and global energy demand rises, energy efficiency labels have emerged as vital tools for promoting sustainable consumption. This chapter explores the evolution, regulatory frameworks, and behavioral impact of labeling systems across regions such as the European Union, India, and the United States. Using theories like the Theory of Planned Behavior and behavioral economics, it examines how consumers interpret and respond to labels. The chapter highlights the role of digital innovations—like QR codes and mobile apps—and emphasizes the importance of public-private collaboration and regulatory enforcement. Drawing from case studies and data analysis, it provides strategic recommendations to ad-

DOI: 10.4018/979-8-3373-3790-6.ch002

dress barriers such as label fatigue and affordability. Energy labels are positioned as key instruments connecting consumer behavior, technological progress, and climate policy.

INTRODUCTION

In today's world, where climate change and environmental issues are becoming more urgent, the need for sustainable consumption is more important than ever. Human activities-like our heavy dependence on fossil fuels, wasteful energy practices, and unsustainable production methods-have ramped up greenhouse gas emissions, leading to global warming and ecological disruption. Tackling these problems calls for not just big changes at the industrial and government levels, but also a shift in how each of us, as consumers, behaves.

One of the key tools we've got to promote eco-friendly consumer choices is energy efficiency labels. You'll often find these labels on household appliances, electronics, and other products that use energy. They offer clear and standardized information about how much energy a product consumes, its performance rating, and the potential savings you could enjoy. By breaking down complex energy data into simple formats-like stars, colors, or grades-these labels make it easier for consumers to consider environmental impact when making purchases (Dötsch & Schmid, 2021).

But energy labels do more than just educate consumers; they also play a crucial role in policy communication, linking climate objectives with what happens in the marketplace. For instance, energy labeling programs are often woven into national energy efficiency strategies, building regulations, and even subsidy programs. This dual role—both educational and regulatory-makes energy efficiency labels powerful tools for connecting public policy goals with the everyday choices consumers make (IEA, 2020).

Energy labels are central to the wider context of sustainable product standards. Far from being simple visual indicators, energy labels act as agents of market change that force manufacturers and retailers toward greater responsibility for product performance. As consumers repeatedly choose energy-efficient products, there is a reinforcing loop-encouraging companies to innovate and governments to enhance regulatory settings. In this way, energy labels act as active policy instruments that speed up the move towards a low-carbon economy.

On a behavioral level, having a label on a product often establishes what behavioral economists call "choice architecture." This means that the energy performance of a product becomes a significant factor in our decision-making process. Studies show that when energy labels are clear, reliable, and standardized, consumers are more likely to choose energy-efficient options-even if they come with a higher initial price tag-because they see the long-term savings and environmental advantages (Thøgersen & Schrader, 2012; OECD, 2022).

In this context, energy efficiency labeling represents a powerful blend of data-driven governance, sustainable consumption, and climate action. It not only empowers consumers to make informed choices but also inspires them to actively participate in environmental protection. Understanding how these labels function, how people interpret them, and how they influence consumer behavior is vital for developing effective sustainability strategies in both public and private sectors.

Key Energy Labeling Programs

1. BEE Star Label (India): Under the 2001 Energy Conservation Act, India's Bureau of Energy Efficiency began a label program that mixes mandatory rules with optional badges (BEE, 2022). Appliances are scored from 1 star (least efficient) to 5 stars, based on standardized national lab tests. With this simple grading, shoppers in big cities and small towns alike can spot fridges, ACs, and fans that will save money over their lifetime.
2. EU Energy Label (European Union): Launched in 1995 and refreshed in 2021, the label's clear A-to-G scale now sits alongside familiar pictograms and a handy QR code (European Commission, 2021). Bright color bars let anyone glance at a washing machine, dishwasher, or bulb and tell at once how energy-saving (and planet-saving) it is.
3. ENERGY STAR (United States): Run by the U.S. Environmental Protection Agency, the ENERGY STAR program is a voluntary label that flags products and practices known for saving energy. To carry the star, an item or system must sit in the top 25 percent for efficiency and meet EPA test rules (EPA, 2020). Consumers and businesses alike easily associate the mark with better-performing appliances, electronics, and commercial buildings.

Energy labels are thus central to public policies that nudge markets toward greener choices. When clear, trusted, and backed by real data, they sway "shoppers, spur companies to innovate, and move nations" closer to their energy and climate goals (IEA, 2020; Thøgersen & Schrader, 2012).

The Role of Informed Consumer Behavior in Sustainable Development

Understanding what an energy label means sits at the heart of every successful labeling scheme. A sticker on a product won't spark greener habits unless people know how to read it, trust the source, and feel confident that what they see matches reality. When shoppers can do all that, they tend to pick energy-saving appliances even if they cost a bit more upfront (OECD, 2022).

Looked at more broadly, these smarter choices help power Sustainable Development Goal 12, which calls for responsible consumption and production. As more households lean toward efficient goods, manufacturers are nudged to innovate, governments are pushed to tighten rules, and whole markets start moving in a greener direction. In that way, energy labels work on many levels—at the checkout counter, in factory design rooms, and in national and global sustainability plans.

Having established the relevance of energy labels for sustainable consumption and supply chains, the next section traces their historical evolution across regions to understand how such programs have developed over time.

Regional Efforts and Worldwide Alignment

In addition to global leaders like the EU, United States, and India, numerous other countries have established their own energy labeling systems, tailored to their national energy priorities, market structures, and consumer behavior. These regional efforts reflect a growing global consensus on the role of labeling in promoting energy efficiency and climate action.

- China introduced its mandatory three-level China Energy Label in 2005, covering major household appliances such as refrigerators, air conditioners, and washing machines. The system classifies products into three efficiency levels and requires clear display of energy consumption figures, playing a critical role in shaping consumer expectations and influencing manufacturer standards.
- Australia and New Zealand adopted comparative star labeling schemes in the 1980s, among the earliest globally. Initially limited to white goods, the program has since expanded to include lighting, commercial refrigeration, and televisions. The two countries have aligned their systems under a joint trans-Tasman energy rating initiative, facilitating regulatory harmonization and cross-border product consistency (OECD, 2022).
- Gulf Cooperation Council (GCC) countries, particularly Saudi Arabia and the United Arab Emirates, have implemented mandatory labeling for appliances such as air conditioners and washing machines. These programs are integral to broader national strategies aimed at improving energy efficiency, managing peak electricity demand, and reducing reliance on fossil fuels.

Collectively, these regional initiatives demonstrate how energy labeling can be adapted to diverse policy environments while contributing to global harmonization efforts. Organizations like the International Energy Agency (IEA) and CLASP continue to support standardization and knowledge sharing to reduce duplication, ease international trade, and elevate minimum energy performance worldwide.

Despite variations in design and scope, these programs share core features: standardized performance metrics, visual simplicity, and regulatory support. Organizations like IPEEC and CLASP have worked toward harmonizing label formats, reducing compliance costs, and promoting cross-border product comparability (CLASP, 2019).

From Static to Smart Labels: Energy labeling is increasingly integrating with smart technologies. QR codes, mobile apps, and IoT systems now enable real-time feedback, personalized savings estimates, and carbon footprint tracking. These digital innovations increase transparency and accessibility, particularly in e-commerce and mobile-first markets (OECD, 2022).

From Awareness to Regulation: The original goal of labels was to raise awareness. Today, most programs have become performance-based regulatory tools that need standardized testing and certification. Products must meet certain energy levels to qualify for higher ratings. Label criteria are updated frequently to reflect new technology and changes in the market (IEA, 2020).

Global Frameworks and Policy Influence: The global adoption of energy labeling is reinforced by international climate frameworks. The Paris Agreement (2015) encourages nations to include energy efficiency in their Nationally Determined Contributions (NDCs), while Sustainable Development Goal 12 promotes information tools like labeling to guide responsible consumption (United Nations, 2015).

Groups such as the IEA and UNEP offer practical assistance, funding, and training to assist nations in setting up and managing effective labeling programs. Energy labels do more than just reveal technical details; they act as instruments for changing behavior, shaping economics, and guiding policy that help turn national and global climate plans into reality. While this historical evolution provides important context, the effectiveness of labeling ultimately depends on the mechanisms through which they shape consumer choices and market outcomes.

Regulatory Frameworks and Compliance Mechanisms

The success of energy efficiency labeling systems relies not only on how informed consumers are but also on having robust regulatory frameworks and effective enforcement mechanisms in place. These frameworks provide the essential legal and institutional support needed to ensure that energy labels are reliable, standardized, and consistently applied. They also enable governments to integrate labeling programs into broader environmental and energy strategies, while making sure that manufacturers are held accountable for adhering to the rules.

Institutional Roles in Energy Labeling Governance

Different national and regional organizations are responsible for designing, implementing, and enforcing energy labeling regulations. Each institution operates within its own jurisdiction, yet they all share common goals related to climate action, energy savings, and market transparency.

India – Bureau of Energy Efficiency (BEE): Founded under the Energy Conservation Act of 2001, the BEE serves as the main agency for energy efficiency initiatives in India. It runs the Standards & Labeling (S&L) Programme, which establishes minimum energy performance standards (MEPS) and star rating criteria for appliances. The BEE also manages testing protocols, public awareness campaigns, and compliance audits (BEE, 2022).

United States – Environmental Protection Agency (EPA): The EPA oversees the ENERGY STAR program in collaboration with the U.S. Department of Energy (DOE). While participation in ENERGY STAR is voluntary, its strong reputation is upheld through rigorous performance criteria and independent third-party verification. The EPA also engages with stakeholders and regularly updates the program (EPA, 2020).

European Union – European Commission (EC): The EC is responsible for developing and enforcing energy labeling and eco-design directives across all EU member states. This framework is legally binding under Regulation (EU) 2017/1369. The EU Commission, in partnership with national market surveillance authorities, supervises label design, MEPS, and compliance efforts.

Challenges in Enforcement and Harmonization

Even with solid institutional frameworks in place, there are still several hurdles to overcome for effective implementation:

Enforcement Gaps and Informal Markets: In many developing nations, the presence of unregulated or informal markets makes it tough to enforce regulations. You often come across products that are sold without proper labels or, worse yet, with misleading ones, especially in rural or low-income areas (OECD, 2022).

Harmonization Across Jurisdictions: The differences in testing standards, performance metrics, and label formats from one country to another make international trade and compliance a real headache for manufacturers. While organizations like CLASP and the IEA are striving for harmonization, the progress is pretty inconsistent.

Manufacturer Accountability: Ensuring that manufacturers accurately represent energy data is no easy feat. Sometimes, self-declaration systems that don't undergo thorough audits can really hurt credibility. To enhance accountability, we need tougher penalties, random product testing, and better ways for consumers to voice their complaints.

Resource Constraints: In many low-income countries, government agencies often face challenges like limited funding, a shortage of trained staff, or inadequate infrastructure to effectively monitor and enforce regulations on a large scale (UNEP, 2020).

Regulatory frameworks and compliance mechanisms are essential for creating credible and effective energy labeling systems. The success of these efforts depends on strong institutional capacity, rigorous standards, and continuous monitoring. As energy efficiency labeling expands globally, improved coordination among national agencies, harmonization of international standards, and the adoption of digital verification tools will be crucial for making a lasting impact.

These mechanisms, however, are not experienced uniformly, as consumer responses vary significantly depending on socioeconomic status, cultural context, and inclusivity factors such as gender, age, and disability.

CONSUMER BEHAVIOR AND LABEL EFFECTIVENESS

Energy efficiency labels aim to sway consumer choices by offering clear, comparable, and trustworthy information. However, the actual influence of these labels on behavior is shaped by a blend of psychological, social, and contextual factors. To truly enhance their effectiveness, it's vital to understand how consumers perceive and respond to label information.

Theory of Planned Behavior (TPB)

The Theory of Planned Behavior, introduced by Ajzen in 1991, provides a robust framework for understanding how consumers make eco-friendly choices. According to TPB, three main components shape our behavior:

Attitudes: These are our beliefs about the outcomes of a behavior, like saving money or helping the environment.

Subjective norms: This refers to the social pressures or cultural expectations we feel, such as the influence from friends to buy green products.

Perceived behavioral control: This is about how capable we feel in performing a behavior, considering factors like affordability or the availability of products.

When it comes to energy labeling, consumers who have a positive outlook on sustainability, believe that their peers value energy efficiency, and feel confident in their ability to choose efficient appliances are more likely to act on the information provided by labels (Thøgersen & Schrader, 2012).

Cross-national & Meta-analytic Studies: Recent research has emphasized the value of integrating evidence in widely different national contexts. Meta-analyses that appeared in Energy Policy and Journal of Cleaner Production show that the effectiveness varies by region, with consumer understanding, willingness-to-pay, and policy compliance found to be important mediators (Waechter et al., 2022; Zhou & Zhang, 2023). Cross-national evidence such as this serves to emphasize that labeling schemes are not uniform but rather need to be tailored closely to local institutional and cultural environments.

Alternative Behavioral Theories (VBN Theory): Although the Theory of Planned Behavior (Ajzen, 1991) has found broad use in the analysis of consumer reactions to energy labels, other frameworks like the Value-Belief-Norm (VBN) theory (Stern, 2000) provide further explanatory capability. VBN highlights the influence of pro-environmental values and moral norms upon behavior and predicts that labeling effectiveness is reinforced where consumers perceive energy-efficient options to be in line with their social and individual norms Combining VBN with TPB yields more meaningful insight into psychological processes underlying sustainable consumption.

Sub-Saharan Africa Perspective: While much of the available literature is centered on the Global North, new evidence from the Global South-specifically Sub-Saharan Africa—is highly insightful. For instance, South Africa's mandatory appliance labelling scheme, introduced in 2015, has faced rural awareness and enforcement challenges but has also recorded quantifiable savings in urban household electricity demand (Mkhize & Thopil, 2021). Drawing in views from Africa underscores the significance of contextual conditions such as affordability limitations, infrastructure shortages, and energy access inequalities, which further broadens the geographical scope of the literature review.

Nudging and Behavioral Economics

"Nudges" are subtle, non-coercive cues that influence decision-making by altering the way choices are presented-without restricting freedom of choice (Thaler & Sunstein, 2008). In the context of energy labeling, nudges operate as informational prompts, helping consumers recognize and compare product efficiency at the point of purchase. Visual elements such as color-coded bars, star ratings, and estimated annual savings serve to simplify technical data, reduce cognitive load, and guide consumers toward more sustainable choices.

Empirical evidence supports the impact of framing effects on consumer decisions. For example, presenting energy costs as "save 500 per year" is more persuasive than "spend 10,000 annually," due to loss aversion-a well-established concept in behavioral economics (OECD, 2022). However, the effectiveness of nudges can vary significantly across product categories, income groups, and levels of energy literacy. For some consumers, nudges may be ignored altogether if labels are perceived as irrelevant or untrustworthy.

In real-world purchasing situations, consumers frequently rely on heuristics—mental shortcuts used to make quick decisions under uncertainty. While heuristics reduce decision fatigue, they can also lead to suboptimal outcomes when they override factual label information. Common heuristics observed in the energy labeling context include:

- Brand heuristic: Choosing well-known brands over unfamiliar ones, even if the latter offer better efficiency ratings.
- Price-quality heuristic: Assuming that a higher price implies better energy performance, despite potential mismatches between cost and efficiency.
- Star-counting heuristic: Relying only on star count, not actual kWh or usage context or context-specific product use.

These tendencies highlight the importance of clear label design, consumer education, and institutional trust. Unless consumers fully understand and trust the meaning behind energy labels, even well-intended nudges may fail to drive meaningful behavior change.

Regarding sociodemographic and contextual factors, there are a few key points to consider.

Income and Education: Typically, people with higher incomes and education levels are more likely to pay attention to energy labels. They tend to understand the information better and are more willing to invest in the higher upfront costs (IEA, 2020). In contrast, those with lower incomes often prioritize the initial price over potential long-term savings, which can make these labels less effective for them.

Label Design and Comprehension: The way labels are designed is super important for how well they work. If a label is cluttered with numbers or filled with complicated jargon, it can leave consumers feeling confused. On the other hand, using color codes (like the EU's green-to-red scale) and visual indicators (like India's 1–5star system) can really enhance usability (European Commission, 2021).

Trust in Institutions: The effectiveness of a label largely depends on how much consumers trust the organization behind it. If a label is viewed as credible and backed by respected entities like government agencies or independent organizations (e.g., BEE, EPA, EU Commission), people are more likely to take it seriously (OECD, 2022). However, trust can quickly fade if there are concerns about fake labels or greenwashing.

Cultural, Social, and Psychological Influences: The way labels are perceived is shaped by cultural values, social norms, and psychological factors. In collectivist cultures, such as India or China, the opinions of others may weigh more heavily than individual cost savings. When making decisions, influences like peer pressure, family advice, and local customs often hold more sway than our own reasoning. Psychological aspects, such as present bias-where we tend to prioritize immediate costs-and information overload can also significantly influence how consumers behave. If people feel swamped by too many options or think that comparing them is too much hassle, they might just overlook the label information entirely.

Gender and Household Decision-Making Dynamics: Energy labeling does not affect all household members equally. Studies indicate that gender is an important factor in sustainable consumption choices, as women tend to be key decision-makers in buying household appliances in most cultures (Senyapar, 2024). Women might focus more on long-term cost savings and well-being within the household, whereas men might be more focused on initial purchase price or technical requirements. In addition, domestic decision-making is often a collective process, and labeling information will engage with family life, bargaining capacity, and inter-generational differences. Recognizing these gendered and relational aspects offers a better explanation of how energy labels influence consumer decisions, especially in complex sociocultural contexts.

Equity and Inclusivity: Income and education are only two of the many factors, including gender, age, disability, and rural residence, that have significant impacts on the access to energy labels. Women, who in many countries determine major household purchases, might read and rank energy information differently. Older and differently-abled consumers could have difficulties with small print, complicated icons, or digital-only presentations like QR codes. Low-income and rural consumers are also hampered by affordability, narrow product offerings, and lower energy literacy. Closing these gaps necessitates goal-directed measures, such as simplified label formats, local public awareness campaigns, low-income household subsidies, and implementing universal design principles to make energy labels usable and understandable for all consumer segments.

Energy labels can be powerful tools for driving behavior change, but their success hinges on a complex mix of individual, social, and institutional factors. To create effective labeling strategies, we need to ground them in behavioral science, taking into account not just how the information is presented, but also how it's perceived and acted upon. It's crucial to tailor labels to resonate with different consumer profiles.

Although consumers remain central to the adoption of energy-efficient products, the broader success of labeling programs is equally dependent on the roles played by governments, manufacturers, and retailers.

DATA-DRIVEN INSIGHTS AND LABEL IMPACT ANALYSIS

To truly understand the impact of energy efficiency labeling in the real world, we need to employ rigorous, data-driven methods. While these labels are crafted to influence behavior, their actual effectiveness relies on how consumers interpret and respond to them in various situations. Over the last decade, a growing number of studies have utilized surveys, controlled experiments, and big data analytics to evaluate the role of energy labels in shaping consumer choices, market trends, and environmental results.

Quantitative Methods for Behavioral and Market Analysis

Surveys and Field Studies: Surveys are one of the most widely used techniques to measure awareness, comprehension, and attitudes towards energy labels. Surveys mainly assess:

Label recognition (such as the ability to recognize ENERGY STAR or BEE labels), Trust in the label-associated organizations, Willingness to pay (WTP) for more efficient products.

For example, EU surveys identified that over 90% of shoppers are able to identify the EU Energy Label, and over 75% do indeed use it when choosing (European Commission, 2021). Likewise, Indian research shows that star-rated products are far more likely to be purchased if customers are made aware of the long-term financial benefits they deliver (BEE, 2022).

Consumer Testing and Behavioral Experiments: Experimental economics has played a vital role in understanding how the presentation of labels affects consumer behavior. Such studies typically use randomized control trials (RCTs), A/B testing, or simulations in a lab.

For instance: Customers who are exposed to easy-to-understand labels (such as color-coded scales or comparison savings) are more likely to select energy-efficient products compared to customers given technical or numerical information (OECD, 2022).

Research shows that framing energy use in terms of losses (like "You'll lose 500/year") is more impactful than framing it in terms of gains (like "You'll save 500/year") due to the concept of loss aversion (Thaler & Sunstein, 2008).

These experiments offer valuable insights into the psychological factors that drive energy-efficient purchasing and help improve label design.

Big Data and Market Analytics: As e-commerce and intelligent appliances expand, big data solutions are becoming the norm for assessing label performance on a mass level. Some of these methodologies are:

Web tracking: Measuring how consumers interact with labeled products online (such as click-throughs and conversion).

Retail scanner data: Examining sales patterns between labeled and non-labeled items.

Smart meter data: Tracking energy consumption post-purchase to validate label effectiveness. purchase behavior insights, regional differences, and long-term usage patterns that conventional surveys often miss.

Measuring Impact: Energy Savings and Market Shifts: Energy Savings Metrics

Quantitative impact assessments have revealed that energy labels play a direct role in achieving substantial electricity savings:

- In the United States, ENERGY STAR-labeled goods saved more than 500 billion kWh of electricity in 2020 alone (EPA, 2020).
- In India, the BEE's labeling scheme reduced total electricity demand by over 70 billion kWh during 2015–2020 (IEA, 2020).

These numbers equate to significant cost savings for consumers and emissions reductions, as labels are both environmentally and economically worthwhile.

Market Share of Labeled Products: Labels not only influence one-off purchases but also induce market change. They impact supply chains over the long term, leading manufacturers to make more efficient products and discontinue inefficient ones.

- In the EU, by 2019, more than 80% of refrigerators and washing machines on sale were A++ or A+++ (prior to the label scale reset) (European Commission, 2021).
- In India, data reported by BEE indicates an expanding share of 4-star and 5-star appliances in the marketplace, particularly in the AC and refrigerator markets (BEE, 2022).

These tendencies demonstrate the influence of labeling on directing consumer markets toward sustainability objectives.

Energy efficiency labels have graduated from mere information tools to quantifiable levers of behavioral and market transformation. As more data become available—from field surveys to moment-by-moment energy consumption-researchers and policymakers can now assess label performance with much greater accuracy. But if future strategies are to have maximum impact, behavioral insights, data analytics, and technological innovation will have to be combined so that labels can continue to be effective across the range of consumer contexts.

Despite these diverse contributions, labeling programs continue to face systemic barriers and unintended consequences that limit their effectiveness, which merit critical evaluation.

Adoption Barriers and Systemic Constraints

Because of their established ability to shape consumer behavior and lower energy consumption, energy labeling schemes encounter various adoption constraints. These constraints arise not only from personal constraints-such as consumer knowledge and cost affordability-but also from systemic factors such as market structure, policy implementation, and information trustworthiness. Knowledge of these limitations is critical in order to improve labeling policies and make them effective in the long run.

Cognitive and Behavioral Barriers

Label Fatigue and Information Overload: Perhaps the most prevalent behavior constraint is label fatigue, in which consumers are faced with too much labeling or too technical information. When faced with many certification schemes (e.g., ENERGY STAR, BEE, eco-labels), consumers can be confused, doubtful, or disconnected (Thøgersen & Schrader, 2012). Likewise, information overload-particularly when labels contain dense numerical data or similar efficiency measures-can cause consumers to tune out label information entirely (OECD, 2022).

Low Awareness in Low-Income and Rural Segments: Label recognition and understanding are often low in most developing nations, especially among rural and low-income consumers. Poor education, internet, or product access increases the lack of awareness of energy efficiency principles (IEA, 2020). In most of these markets, consumers tend to favor fundamental product functionality or price over long-term savings, lessening the power of labels to influence purchasing decisions.

Economic Constraints

Affordability and Price Sensitivity: A major impediment is price sensitivity. Appliances with high efficiency tend to have higher initial prices-even though they provide long-term savings. For price-sensitive customers, particularly in lower-income groups, the immediate difference in cost can be a determining factor. Research conducted in India and Southeast Asia shows that even if consumers appreciate the value of energy efficiency, several are either unwilling or incapable of paying a premium price (BEE, 2022; UNEP, 2020).

Limited Supply of Labeled Alternatives: Efficient or labeled alternatives may not be supplied in smaller markets or isolated locations. Lower-cost, non-compliant appliances may be sold by retailers because of supply chain shortages, manufacturer promotions, or weak enforcement. This will limit further the consumer's ability to respond to label information, even if they wish to.

Institutional and Market-Level Limitations

Misuse of Labels and Greenwashing: If not monitored strictly, labels may be misused or falsely represented. Labels are used by some producers or retailers on non-conforming products, or overly ambitious efficiency claims are made. In weakly regulated markets, these practices, as they are called greenwashing, undermine public confidence in labeling schemes (OECD, 2022). This is especially an issue in second-hand or informal appliance markets, where regulatory control is limited.

Weak Monitoring and Enforcement: Successful labeling schemes necessitate frequent checks for compliance, such as random product testing, audits, and sanctions for transgressions. Yet most national schemes-particularly in developing economies-do not have the institutional capacity or resources to implement far-reaching enforcement (UNEP, 2020). In the absence of serious consequences for non-compliance, labeling is symbolic and ineffective.

Shortage of Manufacturer Incentives: Manufacturers might not have enough incentives to make investments in energy-efficient technologies or apply for certification labels. The procedure is usually expensive, involving testing, redesign, and regulatory clearance, which can discourage small- and medium-sized manufacturers from doing so. In the absence of specific subsidies, tax credits, or procurement incentives, their engagement in labeling schemes is limited (IEA, 2020).

Barriers to energy labeling are multi-faceted: Ranging from consumer-level behavioral gaps through to systemic market frailties. Overcoming these challenges will require a combination of efforts: streamlining label design, increasing consumer education, encouraging manufacturers, and boosting regulatory enforcement. By accepting and addressing these limitations, only then can energy labeling programs reach their full potential for sustaining consumption.

Critical Assessment of Labeling Issues: Whereas energy efficiency labeling schemes have yielded quantifiable environmental and economic savings, their potential is frequently limited by systemic weaknesses. In reality, a number of schemes threaten to be cosmetic rather than revolutionary. For instance, poor enforcement in informal markets allows the presence of appliances with counterfeit or deceptive labels on the market, eroding consumer confidence and **diluted policy impact**. Likewise, labeling programs can inadvertently leave behind vulnerable groups-rural or poor households-who do not have access to high-efficiency appliances or digital means (e.g., QR-based smart labels). Another is the rebound effect in which consumers who buy efficient appliances will end up consuming more energy overall by using them more intensively, thus negating possible savings. In addition, most national initiatives focus on consumer awareness campaigns but allocate lesser resources to removing supply-side impediments, including manufacturer incentives and cross-border regulatory harmonization. Without addressing such underlying structural challenges, labeling policies could easily degenerate into compliance checklists, not effective drivers of sustainable supply chain change. To better illustrate how these challenges play out in practice, the following section examines case studies from leading countries and compares their policy outcomes.

Adoption Barriers and Systemic Constraints

Microeconomic Facets: Though awareness and trust receive the bulk of the focus, the economic aspect of consumer behavior is as important. Research indicates that price elasticity of demand has a predominant impact on the adoption of efficient appliances, and people with lower incomes are more price-sensitive in terms of initial buying price compared to future savings (Guo et al., 2021). Willingness-to-pay (WTP) studies also indicate that consumer segments vary in their valuation of efficiency attributes, implying labeling must be backed by targeted subsidies or financing mechanisms to bridge affordability gaps.

Interaction with Fiscal Incentives: Energy labeling also closely interacts with fiscal instruments like subsidies, tax credits, or rebates. Labels enable governments to determine which products are eligible for financial incentives, and incentives complement the behavioral impact of labeling by compensating for higher initial costs. Yet ill-designed subsidy schemes tend to distort consumers' decisions or induce short-term conformity over long-term efficiency improvements. An optimal mix that combines labeling with fiscal incentives can achieve maximum policy impact.

Challenges for SMEs: Small and medium-scale enterprises (SMEs), especially in developing economies, usually have out-of-proportion difficulties in accessing labeling and certification. Cost of compliance, insufficient technical capacity, and inaccessibility to testing facilities may lock SMEs out of mainstream markets. Not only does this decrease competition, but it also limits consumer choice in local markets. Policy frameworks should therefore prioritize capacity building, financial support, and simplified procedures to ensure that SMEs can comply with energy labeling standards without excessive burden.

CASE STUDIES AND BEST PRACTICES

Case studies from real-world scenarios provide important information on how energy labeling programs are developed, executed, and developed further. Through the examination of both voluntary and mandatory schemes, this section points to the pragmatic solutions that have made labels more influential in inducing energy-saving decisions. The ensuing examples from the European Union and India demonstrate how regulatory innovation, public participation, and online integration can greatly improve the performance of labeling schemes.

Case Study 1: European Union – Label Redesign and Digital Transparency

The EU Energy Label, introduced in 1995, is one of the most well-known and successful voluntary energy labeling programs in the world. Its A–G scale, together with simple color coding, made it convenient for customers to recognize high-efficiency products. Nevertheless, as years passed, the improvement of technology resulted in a flooding of products within the A+, A++, and A+++ categories, which resulted in consumer confusion and de-escalation of the effectiveness of the label (European Commission, 2021).

In turn, the EU made a significant redesign in 2021, going back to a streamlined A–G scale and dropping the "plus" grades. The revised label features:

- A QR code connected to the EPREL database, enabling consumers to view detailed product data, including noise, size, and particular energy consumption.
- New testing procedures and revised performance thresholds to guarantee the accuracy and validity of label grades.

This redesign was based on consumer research and behavioral studies that revealed, clear and easy-to-understand images and digital tools enhance trustworthiness and ease of use (OECD, 2022).

Case Study 2: India – Public Engagement and Compliance Innovation

India's Bureau of Energy Efficiency (BEE) has had notable progress in public engagement as well as regulatory innovation via its Standards & Labeling Programme, introduced in 2006. The program commenced with voluntary labels on air conditioners and refrigerators and extended to mandatory application on major appliances by 2010 (BEE, 2022).

Main features of the BEE program include:

(i) Mass communication campaigns: BEE employs television, radio, print media, and social media in order to enhance label exposure and inform consumers about long-term cost savings from high-efficiency products.
(ii) theBEE mobile app: Offers a searchable database of star-rated appliances and enables consumers to compare levels of efficiency and cost.
(iii)Online compliance tools: The Star Labeling Portal allows manufacturers to register products, submit test results, and get approved online. The system facilitates greater openness, less paperwork, and faster compliance verification.

BEE's model illustrates how a developing nation can rise above affordability and awareness barriers through focused public outreach and regulatory creativity. IEA (2020) states that BEE-marked appliances have saved around 70 billion kWh between 2015-2020, in addition to improving the market share of 4- and 5-star products.

Whereas voluntary initiatives such as ENERGY STAR encourage early innovation and brand distinction, mandatory programs create wider, more inclusive effects, particularly in low-literacy or price-sensitive markets. Optimal outcomes generally result when programs marry legal requirements with effective public awareness, frequent updates, and available online resources.

The case studies of the EU and India demonstrate that well-designed labeling systems, when supported by digital tools and proactive communication, can substantially influence consumer behavior and market dynamics. The evolution of these programs reflects a shift toward consumer-centered, data-enabled, and policy-integrated approaches. Future best practices will likely continue blending behavioral insights, regulatory frameworks, and technology platforms to scale energy efficiency globally. Insights from these case studies highlight both achievements and limitations, which inform the need for an integrated conceptual framework and forward-looking policy recommendations.

FUTURE TRENDS AND POLICY RECOMMENDATIONS

As energy labeling systems become more advanced, their embedding within digital technology, behavioral systems, and policy environments will determine their future success. Breakthroughs in product labeling are reorienting focus away from static physical stickers and toward interactive, real-time, and personalized consumer experiences. These trends create compelling opportunities to improve user engagement and regulatory performance.

Smart Labels, Mobile Apps, and IoT Integration: Future energy labels will move beyond bar graphs and star ratings. Smart labels-equipped with QR codes, NFC chips, or IoT connectivity-can share real-time information on consumption, maintenance requirements, or price comparisons. For instance:

- In the EU, EPREL database enables users to scan a QR code and view detailed technical specifications (European Commission, 2021).
- In India, the BEE app already allows consumers to look and compare labeled products through smartphones (BEE, 2022).

With smart appliances becoming increasingly common, labels can be linked to domestic energy management systems, providing dynamic feedback and optimizing behavior.

Strengthening Public-Private Partnerships: Labeling programs cannot be made successful solely by government action. Manufacturer, retailer, and online platform partnerships are needed to scale up compliance and consumer education. Retailers are frequently the last store of consumer contact, but few are trained in how to effectively describe label features or benefits.

Policy Recommendations: Strengthening Stakeholder Engagement

To enhance the real-world effectiveness of energy labeling programs, policy must go beyond technical standards and actively engage key market actors-particularly retailers, manufacturers, and digital platforms. The following strategies are essential:

Incentivize Retailers to Market Efficient Products: Retailers are the final link between product and consumer. By offering financial incentives, training, or recognition programs, governments can encourage retailers to prioritize and actively promote energy-efficient products. Trained sales personnel can explain the long-term cost benefits and environmental impact of high-efficiency appliances, making labels more meaningful at the point of sale.

Promote Manufacturer Co-Branding Linked to Label Visibility: Policies should support co-branding initiatives, where energy-efficient product lines are jointly marketed by manufacturers and labeling authorities (e.g., BEE, ENERGY STAR). This strategy boosts label recognition and consumer trust, especially when paired with clear messaging around cost savings and environmental impact. Co-branding can also encourage manufacturers to improve energy performance to align with trusted labeling programs.

Partner with Online Platforms for Label Standardization: With the rapid growth of e-commerce, energy labels must be clearly displayed in digital product listings. Governments and regulatory bodies should collaborate with online retailers and marketplaces to ensure that:

- Label formats are standardized and visible.
- Product descriptions include label grades and energy usage.
- Filters allow consumers to search or sort by efficiency rating.

Such partnerships will extend the visibility and accessibility of labels beyond physical stores, reaching digital-first consumers.

Multi-Stakeholder Collaboration as a Bridge: Bringing together regulators, manufacturers, retailers, and digital platforms creates a multi-stakeholder ecosystem that can improve both the credibility and usability of energy labels. While regulations ensure compliance, collaboration ensures that labels are promoted, understood, and trusted by the public.

Proposed Conceptual Framework

This model depicts the interplay among technology adoption, renewable energy integration, and energy efficiency labeling in the formation of sustainable supply chains. Technologies like IoT, blockchain, and artificial intelligence, if coupled with renewable energy adoption, are sustainability enablers. Energy labels play the role of a mediator by converting technical content into consumer-oriented signals, thus guiding buying behavior and market demand. This, in turn, encourages producers and retailers to be greener and promotes government regulation. Ultimately, the overall effect is increased transparency, efficiency, and sustainability in supply chains.

Future Trends & Policy Recommendations

Circular Economy and Extended Producer Responsibility (EPR): Energy labeling is being put more and more into the larger sustainability context of the circular economy and extended producer responsibility (EPR). Through the identification of life-cycle energy performance of appliances, labels can be used to support policies requiring recycling, repairability, and material recovery. Combining labeling with circular economy policy means that consumer choices not only consider operating energy efficiency but also help minimize waste and resource use throughout the product life cycle.

Artificial Intelligence and Label Personalization: The future developments in artificial intelligence (AI) and machine learning (ML) can potentially make energy labeling more personalized. AI algorithms through digital platforms and smart meters can correlate product data with household consumption habits, income levels, or environmental priorities, thus providing customized recommendations. This personalization can increase consumer involvement, bypass information overload, and make purchasing habits closer to energy-saving objectives.

Geopolitical and Trade Harmonization Challenges: Notwithstanding the international spread of energy labeling, there are still major obstacles to harmonizing national standards. Geopolitical tensions, differing regulatory agendas, and protectionist trade practices have hampered efforts to harmonize. For instance, having three competing star-rating, A–G, and numerical kWh formats makes cross-border trade in appliances more difficult and increases manufacturers' compliance costs. Converging to internationally harmonized labeling schemes would lower transaction costs, facilitate the role of SMEs, and enhance the comparability of energy efficiency information for worldwide consumers.

Empowering Consumers with Interactive and Digital Tools

Modern consumers are increasingly reliant on digital platforms for product research and decision-making. To maintain the relevance and effectiveness of energy labels in this environment, policymakers and program designers must adopt interactive, digital-first approaches that actively engage users and simplify energy efficiency information.

Interactive learning tools-such as energy savings calculators, gamified apps, and virtual product demonstrations-translate complex label data into tangible benefits. These tools help consumers understand how energy ratings relate to long-term cost savings and environmental impact, enhancing both awareness and motivation (OECD, 2022).

Key innovations include:

- Augmented Reality (AR) Applications: AR tools can provide immersive experiences, allowing consumers to visualize the lifetime energy costs of appliances in their own homes. For example, scanning a product with a smartphone could display potential annual savings compared to less efficient models, reinforcing the value of higher-rated products.
- Digital Nudging on E-commerce Platforms: E-commerce websites can embed digital nudges-subtle prompts that encourage energy-conscious choices. These may include:
- Highlighting the most efficient product in a category.
- Recommending alternatives with lower energy use.
- Tagging energy-saving products during seasonal sales.

Such nudges reduce decision fatigue and guide consumers toward more sustainable options without limiting their freedom of choice.

Gamified and Personalized Content: Gamification strategies-like scoreboards for household energy use or interactive quizzes-can boost engagement, especially among younger audiences. Personalization features (e.g., suggesting appliances based on usage patterns) make label content more meaningful to individual consumers.

Together, these digital approaches not only increase label visibility but also empower consumers to take control of their energy choices. In a fast-evolving, tech-driven marketplace, such innovations are essential for ensuring that energy labels remain relevant, actionable, and widely adopted. Taken together, these future-oriented perspectives underline the transformative potential of energy labels when combined with technology, renewable energy, and sustainable supply chain practices, leading to the concluding reflections.

CONCLUSION

Energy efficiency labeling is a key nexus between consumer action, technological innovation, and green supply chain management. By making what is invisible energy performance visible, labels give consumers power to make conscious choices, encourage manufacturers' innovation, and assist governments in promoting climate and energy policy objectives.

The chapter has demonstrated that labeling is not merely conducive to SDG 12 (Responsible Consumption and Production) but also drives directly towards SDG 7 (Affordable and Clean Energy) by facilitating access to efficient technology and decreasing energy costs for households. Enhancing these synergies guarantees that labeling policies deal with both sustainability and equity aspects.

The analysis also highlights that labels cannot act independently. Their use is mediated by demographic and cultural drivers, fiscal benefits, technology developments, and supply-side preparedness. Fundamental challenges-such as consumer fatigue, rebound effects, SME compliance obstacles, and cross-border harmonization problems-require joined-up solutions. Emerging technologies like digital platforms, AI-powered personalization, and circular economy integration emphasize the changing function of labels in creating sustainable futures.

In general, energy labeling strength is its interdisciplinary reach: it brings behavioral science, policy making, and technological innovation together with managerial relevance. Strategically integrated in renewable energy transitions and supply chain structures, energy labeling is more than a policy tool; it is an accelerator of transformation toward a low-carbon, equitable, and resilient economy.

REFERENCES

Ajzen, I. (1991). The theory of planned behavior. *Organizational Behavior and Human Decision Processes, 50*(2), 179–211. DOI: 10.1016/0749-5978(91)90020-T

Bureau of Energy Efficiency. (2022). *Annual report 2021–2022*. https://beeindia.gov.in/

Bureau of Energy Efficiency. (2022). *Energy efficiency labelling programmes*. https://beeindia.gov.in/

CLASP. (2019). *Opportunities for international collaboration on energy efficiency labels*. https://www.clasp.ngo/

Dötsch, J., & Schmid, N. (2021). The effectiveness of energy efficiency labels in influencing consumer choice: A meta-analysis. *Journal of Cleaner Production, 279*, 123456.

Environmental Protection Agency. (2020). *About ENERGY STAR*. https://www.energystar.gov/

Environmental Protection Agency. (2020). *ENERGY STAR overview*. https://www.energystar.gov/

EPA. (2020). *ENERGY STAR Overview*. https://www.energystar.gov/

European Commission. (2021). *Consumer research on the effectiveness of the EU energy label*. https://ec.europa.eu/

European Commission. (2021). *EU Energy Label*. https://ec.europa.eu/energy/topics/energy-efficiency/energy-labelling_en

European Commission. (2021). *The new energy label explained*. https://ec.europa.eu/info/energy-climate-change-environment/standards-tools-and-labels

Guo, S., Li, M., & Chen, X. (2021). Influence mechanism of energy efficiency label on consumers' purchasing behavior of energy-saving household appliances. *Frontiers in Psychology, 12*, 711854. DOI: 10.3389/fpsyg.2021.711854 PMID: 34733202

International Energy Agency. (2020).. . *Energy Efficiency, 2020,* •••. https://www.iea.org/reports/energy-efficiency-2020

International Energy Agency. (2022). *Solar PV global supply chains*. International Energy Agency. https://www.iea.org/reports/solar-pv-global-supply-chains

Mkhize, S., & Thopil, S. (2021). Evaluating the effectiveness of appliance energy labeling in South Africa: Challenges and outcomes. *Energy Research & Social Science*, *75*, 102052. DOI: 10.1016/j.erss.2021.102052

OECD. (2022). *Consumer behavior and energy efficiency: Insights from behavioral economics*. https://www.oecd.org/environment/

Organisation for Economic Co-operation and Development. (2022). *Behavioural insights for energy labelling*. https://www.oecd.org/environment/

Senyapar, H. N. D. (2024). Energy labels and consumer attitudes: A study among Gazi University staff. *Sustainability (Basel)*, *16*(5), 1754. DOI: 10.3390/su16051754

Stern, P. C. (2000). Toward a coherent theory of environmentally significant behavior. *The Journal of Social Issues*, *56*(3), 407–424. DOI: 10.1111/0022-4537.00175

Thaler, R. H., & Sunstein, C. R. (2008). *Nudge: Improving decisions about health, wealth, and happiness*. Yale University Press.

Thøgersen, J., & Schrader, U. (2012). From knowledge to action—New paths towards sustainable consumption. *Journal of Cleaner Production*, *29–30*, 1–3. DOI: 10.1016/j.jclepro.2012.02.023

United Nations. (2015). *Transforming our world: The 2030 agenda for sustainable development*. https://sdgs.un.org/goals

United Nations Environment Programme. (2020). *Monitoring, verification, and enforcement for appliance energy efficiency*. https://www.unep.org/

Waechter, S., Klingenberg, H., & Meyer, R. (2022). Cross-national evidence on energy label effectiveness: A meta-analytic review. *Energy Policy*, *164*, 112891. DOI: 10.1016/j.enpol.2022.112891

Zhou, Y., & Zhang, L. (2023). Global consumer response to energy efficiency labels: Meta-analysis and policy implications. *Journal of Cleaner Production*, *389*, 136262. DOI: 10.1016/j.jclepro.2023.136262

Chapter 3
Fueling the Future With SAF:
How Airlines Are Reshaping Value Chains for a Decarbonized Aviation

Gezia Damergy
https://orcid.org/0009-0001-2522-7135
Université de Paris 1 - Panthéon Sorbonne, France

ABSTRACT

The race to decarbonize aviation is no longer a distant ambition, it is an immediate necessity. Among the limited options available, Sustainable Aviation Fuel (SAF) has emerged as the only scalable lever capable of reshaping the industry's carbon trajectory. Yet SAF is more than a new fuel: it overturns decades of reliance on a uniform, fossil based Jet A 1 system, replacing it with costly, scarce, and traceable alternatives. This shift forces airlines to abandon their traditional role as passive fuel buyers and instead act as strategic players—securing supply through long term commitments, partnerships, and vertical integration to safeguard their future in a low carbon world.

DOI: 10.4018/979-8-3373-3790-6.ch003

I) DECARBONIZING AIR TRANSPORT AND THE ROLE OF SUSTAINABLE AVIATION FUELS

I.a) The Decarbonization of Air Transport: an Urgent Priority

Although it is not the largest emitter in absolute terms, air transport remains one of the most greenhouse gas-intensive modes of travel per passenger-kilometer. An average flight generates between 80 and over 120 grams of CO_2 per passenger-kilometer, compared to around 10 to 30 grams for electric rail or 30 to 50 grams for intercity buses (International Council on Clean Transportation (ICCT), 2022; Lee et al., 2021). According to the International Energy Agency (IEA, 2020), aviation accounted for approximately 2.5% of global direct CO_2 emissions and around 13% of transport-related emissions (Environmental Protection Agency, 2022 ; International Transport Forum, 2019). Already concerning, these figures actually underestimate the aviation sector's true climate impact. The additional radiative forcing caused by contrails (Ansel, 2023), induced cirrus clouds, nitrogen oxide (NOx) emissions, and fine particulate matter amplifies the warming effect by a factor of two to three compared to CO_2 emissions alone (Nakano et al., 2022). When these non-CO_2 effects are taken into account, aviation may contribute up to 5% of total anthropogenic climate impact (Lee et al., 2021). This situation is all the more critical given that the sector remains heavily reliant on liquid fossil fuels, with average consumption estimated at nearly 350 million tonnes of oil equivalent (Mtoe) per year before the COVID-19 crisis (IEA, 2020), and around 320 Mtoe in 2023 (International Air Transport Association (IATA), 2024).

Figure 1. The increasing contribution of air transport to carbon emissions

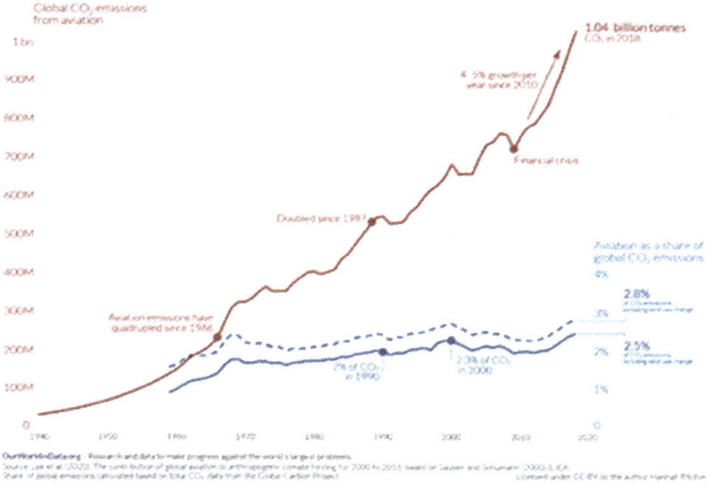

Source: *Lee et al. (2020)*

Unlike other modes of transport, aviation has fewer viable technological alternatives to combustion engines, particularly for long-haul flights. This structural dependence on fossil fuels makes emissions reduction especially difficult. While significant progress has been made over the years in improving engine efficiency and aircraft aerodynamics, the resulting gains in energy performance per passenger-kilometer have not been enough to counterbalance the surge in global air traffic. Between 1980 and the onset of the covid-19 crisis in 2019, air traffic grew from under 2,000 billion passenger-kilometers to over 8,000 billion (International Civil Aviation Organization (ICAO, 2023 ; IATA, 2023). These growth projections are a further cause for concern. IATA estimate that the number of annual air passengers could exceed 10 billion by 2050, compared to around 4.9 billion in 2024 (IATA, 2025). This projected growth, driven by globalization and rising incomes in emerging economies, risks negating or even surpassing the efficiency gains made through technological or operational innovation (IEA, 2023), and would contribute to an even heavier carbon footprint for the sector (IATA, 2023; ICAO, 2023). This concern is widely echoed in academic literature, which highlights the existence of a rebound effect: improvements in energy efficiency tend to reduce emissions per passenger-kilometer but are often offset by increases in overall demand (Peeters et al., 2016; Gössling & Humpe, 2020).

The Intergovernmental Panel on Climate Change (IPCC), in its special report on global warming, emphasizes the urgency of reducing global greenhouse gas emissions by 84% by 2050 compared to 2019 levels. The International Council on

Clean Transportation (ICCT) has underscored the need to reach net-zero emissions by mid-century in order to align with the Paris Agreement goal of limiting global warming to 1.5°C above pre-industrial levels (ICCT, 2022). Achieving this trajectory will require major, rapid transformations—even in sectors traditionally considered hard to decarbonize, such as aviation (IPCC, 2022). The challenge is so significant that many institutions now call for the full decarbonization of aviation by 2050. Without it, the sector's absolute CO_2 emissions could continue to rise, making it incompatible with international climate goals.

I.b) Measures to Decarbonize Air Transport

The alarm has been sounded, urging all stakeholders to take decisive action toward aviation decarbonization. Achieving carbon neutrality by 2050 will require mobilizing a range of levers, many of which are already partially in place but still need to be significantly refined and scaled up.

1. Public policy instruments and market-based mechanisms -such as emissions trading systems and carbon offset and reduction programs like CORSIA (Carbon Offsetting and Reduction Scheme for International Aviation)- aim to limit net emissions growth by enabling the purchase of carbon credits or the funding of low-carbon initiatives (ICAO, 2022). However, these mechanisms remain a work in progress and have come under criticism for lacking sufficient ambition.
2. Technological improvements in aircraft and engine performance are another critical focus (Air Transport Action Group, 2022). This includes using lighter materials and developing innovations that lower fuel consumption per passenger-kilometer. In parallel, enhanced air traffic management can reduce fuel use through modernized control systems and optimized flight paths (Eurocontrol, 2021). Operational practices such as single-engine taxiing or minimizing in-flight and ground delays are also being promoted (Air Transport Action Group, 2022). Moreover, increasing reliance on artificial intelligence and digital systems for trajectory planning and delay minimization offers additional efficiency gains (McKinsey & Company, 2020).
3. Fleet renewal—with the introduction of next-generation[1], more fuel-efficient aircraft—is a vital strategy, though its impact is constrained by the long service life of commercial airliners (ICCT, 2022). It is widely recognized that current technological and operational improvements alone will not suffice to achieve net-zero emissions by 2050. In the longer term, breakthrough technologies could enable more substantial progress. Electric propulsion, and especially hydrogen-powered flight, hold considerable promise. Electric aircraft are particularly well suited to regional routes; prototypes are already flying, though current models

remain limited in both range and capacity (Bullerdiek, Neuling & Kaltschmitt, 2021). Despite their potential, these technologies still face major hurdles in storage, airport infrastructure, aircraft design, and the large-scale production of low-carbon hydrogen (McKinsey & Company, 2020; IEA, 2023).

4. According to the Clean Skies for Tomorrow coalition, led by the World Economic Forum, the most immediate and impactful lever for achieving carbon-neutral aviation is to rapidly increase investment in and production of sustainable aviation fuels (SAF), which would gradually replace fossil-based jet fuel. With a sufficient supply of SAF, global airlines could become nearly carbon-neutral almost immediately, as their energy demands would no longer require new fossil fuel extraction. Sustainable aviation fuels (SAF) can reduce CO_2 emissions by up to 80% compared to conventional kerosene[2]. Classified as drop-in fuels, they require no modifications to existing aircraft engines or airport infrastructure—a decisive advantage given that most airports manage fuel through centralized systems where incoming batches are blended before distribution. Without this compatibility, airports would need to invest in separate storage and distribution facilities, a costly undertaking for which most lack both space and funding. To ensure uninterrupted operations, SAF must comply fully with existing certifications, covering aircraft systems such as engines, fuel tanks, pipelines, and monitoring equipment. This guarantees that airlines can continue using the same aircraft, with the same capacity, routes, and cruising speeds, while maintaining punctual schedules even at congested airports—without delays or payload restrictions linked to the fuel type.

The figure below illustrates the mid-range scenario studied by ATAG. It highlights the four main pillars of the decarbonization strategy: carbon offsetting, technological development, improvements in operations and infrastructure, and SAF. Among these, SAF have now become the central focus of the industry's efforts, as they represent the most practical pathway currently available to achieving net-zero carbon emissions by 2050.

Figure 2. Contribution of technology, operations and infrastructure, sustainable aviation fuels, and market-based measures to achieving net-zero carbon emissions in air transport by 2050 (mid-range scenario).

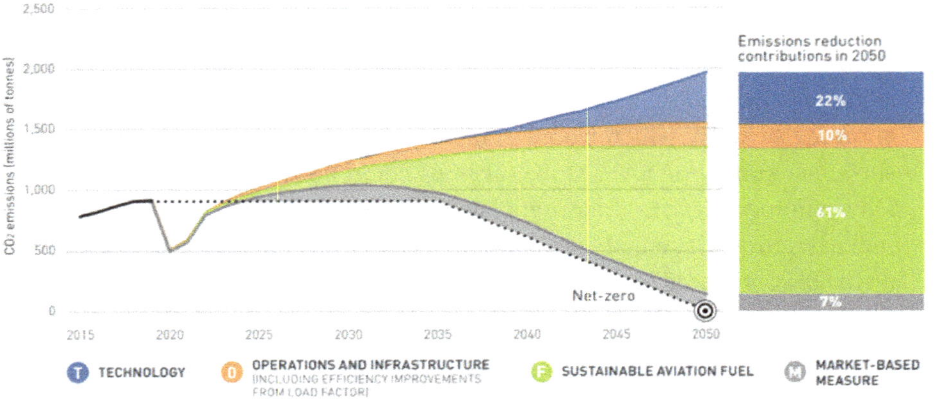

Source: *Air Transport Action Group, Waypoint 2050, 2nd edition, September 2021.*

I.c) Sustainable Aviation Fuels

Air transport requires a highly specific fuel (called Jet A-1fuel) that meets strict criteria: (1) high energy density to keep fuel weight within reasonable limits, (2) a freezing point of −47°C for flight at altitudes up to 40,000 feet, (3) a high flash point combined with effective vaporization in the combustion chamber, and (4) worldwide availability. For any substitute to be viable, it must closely match the physical and chemical properties of fossil-based Jet A-1, enabling seamless integration into existing supply chains. This includes meeting the same exacting standards for storage, handling, transport, and aircraft operations, without compromising safety or performance. SAF fulfills these conditions and it is a certified liquid fuel under ASTM D7566, a standard regularly updated to incorporate newly validated production methods and ensure consistent global quality. Together with the ICAO CORSIA framework and the ReFuelEU Aviation regulation, ASTM certification confirms that SAF -whether bio-based or synthetic- can be safely used in current aircraft and airport systems without modifications. Until recently, aircraft relied exclusively on kerosene derived from crude oil. Today, however, a growing number of flights[3] are using sustainable aviation fuels typically blended with conventional jet fuel as a low-carbon alternative.

Although often referred to in the singular, sustainable aviation fuel actually encompasses a variety of categories, distinguished by their feedstocks, production methods, technological maturity, scalability, and cost. Despite these differences, they share a common objective: to significantly reduce aviation's greenhouse gas emissions by progressively replacing fossil-based kerosene. Biofuels are derived from biological materials of plant or animal origin. As of 2024, they represent 90.1% of global SAF production (Global Market Insights, 2024). Biofuels are categorized into three generations, based on the type of feedstocks used and the maturity of the production technologies. First-generation biofuels are produced from food crops such as rapeseed, soybean, corn, and sugarcane (IEA, 2023 ; Padder et al., 2024). These crops are processed into liquid fuels using established industrial methods, particularly the hydroprocessing of vegetable oils -a certified and widely adopted technique for producing aviation fuel (ASTM International, Standard D7566, latest revision 2023). Second-generation biofuels use feedstocks that do not compete with food production, such as organic waste, used cooking oil, inedible animal fats, and agricultural or forestry residues (IFPEN, 2020; IEA, 2023). These fuels can reduce CO_2 emissions by approximately 80% or more over their full lifecycle[4]. Third-generation biofuels are based on the cultivation of algae, which have the advantage of not requiring arable land and can grow in water unsuitable for human consumption (IFPEN, 2020).

The second main category of sustainable aviation fuels consists of synthetic fuels, also known as e-fuels or Power-to-Liquid (PtL) fuels. These are liquid fuels produced without biomass, using green hydrogen and captured carbon dioxide. Hydrogen is generated through electrolysis, a process that splits water molecules into hydrogen and oxygen using electricity from renewable sources such as solar, wind, or hydro. When powered by carbon-free electricity, this method generates no direct CO_2 emissions. The carbon dioxide can be captured either from industrial facilities or directly from the atmosphere using direct air capture (DAC) technologies. Once both components are available, they are combined in a chemical reaction—most commonly the Fischer-Tropsch synthesis. This process converts a mixture of carbon monoxide (CO) and hydrogen (H_2), known as syngas, into liquid hydrocarbons using metal catalysts (typically cobalt or iron) under high temperatures and pressures. The resulting fuel is then refined to meet aviation kerosene specifications, making it chemically identical—or nearly so—to fossil-based Jet A-1. As a result, it can be used in existing aircraft without any technical modifications, qualifying as a drop-in fuel. This type of fuel offers several advantages: it does not depend on agricultural or forest-based resources, is fully compatible with current engines and infrastructure, and has the potential to be carbon-neutral—provided the hydrogen is produced using renewable electricity and the CO_2 is captured and stored sustainably (Nemmour et al., 2023), However, production remains extremely costly and energy-intensive due to low overall process efficiency and the early stage of infrastructure development[5].

Conclusion – Section I

While sustainable aviation fuels (SAF) have emerged as the most promising pathway to decarbonize air transport, their effective deployment remains constrained by economic, technical, and structural barriers. The diversity of feedstocks and production technologies reveals both the dynamism and the complexity of this emerging sector. Yet, despite their potential, current production volumes and cost structures fall far short of what is needed to achieve large-scale substitution. Understanding these limitations—rooted in feedstock scarcity, high production costs, and slow industrial scaling—is therefore essential. The following section examines these challenges in detail and explores how they shape the future trajectory of SAF development and deployment.

II) SAF: SCARCITY, COST BARRIERS, AND THE CHALLENGE OF MARKET SCALING

While it is widely recognized that the aviation sector must decarbonize urgently and that large-scale adoption of sustainable aviation fuels is the most viable path forward, current production capacity remains far from sufficient to meet today's global kerosene demand—let alone the needs of the future. « SAF production encounters numerous difficulties, including the limited availability of raw materials such as waste oils and fats, and competition from other sectors for the same resources. Furthermore, the development of new production technologies, such as power-to-liquid (PtL), requires significant investment and time to reach commercial scale. Additionally, airport infrastructure must be adapted to handle and distribute SAFs, which entails further costs and logistical challenges » (Klimczyk et al., 2025). Placing such a central responsibility on SAF comes with significant challenges. The expectation that they will carry much of the burden of achieving net-zero emissions by 2050 is daunting, given that the industry is still far from ready to deliver on this scale. At present, SAF account for only about 0.1% of total aviation fuel consumption, and the sector is struggling to scale up. The challenges are significant. Investments are costly, and uncertainty about the long-term stability of demand remains high. Airlines, for which fuel represents nearly 30% of operating costs, are still reluctant to commit to SAF purchases as long as prices remain four to eight times higher than fossil kerosene. Investors also face difficulties projecting into this market, as the different types of SAF vary considerably in terms of feedstocks and technological maturity. Some rely on abundant resources whose large-scale use raises sustainability concerns, while others depend on scarcer or more technically demanding inputs. Likewise, production technologies are at different stages of industrial development:

a few are already proven and approaching large-scale deployment, whereas others remain at the experimental stage. These disparities directly affect production costs and fuel much of the uncertainty surrounding SAF.

Understanding these specificities and the obstacles to SAF development is an essential prerequisite to analyzing recent commitments by airlines in their supply chains (see section III) as part of the sector's decarbonization effort. Yet the objective remains daunting. Current production volumes and supply chains fall far short of projected demand, and bridging this gap will require an unprecedented acceleration in investment, innovation, and policy support. Without a substantial shift in both industry practices and government frameworks, achieving net-zero emissions by 2050 will be extremely difficult.

II.a) SAF Challenges: Technological Diversity, Resource Scarcity and High Costs,

First-generation biofuels are currently the most technologically mature and industrially developed. They remain the only SAF available at commercial scale, representing around two-thirds of global biofuel output in 2020 (IFPEN, 2020). However, they face strong criticism: by using arable land, they may compete with food crops, drive deforestation, and contribute to higher food prices. For these reasons, their development is strictly regulated within the European Union. Production costs remain high—two to four times that of fossil-based kerosene. Beyond these economic and social constraints, an additional methodological challenge concerns how their climate impact is assessed. Traditional life-cycle analyses (LCA) often account only for "core" emissions linked to cultivation, conversion, and combustion, but tend to overlook indirect effects such as induced land use change (ILUC). As highlighted by Zhao et al. (2021), excluding ILUC from these assessments can substantially distort the actual mitigation potential of SAF pathways. When such indirect land use effects are included, results vary widely—from significant net benefits to even positive carbon balances, depending on the feedstock and region. In other words, some feedstocks such as miscanthus can generate a net carbon benefit—because they store more CO_2 in soils than they emit over their life cycle—whereas others like palm oil can lead to additional emissions due to deforestation and peatland oxidation.

Second-generation biofuels are technically mature and certified for aviation use. However, their industrial scale-up remains constrained by the limited and dispersed availability of suitable feedstocks. This scarcity is a key factor behind their high production costs—typically three to five times that of fossil kerosene. Although these advanced pathways alleviate pressure on arable land, uncertainties persist regarding the indirect effects of land conversion and the overall carbon payback time, both of which are still poorly represented in conventional LCA models.

Third-generation biofuels, derived from algae cultivation, are of particular interest due to their high yield potential and lack of competition with agricultural land. However, they remain at an early stage of development. Significant technical and economic challenges persist in cultivating, harvesting, and processing algae at competitive cost levels. At present, there is no meaningful industrial production—only pilot projects are operational—and current costs can reach up to ten times those of fossil kerosene. Nevertheless, from a carbon accounting perspective, these algal pathways may offer a more resilient long-term solution, as their emissions are less sensitive to ILUC-related uncertainties that continue to undermine the credibility of first- and second-generation biofuel assessments.

Synthetic fuels—or electrofuels—are also largely in the R&D phase. While some process steps, such as water electrolysis, are already industrially viable, others are still emerging, particularly direct air capture (DAC) of atmospheric CO_2 and the final fuel synthesis (Nemmour et al., 2023). Currently, production is limited to demonstration and pilot-scale facilities. Costs remain prohibitive, at five to ten times that of fossil kerosene (Air Transport Action Group, 2021; European Commission, 2022). However, industry projections suggest that more significant commercial-scale production could emerge after 2030, driven by falling costs for renewable electricity and electrolyzers, along with advances in CO_2 capture technologies (World Economic Forum, 2023).

There are also hybrid approaches, known as electro-biofuels, which combine biomass with renewable electricity to improve energy efficiency. These pathways are still at the experimental stage (IFPEN, 2020). Today, synthetic fuels cost roughly ten times more than fossil kerosene and are expected to remain two to three times more expensive over the long term.

Figure 3. Estimated prices and production costs in 2023 for ReFuelEU Aviation eligible

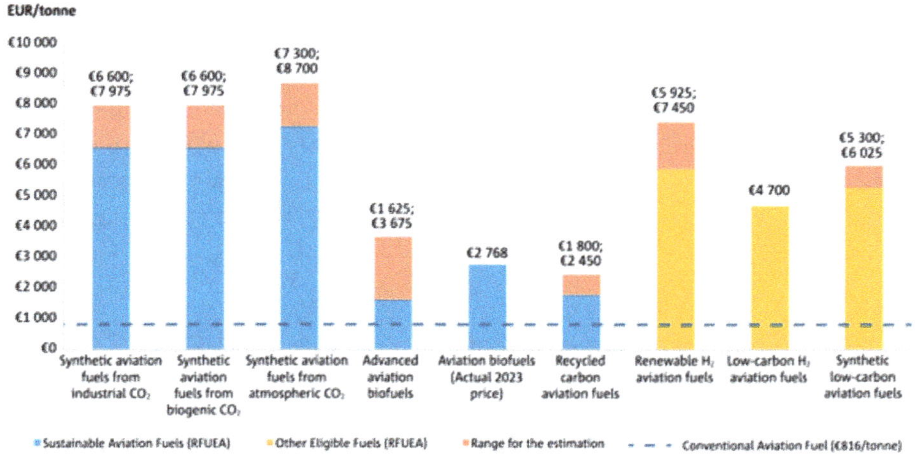

Source: *EASA: European aviation environmental report, 2025.*

The graph above compares the estimated production costs (in €/tonne) of various sustainable aviation fuels (SAF) and other eligible low-carbon fuels, relative to conventional fossil-based kerosene (represented by the dashed line at €816/tonne). The analysis clearly highlights the significant cost gap between traditional jet fuel and its sustainable alternatives. Synthetic fuels produced from captured CO_2—whether from industrial sources (€6,600–€7,975/tonne), biogenic sources (€6,600-€7,975/tonne), or atmospheric CO_2 (€7,300–€8,700/tonne) are among the most expensive options, costing up to ten times more than fossil kerosene. Advanced biofuels are comparatively more affordable, with estimated costs ranging from €1,625 to €3,675/tonne, while current-generation aviation biofuels (as of 2023) average around €2,768/tonne. Recycled carbon aviation fuels, combining waste feedstocks and established processing methods, show a slightly lower cost range (€1,800–€2,450/tonne) but still remain well above conventional levels. Hydrogen-based options also display high cost variability: renewable hydrogen aviation fuels range from €5,925 to €7,450/tonne, while synthetic low-carbon hydrogen fuels fall between €5,300 and €6,025/tonne. Even low-carbon hydrogen fuels—produced using less carbon-intensive methods—are estimated at around €4,700/tonne[6].

These figures underscore the urgency of targeted investment, policy support, and industrial scaling to bridge the cost gap and ensure SAF becomes a viable and competitive solution for decarbonizing aviation.

II.b) Overcoming Structural Barriers to SAF Production and Deployment

The large-scale production of sustainable aviation fuel (SAF) faces several obstacles: high production costs, the difficulty for new entrants to penetrate a market dominated by major oil companies, a lack of financial incentives, and insufficient targeted support. As a result, there is growing uncertainty over whether global SAF output can keep pace with current and future demand. Enormous expectations rest on SAF to help decarbonize the aviation sector by 2050, yet the production ecosystem remains nascent.

Table 1. Evolution of SAF production versus fossil fuel consumption (in millions of gallons)

Millions of gallons

Year	SAF produced	Jet fuel consumed by major U.S. airlines
2016	1.9	17,138
2017	1.7	17,662
2018	1.8	18,325
2019	2.4	18,746
2020	4.6	11,067
2021	5.1	14,617
2022	15.8	17,510

Source: GAO, 2023

This table underscores the enormous gap between the production of sustainable aviation fuel (SAF) and the total jet fuel demand of major U.S. airlines. In 2022, only 15.8 million gallons of SAF were produced, compared to 17.5 billion gallons of jet fuel consumed—a share of less than 0.1%. This disproportion, which has persisted for several years, highlights the urgent need to expand SAF production capacity and distribution infrastructure if the aviation sector is to meet its climate goals. This mismatch is not unique to the United States; it reflects a broader global trend. While SAF production has increased by 1,150% worldwide over the past three years, this growth has occurred from a very low starting point. In absolute terms, the situation remains highly constrained: by 2024, SAF still represented just 0.3% of total global jet fuel consumption (IATA, 2025). Althvalleyough recent developments in 2024 and 2025 suggest a renewed momentum, current projections indicate that the International Energy Agency's global production targets for 2030 are unlikely to be met unless key market dynamics -particularly regarding investment, demand certainty, and production capacity- shift more decisively. « The infrastructure and supply chains for efficient large-scale production of SAF are still in their infancy » (Braun et al., 2024).

Against this backdrop of limited progress in scaling up production, policymakers have become increasingly critical, with regulatory frameworks and economic incentives now standing out as essential tools to foster SAF deployment. Governments -though often criticized for a lack of ambition- are increasingly combining regulatory instruments with economic incentives. Since 2012, countries like Norway and Sweden have introduced binding or incentive-based policies, such as blending mandates and subsidies to encourage SAF use in aviation. These measures aim to shift both producers and airlines away from fossil-based kerosene (JET A-1) toward SAF, rather than continuing to rely on carbon offsetting, lowest-cost alternative. In the U.S., the Inflation Reduction Act[7] provides subsidies and tax credits designed to reduce SAF's cost disadvantage (U.S. Department of Energy, 2022, U.S. Department of Energy, 2025).

Table 2. SAF Blending Mandate Timeline – ReFuelEU Aviation Regulation

Year	Mandatory SAF Share	Minimum Share of Synthetic Fuels
2025	2 % SAF	–
2030	6 % SAF	1,2 % e-fuels
2035	20 % SAF	5 % e-fuels
2040	34 % SAF	10 % e-fuels
2045	42 % SAF	15 % e-fuels
2050	70 % SAF	35 % e-fuels

Source: Official Journal of the European Union, Regulation (EU) 2023

In the EU, the ReFuelEU Aviation Regulation (European Commission, 2023) mandates a gradual increase in SAF blending. Under this framework, fuel suppliers operating at EU airports must ensure that a minimum percentage of their total aviation fuel volume—blended to meet technical standards—consists of SAF. These minimum shares must reach 2% by the end of the first compliance year, and increase progressively to 6%, 20%, 34%, 42%, and ultimately 70% (see Table N°2). These quotas apply as average annual targets across all airports served within the EU. The chart below illustrates the projected shift from fossil jet fuel to sustainable aviation fuels (SAF) by 2050. While fossil fuel remains dominant in 2030 at 95%, its share decreases significantly to 68% in 2040 and 37% in 2050. At the same time, various SAF pathways—particularly Power-to-Liquid (PtL), Alcohol-to-Jet (AtJ), and Gasification—gain importance, with PtL accounting for the largest share of SAF by mid-century.

Figure 4. Predicted share of EU aviation fuel from different technology pathways

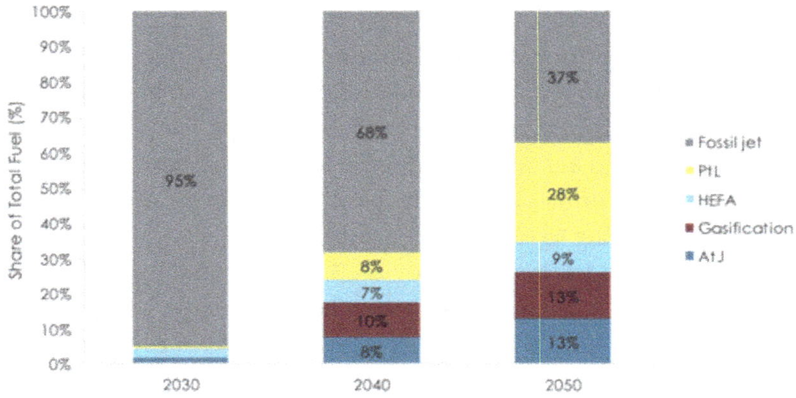

Source: *Intergovernmental Panel on Climate Change, 2022.*

While policy momentum has been strongest in the United States and the European Union, other regions have also begun to engage in SAF deployment—albeit through diverse policy instruments and at different levels of ambition. A growing number of governments in Latin America, Africa, and Asia have recently adopted measures to stimulate SAF production and consumption, either through binding mandates or non-binding, incentive-based frameworks. In Latin America, Brazil stands out as a first mover: the government has developed a national SAF program that transitions from voluntary to mandatory blending phases in the late 2020s, supported by the state-owned oil company Petrobras and the national development bank. This framework aims to create a domestic market for bio-based jet fuel while leveraging the country's agricultural expertise. South Korea and Thailand in Asia have announced compulsory blending mandates starting around 2026–2027, with minimum shares set at 1% for international flights departing from national airports. These policies mark a significant shift toward binding frameworks in regions previously focused on research and demonstration.

By contrast, several countries have chosen non-binding or incentive-led approaches. In Colombia, public investment by Ecopetrol in commercial-scale SAF facilities reflects a strong industrial policy but without a formal blending requirement. In the United Arab Emirates, the government has set national production targets and provided strategic funding for SAF plants as part of its broader decarbonization agenda, yet adoption remains voluntary for airlines. Australia has followed a similar path, launching federal and state-level roadmaps, funding programs, and public–private partnerships to accelerate SAF production, but without enacting a binding national quota. In India, feasibility studies and pilot projects are supported by government incentives and targets, although no legal blending obligation has yet been implemented.

These distinctions are crucial for understanding the global policy landscape. Mandatory policies, such as blending mandates (for example, requiring at least 1% SAF in jet fuel from a specific year), directly reshape market demand. They compel airlines and fuel suppliers to purchase SAF, creating an immediate and structural demand signal. This can generate broader economic ripple effects, including upward pressure on agricultural commodity prices, increased competition for land, and the mobilization of specific agricultural or forestry sectors for energy purposes. In contrast, incentive-based policies primarily act on the supply side. They support the construction of production plants, research and development, domestic feedstock supply, and industrial partnerships, but without imposing immediate consumption obligations. As a result, these measures mainly influence production capacity and investment risk, rather than directly affecting land use or agricultural resource allocation.

While the U.S. and EU continue to lead, emerging economies are experimenting with diverse SAF strategies. Yet, despite these efforts, adoption remains uneven, and the pace of change slow. This raises broader questions: why do some initiatives succeed while others lag, and how do technological, industrial, and institutional factors interact to shape the transition? Answering these questions requires not only examining policies, but also considering insights from innovation management, institutional economics, and theories of technological change—approaches that can shed light on the governance of SAF and the transformation of aviation fuel value chains. Such a perspective will be developed in the following section.

II.c) Interpreting the SAF Transition Through an Evolutionary and Socio-Technical Lens

The emergence of Sustainable Aviation Fuels (SAF) can be interpreted through the analytical framework of socio-technical transitions developed by Geels (2002). In this perspective, technological change is not only a matter of technical substitution, but the outcome of interactions between multiple levels: niche innovations, socio-technical regimes, and the broader socio-technical landscape. This multi-level perspective (MLP) helps explain why the SAF transition remains gradual, fragmented, and uneven, despite growing policy and corporate support.

At the landscape level, powerful exogenous forces such as global climate imperatives, the decarbonization of transport systems, and new policy frameworks like *ReFuelEU Aviation* and the *Inflation Reduction Act* are reshaping the aviation sector. These macro-level dynamics create strong external pressures on the incumbent fossil fuel regime and open new avenues for alternative technologies. Yet, as evolutionary economists remind us, such openings only lead to genuine transitions when actors at lower levels can seize them through coordinated innovation and institutional adaptation.

At the regime level, the aviation fuel system remains characterized by deep structural lock-in. Industrial actors, infrastructures, maintenance systems, and regulatory standards are tightly aligned with fossil kerosene. Radically new technologies therefore face major barriers to diffusion, as they must adapt to a network of practices and norms built around the incumbent system. This explains why SAF, though technically certified and politically supported, still represents less than one percent of global jet fuel consumption. The persistence of high costs, feedstock scarcity, and fragmented production capacity reflects this regime stability and the slow pace at which established systems reconfigure.

Within this context, SAF innovation still occupies what Geels (2002) calls *niche spaces*—protected arenas in which radical technologies can develop under specific market, regulatory, or policy shields. Demonstration plants, voluntary purchase agreements, and national blending mandates serve as these experimental niches, providing opportunities for learning, cost reduction, and the formation of specialized actor networks. Yet these niches remain fragile. Feedstock availability is limited, investment uncertain, and the coordination between supply and demand incomplete—features typical of what Mokyr (1990) described as "hopeful monstrosities," early-stage innovations that are cumbersome, costly, and not yet competitive with incumbent systems. One of the central mechanisms in such transitions is technological hybridization—the phase in which new and old systems coexist in partial symbiosis.The SAF sector exemplifies this mechanism. Rather than replacing fossil kerosene outright, SAF is blended with it, enabling a progressive integration within the existing fuel infrastructure. This hybridization reduces the risks of disruption and allows for cumulative learning by both industry and regulators. Over the next three decades, such coexistence between fossil and sustainable fuels is expected to persist, gradually shifting the composition of the global fuel mix rather than triggering a sudden rupture. This process aligns with the evolutionary view that transitions emerge not from revolution, but from successive adaptations and gradual reconfigurations (Van den Ende and Kemp, 1999).

As Kemp et al. (2001, p. 277-278) emphasize, "it is the alignment of developments—successful processes within the niche reinforced by changes at regime level and at the level of the socio-technical landscape—which determine if a regime shift will occur." The SAF transition illustrates this interdependence clearly. Progress at the niche level must coincide with regime-level adjustments and sustained landscape pressures (see Figure below). Only their convergence can unlock a structural shift from the fossil fuel regime to a low-carbon aviation fuel regime.

Figure 5. Constructing transition paths from fossile to SAF

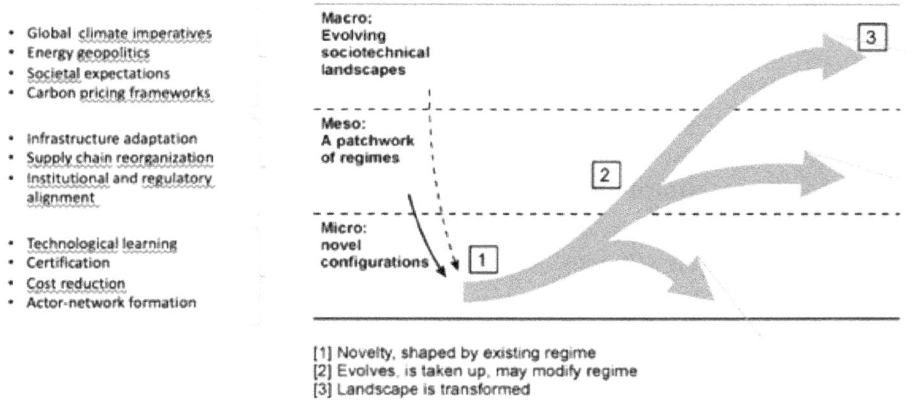

(Kemp et al., 2001; Geels, 2002)

The SAF trajectory embodies the logic of socio-technical and evolutionary theories of change. It demonstrates that transitions are long-term, non-linear processes shaped by institutional inertia, learning dynamics, and multi-level coordination. The current coexistence of SAF and fossil kerosene is not a sign of stagnation, but rather the hallmark of a transition in progress—one where technological innovation, industrial adaptation, and policy evolution gradually reinforce one another toward a new regime of sustainable aviation.

Conclusion – Section II

The dynamics surrounding SAF development reveal a sector caught between technological promise and structural inertia. Progress across diverse fuel pathways is evident, yet persistent cost barriers, resource constraints, and institutional lock-ins continue to slow large-scale deployment. Rather than a rapid substitution, the coming decades will likely see a gradual reconfiguration, where fossil and sustainable fuels coexist. The extent to which SAF can drive a genuine transformation will ultimately depend on the collective alignment of technological innovation, industrial adaptation, and public policy.

III) THE DISRUPTIVE DYNAMICS INTRODUCED BY SAF: A NEW RELATIONSHIP PARADIGM BETWEEN FUEL SUPPLIERS AND AIRLINES

While key stakeholders in the aviation sector, most notably governments and aircraft manufacturers[8], have supported SAF production initiatives, airlines have long remained on the sidelines, adopting a cautious, sometimes wait-and-see approach. Their contribution to climate action has historically focused on improving operational efficiency, both on the ground and in flight, and modernizing fleets with more fuel-efficient aircraft, a strategy pursued for over fifty years. Today, airlines increasingly recognize that securing access to SAF is, and will continue to be, a critical factor in maintaining operational continuity and meeting environmental commitments. The SAF supply chain differs fundamentally from that of fossil-based kerosene in terms of feedstock origin, production technology, logistics, actors involved, and regulatory requirements. In this context, airlines can no longer afford to act solely as passive fuel buyers. Faced with uncertainties around supply availability and pricing, they are taking a more proactive role further upstream in the SAF value chain to secure access to these scarce and costly fuels. Although they remain end-users in both cases, their role in the SAF supply chain differs markedly from their traditional position in the fossil fuel market. This shift is reflected in long-term offtake agreements, industrial partnerships, and in some cases, direct equity investments in SAF production projects.

III.a) Supply chain JET-A-1 Fossile versus supply chain JET-A-1 SAF

The supply chain encompasses all interconnected and interdependent organizations that collaborate to manage and optimize the flow of materials, information, and resources from upstream suppliers to end users[9]. In the case of fossil-based Jet-A1, the supply chain follows a traditional, standardized, and vertically structured model that has been firmly established for several decades and demonstrates reliability and maintained steady availability. It is designed around principles of large-scale production and mass distribution. « Petroleum products are referred to as "commodities" because their product characteristics are uniformly standardised by specifications to ensure that products from different producers do not differ in quality and can be mixed and stored together in the same tank (commingled storage). The only difference between commodities is the price at which they are sold » (Buse, 2024). The process begins with crude oil extraction in regions where production remains profitable, such as the Middle East, Russia, or North America. The crude is transported to refineries where it undergoes a series of treatments (distillation, hydrocracking, desulfurization) to yield various petroleum fractions, including the kerosene cut used in aviation. This product is further refined to meet the technical and environmental specifications defined by ASTM Standard D1655, ensuring compatibility with modern aircraft engines and airport fueling infrastructures. Once processed, Jet-A1 is delivered to airport storage terminals and then supplied to aircraft via tanker trucks or underground hydrant systems, enabling secure and efficient refueling operations.

This supply chain is marked by a high degree of vertical integration and market concentration. A small number of suppliers provide oil products to a highly fragmented consumer base, where numerous individual buyers have no significant influence over prices or demand. In such a polypolistic market, consumers lack market power. A small number of multinational oil companies -such as ExxonMobil, Shell, BP, Chevron, and TotalEnergies- control most stages of the chain, from upstream extraction through midstream refining and transport, to downstream airport distribution. These firms operate major refining complexes (e.g., Shell's Pernis facility in the Netherlands, ExxonMobil's Baytown and Baton Rouge refineries in the United States), manage international logistics, and maintain fueling operations at hundreds of airports. For example, Shell Aviation supplies more than 850 airports worldwide, and Air BP delivers over 8 billion gallons annually to 600 airports across 55 countries. In some markets, such as China, the concentration is even more pronounced: China National Aviation Fuel Group (CNAF) controls over 90% of domestic jet fuel imports. Within this vertically integrated and oligopolistic configuration, airlines play a limited and passive role. They act as final customers, procuring fuel through standardized contracts, joint purchasing consortia, or competitive bidding processes. Their involvement remains primarily commercial and logistical; they exert no control over upstream production or infrastructure. Strategic priorities for airlines are generally limited to cost optimization, securing supply through collective procurement mechanisms, and managing price volatility via financial hedging. The sole notable exception is Delta Air Lines, which acquired a refinery in Pennsylvania in 2012 that was being prepared for closure, a rare case of partial vertical integration. Overall, airlines remain structurally dependent on a small number of powerful suppliers and have little capacity to influence the architecture or dynamics of the fossil jet fuel supply chain.

In contrast to the vertically integrated and oligopolistic structure of the fossil jet fuel supply chain, the SAF sector is more fragmented, technologically diverse, and organizationally complex. It is built around a wide range of actors -including renewable energy firms, agricultural processors, biotech startups, and local utilities-rather than being dominated by traditional oil majors. While large traditional energy companies like BP, TotalEnergies, and Shell have recently entered the SAF market, recognizing the potential of SAF to complement their renewable energy strategies, many key players remain independent or specialized. This plurality reflects not only the diversity of feedstocks but also the multiplicity of technological pathways recognized under ASTM Standard D7566, each requiring different conversion techniques, infrastructure, and input logistics.

Table 3. Global SAF producers

SAF Producers	Estimated SAF Market Share	Oil Company	Key Notes
Neste (Finland)	35% – 45%	NO	Global leader in bio-SAF (HEFA)
World Energy (USA)	15% – 20%	NO	Bio-SAF specialist, partners with major airlines
TotalEnergies (France)	12% – 18%	YES	Rapidly expanding in SAF, including e-SAF
LanzaJet / LanzaTech (USA)	10% – 15%	NO	ATJ (alcohol-to-jet) technology, industrial partnerships
Gevo Inc. (USA)	~4%	NO	Isobutanol-based SAF, bio-SAF
Shell (UK)	<5%	YES	Investing in SAF and PtL, catching up
BP (UK)	<5%	YES	Various SAF projects, especially in Europe
OMV (Austria)	<3%	YES	Growing HEFA-SAF project
Aemetis (USA)	<2%	NO	Bio-refining, scaling up plans
SkyNRG (Netherlands)	<1%	NO	SAF logistics specialist, not a refiner
Fulcrum BioEnergy (USA)	<1%	NO	Municipal solid waste-based SAF
Valero + Darling (USA)	<5%	YES	HEFA production via joint-venture
Alder Fuels (USA)	<1%	NO	Innovative bio-SAF from residual wood, pilot stage
Sasol (South Africa)	<2%	YES	Synthetic SAF, Fischer-Tropsch process

Source: Specialist publications and company websites' (author)

As Martinez-Valencia (2021, p.1) explains, the SAF's supply chain structure « can be divided into three main components: (1) upstream, which includes feedstock production, transportation, storage, and preprocessing, (2) midstream, the conversion of the feedstock to fuel, and (3) downstream, the fuel distribution, storage, and final use. Despite the apparent simplicity and linearity of the renewable fuel supply chain, its configuration, design, and analysis is a complex process due to the type, quantity, quality, seasonality, and dispersion of the feedstock; transportation constraints; selection of conversion technology; size and location of facilities; distribution or integration of the supply chain structure and the impact of these elements in the final product price". Unlike conventional Jet-A1, SAF must undergo a dual certification process. First, it is certified under ASTM D7566 depending on the production pathway -such as HEFA, AtJ, or PtL- before being blended (up to 50% under current standards) with fossil Jet-A1 to meet ASTM D1655, ensuring full operational equivalence with conventional kerosene. This blending requirement and the reliance on mass-balance accounting systems add further complexity to logistics, traceability, and certification, especially in airports not yet equipped for SAF handling. As noted by the U.S. SAF Grand Challenge Roadmap: « Because current fuel certifications currently require SAF to be blended with conventional fuels, the SAF supply chain also requires coordination with conventional jet fuel industries. As SAF production is a nascent industry, SAF supply chains are immature ». Neste, the current leader, produced around 500,000 tonnes of SAF, with others such as World Energy, SkyNRG, and Gevo contributing smaller volumes. By contrast with fossil jet fuel (where the top five suppliers control over 90% of the market), the SAF sector is less concentrated: its five largest producers account for under 60% of global output (ICCT, 2023). National contexts also vary: in China, production is still strongly state-led, while in the EU and U.S., SAF development involves diverse partnerships driven by policy incentives and regulatory mandates (e.g., ReFuelEU and the U.S. Inflation Reduction Act).

This decentralization is transforming airline–fuel supplier dynamics. More than a purchasing decision, SAF procurement is becoming a strategic endeavor for airlines -a shift from passive dependence toward active supply chain participation. « SAF, with its completely different production pathways compared to crude oil refining, could become a game changer for aviation. New technologies, new feedstocks and new players will provide a unique opportunity for airlines to start vertically integrating their jet fuel supply chain as long as the new SAF market is open to new supply models » (Buse, 2024). While fossil fuel procurement was historically a price and volume based negotiation with a handful of powerful suppliers, SAF sourcing requires proactive airline involvement in upstream decisions.

The SAF sector is locked in a standoff between airlines, producers, and investors. Airlines need SAF to meet climate targets, yet limited supply and prices up to eight times higher than fossil fuel threaten their margins. Producers require major upfront investment to scale production but cannot secure it without long-term purchase agreements. Investors, in turn, hesitate to provide funding without clear commercial guarantees. This deadlock is blocking large-scale deployment. Overcoming these barriers is critical. SAF growth depends on removing persistent financing obstacles, most notably the "valley of death" (VoD). As Hartley and Medlock (2017) highlight, the VoD has long slowed the adoption of renewable energy technologies in Europe. Indeed, as Muscio et al (2023) observes, "The VoD could hamper the effectiveness of public or private funding to research and development (R&D) activities, as new technologies developed in labs are unable to compete, before being deployed at a scale, with technologies that are already used and diffused in the economic system, for instance, because they are still too expensive or they do not perform yet as efficiently as more mature technology ». This observation is particularly relevant in the case of SAF, where bridging the gap from pilot or demonstration units to full-scale industrial production is especially demanding. Estimates[10] suggest that scaling up can require several hundred million dollars -often exceeding one billion for standalone facilities- and may take five to ten years before becoming operational. Without clear visibility on long-term demand, producers face a heightened risk that financing will be depleted before projects become profitable. To break this impasse, coordinated action across the value chain is essential. In particular, establishing predictable demand through long-term offtake agreements would enhance the bankability of SAF projects, thereby attracting the capital required to expand production capacity. As capacity increases and technologies mature, economies of scale would drive down unit costs, creating a virtuous cycle that would reinforce demand, encourage further investment, and accelerate the diffusion of SAF technologies. Ultimately, such a transformation is crucial if the aviation sector is to meet its decarbonization objectives while ensuring the long-term viability of SAF production.

III.b) Competing Pathways and the Risk of Technological Lock-in: Hydrogen, Electrification, and Hybrid Systems

While Sustainable Aviation Fuels represent the only immediately deployable solution compatible with existing fleets and infrastructures, they do not constitute a definitive technological endpoint. Rather, SAF should be understood as a transitional pathway within a broader landscape of aviation decarbonization technologies. In parallel, other disruptive innovations — notably hydrogen propulsion, battery-electric aircraft, and hybrid-electric systems — are progressing rapidly, albeit on longer time horizons.

Hydrogen aircraft, which entail a deep technological rupture requiring new airframe designs, cryogenic storage systems, and dedicated refueling infrastructures, are expected to reach market readiness between 2035 and 2045. By contrast, battery-electric propulsion remains constrained to short-haul or regional aviation, limited by the low energy density of current battery chemistries. Hybrid-electric concepts, combining electric drive with conventional or SAF-based combustion, offer intermediate solutions that could enable partial decarbonization of regional segments before full electrification or hydrogen adoption becomes viable.

This technological diversity creates a strategic dilemma for both policymakers and investors. Over-concentration of resources and regulatory incentives on SAF risks generating a « technological lock-in », where the sunk costs of infrastructure, certification systems, and industrial partnerships hinder the later diffusion of more radical technologies such as hydrogen. Conversely, insufficient investment in SAF could jeopardize near-term decarbonization objectives, leaving the sector overly dependent on future breakthroughs that may take decades to scale. In this regard, the « valley of death » described earlier applies not only to SAF producers but to the entire aviation innovation ecosystem. Each technological pathway faces its own financing and diffusion barriers, yet with distinct risk profiles and maturity timelines. To ensure a balanced and adaptive transition, governance mechanisms must therefore accommodate a portfolio approach — one that combines short-term deployment of SAF with long-term investments in hydrogen and electric technologies.

A comparative SWOT analysis (Table 3) helps visualize these complementarities and trade-offs: SAF's immediate scalability contrasts with hydrogen's disruptive potential and electricity's operational simplicity. From a governance perspective, such pluralism mitigates systemic risk by diversifying technological bets rather than locking the sector into a single trajectory.

Table 4. Comparative SWOT Analysis of Aviation Decarbonization Technologies (SAF, Hydrogen, Electric, and Hybrid-Electric)

Technology	Strengths	Weaknesses	Opportunities	Threats
SAF (Sustainable Aviation Fuels)	• Drop-in compatible with existing aircraft and infrastructure • Up to 80% CO_2 reduction over the lifecycle • Commercially available today (up to 50% blends, 100% by 2030)	• 2–5× more expensive than fossil jet fuel • Limited sustainable feedstock (biomass, waste oils) • Current production <1% of aviation fuel demand	• Rapid transition without redesigning aircraft • Massive global investment (EU, US, IATA targets) • Growth of synthetic e-fuels using green hydrogen + captured CO_2	• Feedstock competition with other sectors (shipping, road transport) • Risk of greenwashing if lifecycle emissions not well verified
Hydrogen (H_2)	• Zero direct CO_2 emissions in flight • Higher efficiency with fuel cells • Long-term potential for regional and medium-haul aircraft	• Complex storage (cryogenic, high volume) • Requires entirely new aircraft and airport infrastructure • Green hydrogen production still costly and energy-intensive	• Enables new aircraft designs (blended-wing, distributed propulsion) • Can also produce e-fuels (intermediate solution)	• Limited availability of green H_2 • Safety and certification challenges • Technology readiness beyond 2035
Electric (battery-powered)	• Zero direct emissions, no NOx or contrails • Very high energy efficiency (~90%) • Quiet, low maintenance	• Battery energy density far below jet fuel (~1/50) • Heavy mass limits range and payload • Battery production has significant environmental footprint	• Ideal for short-range, regional, and urban air mobility • Strong "zero-emission" public image • Synergy with ground EV technologies	• Slow progress in battery energy density • Supply chain risks (lithium, cobalt) • Limited market scope (short-range only)
Hybrid-electric	• Combines thermal and electric energy for higher overall efficiency • Potential 30–50% fuel reduction • Smooth transition technology between conventional and electric	• Technically complex, heavier systems • Efficiency gains depend on flight profile • Still partially reliant on fossil or SAF fuel	• Bridge toward full hydrogen/electric • Compatible with SAF for near-term cuts • Builds learning for future systems	• Limited long-term potential • High cost and complexity • Risk of tech lock-in

This comparative analysis highlights that each technology occupies a distinct temporal and functional niche within the aviation decarbonization trajectory. SAF offers short-term compatibility and scalability, but at the cost of high feedstock dependence and uncertain life-cycle performance. Hydrogen and electric propulsion represent long-term disruptive options with transformative potential, yet face formidable technological and infrastructural barriers before 2040. Hybrid-electric systems stand as bridging solutions that enable progressive efficiency gains and learning effects across both domains. The coexistence of these pathways suggests that the sector's transition will not be linear but multi-modal, requiring coordinated governance to avoid premature lock-in and stranded assets while sustaining innovation diversity.

Major industrial actors (Airbus, Rolls-Royce, Safran) already follow this diversification logic by simultaneously investing in SAF, hydrogen propulsion, and hybrid-electric demonstrators. European and international initiatives such as Clean Aviation and the EU Emissions Trading System explicitly promote such technology-neutral strategies, encouraging firms to pursue sequential rather than purely competitive transitions. In sum, the coexistence of multiple innovation pathways should not be seen as a sign of fragmentation, but as an adaptive response to uncertainty. SAF provides the bridge technology enabling near-term emission reductions, while hydrogen and electric systems define the long-term transformation horizon. The strategic challenge for both industry and policymakers is to orchestrate this temporal layering without inducing a structural lock-in.

In this context, investment and contracting strategies become crucial levers to balance short-term commitments with long-term flexibility. Airlines and producers must secure SAF supply today while preserving the capacity to adapt as hydrogen and electric solutions mature—a tension that underpins the growing importance of long-term offtake agreements discussed in the next section.

III. c) Long-Term Offtake Agreements

Regarding fossil fuels, airlines were chiefly concerned with price volatility and the need to shield themselves from abrupt spikes in kerosene costs through forward hedging contracts. In the case of sustainable aviation fuels, the concern is of an entirely different order, and the nature of the transaction itself changes. It is no longer just a matter of price. The real challenge now lies in securing long-term commitments, mobilizing large-scale investment, and overcoming the structural barriers that keep the SAF sector from expanding. As outlined in the previous section, a vicious cycle binds SAF buyers, producers, and investors. Transaction cost theory offers a useful framework for understanding what kind of contractual arrangements might allow this exchange to take place and shift the cycle in a more virtuous direction. It shows that when uncertainty and high asset specificity are involved, transactions can no longer be anonymous or instantaneous. As Oliver Williamson (1989) explains, "Asset specificity has reference to the degree to which an asset can be redeployed to alternative uses and by alternative users without sacrifice of productive value. This has a relation to the notion of sunk cost. But the full ramifications of asset specificity become evident only in the context of incomplete contracting." This situation gives rise to particular vulnerabilities in the transaction: "When a transaction entails one party committing capital that has little value for other uses, the other party has a strong incentive to appropriate the rents arising from the relationship through opportunistic behavior. Anticipating this risk, also called the 'hold-up' problem, buyers and sellers sign long-term contracts" (Creti and Villeneuve, 2004). These capital-intensive investments, being difficult to redeploy, create bilateral dependencies between parties, thereby increasing the risk of opportunistic behavior.

This mutual dependency highlights the need for governance mechanisms—such as long-term contracts—that help mitigate opportunism and enable credible commitments between parties. To secure both production volumes on the one hand, and reliable SAF supply on the other, producers and airlines often enter into offtake agreements well before production facilities are even operational. These long-term contracts, which typically span between five and twenty years, commit the airline to purchasing specified quantities of SAF in advance, provided certain conditions are met -such as the fuel being available at an agreed price. The agreements may involve fixed or flexible volumes (see the table below), incorporate various price adjustment mechanisms (fixed, indexed, or revisable), and include clauses allowing delivery flexibility. Their core purpose is to provide the producer with the financial visibility needed to unlock the capital investment required for production. Investment risks present major barriers to entry. In the context of sustainable aviation fuels, these agreements play a pivotal role in enabling project financing: they offer producers a way to demonstrate predictable revenue streams from facilities under development or expansion, thus reducing perceived risk for investors. Crucially, firm long-term demand helps to de-risk projects, making it easier to mobilize capital and expand production capacity. The visibility and stability provided by these contracts are often prerequisites for securing financing, especially when facilities are not yet operational. As one study observes, "In many cases, (...) facilities [. . .] have been constructed and operated on the basis of preferential arrangements secured by long-term contracts as a condition for making large investments" (Cameron, 2006). Beyond bilateral agreements, the industry could also benefit from the emergence of a more structured marketplace—one capable of consolidating demand across a broad spectrum of stakeholders. Such coordination would send stronger market signals, reinforce investor confidence, and help accelerate project timelines, thereby supporting the sector in reaching critical mass. By way of illustration, Air France-KLM entered into a long-term offtake agreement with TotalEnergies for the supply of 1.5 million tons of Sustainable Aviation Fuel (produced from organic residues and non-food waste) over a ten-year period, covering flights departing from several EU member states, most notably France and the Netherlands. The terms of the contract were renegotiated in 2024 to increase the committed volumes, in particular as a response to the tightening regulatory requirements for SAF blending within the European Union.

Table 5. Offtake Agreements for SAF

SAF Producer	Airline Customer	Signed (Date)	Duration (years)	Annual Volume (t)	Contract Type
TotalEnergies	Air France-KLM	2024	10	150 000	Long-term offtake
DG Fuels	Air France-KLM	2022	10	60 000	Binding long-term
Neste	Air France-KLM	2022	8	100 000	Binding long-term
OMV Group	Ryanair	2022	8	20 000	MoU off-take option
OMV Group	Lufthansa	2022	7	157 000	Long-term offtake
Gevo Inc.	American Airlines	2022	5	228 000	Strategic partnership
Gevo Inc.	Finnair	2022	5	24 000	Long-term offtake
Gevo Inc.	Aer Lingus	2022	5	19 000	Long-term offtake
Gevo Inc.	Alaska Airlines	2022	5	112 000	Long-term offtake
Gevo Inc.	Qatar Airways	2022	5	19 000	Long-term offtake
easyJet / Moeve	easyJet	2024	6	20 000	MoU supply agreement
AMANA (via Shell)	Cebu Pacific	2023	5	25 000	Long-term supply

Source: Specialist publications and company websites' (author)

A take-or-pay clause[11] is sometimes included in long-term offtake agreements. Under such a provision, the buyer commits to paying for a fixed or minimum quantity of product, regardless of actual consumption and even if physical delivery does not occur—provided that the supplier has fulfilled its obligation to make the product available at the agreed times. Compared to standard offtake contracts, this structure is more rigid, as it transfers a significant portion of risk to the buyer, who may be required to pay for volumes not ultimately needed or received. The primary purpose is to secure a stable and predictable revenue stream for the producer—an essential condition for project finance arrangements, where future cash flows must be contractually demonstrated to reassure lenders. Take-or-pay[12] clauses are commonly used in capital-intensive sectors such as energy, where they enhance the bankability of projects by guaranteeing long-term income streams. For buyers, such commitments mitigate supply risk by ensuring that the product will be available to meet downstream obligations. In practical terms, this is linked to a corresponding obligation on the supplier to make the fuel available at pre-agreed intervals. In most cases, the buyer agrees to purchase a minimum share—say, 80%—of the contracted volume, regardless of actual demand during the delivery period, while the supplier undertakes to provide that same minimum quantity on the specified dates. This mutual commitment creates a balanced allocation of risks: the buyer assumes exposure to demand fluctuations, while the producer bears the risk of market price movements, since the price is usually fixed in the contract. Such provisions serve the dual function of securing both physical offtake and financial flows. They are often paired with infrastructure contracts such as build-own-operate-transfer (BOT) schemes, linked to the construction of production or distribution facilities, whose financing is typically provided by investors expecting reliable returns. Once such specific, capital-intensive investments are made, buyers and sellers become economically interdependent. The high costs of exiting the relationship result in a mutual lock-in, which further reinforces the need for contractual rigidity and predictability throughout the project's lifecycle. Velocys Renewables LLC has secured long-term agreements with Southwest Airlines and IAG for the supply of SAF, covering the full output of its Bayou Fuels plant starting in 2026. These contracts, spanning 10 to 15 years and involving hundreds of millions of gallons, include fixed pricing and provisions linked to carbon credit values, providing stable revenues, essential for financing. Although the term *take-or-pay* is not explicitly used, this agreement functions similarly by ensuring payment for agreed volumes regardless of actual delivery or consumption. This structure shifts significant commercial risk to buyers while guaranteeing producers predictable cash flows, thereby supporting the viability of capital-intensive SAF projects.

III.d) From Buyer to Builder: The Vertical Turn in Airlines' SAF Strategies

The aviation industry is characterized by a significant degree of vertical disintegration, with airlines progressively limiting their ownership stakes in sectors beyond their core business (Tretheway and Markhvid, 2014). Historically, some carriers had pursued diversification strategies, seeking to exert influence over upstream and downstream segments of the value chain, such as aircraft manufacturing, computer reservation systems, maintenance and repair operations, or hotel chains. Over time, however, the mounting pressure of global competition and the persistent volatility of airline profitability compelled carriers to concentrate resources on their primary function of transporting passengers and cargo. Scholars have noted that vertical disintegration may in fact yield benefits, as it can stimulate competition both upstream and downstream, thereby improving efficiency and innovation (Carvalho and Marques, 2014). In practice, divestitures in areas such as hotel chains or maintenance subsidiaries not only reduced exposure to risks in non-core markets but also freed capital for reinvestment into operational efficiency. A prominent example is the widespread divestment from computer reservation systems, once owned by airlines seeking to secure distribution channel[13]s, but later sold under regulatory pressure and in recognition that independent, technology-driven providers could better manage these platforms.

Yet this withdrawal from ownership in certain domains has been accompanied by carefully targeted investments in others, reflecting a strategic imperative to secure control over critical inputs and to reinforce the overall customer experience. Unlike simple contractual partnerships, these investments take the form of equity participation or the creation of wholly owned subsidiaries, particularly in areas such as ground handling, catering, and airport services. Control over these activities allows airlines to mitigate cost fluctuations, ensure operational reliability, and differentiate their service offering in an industry where margins are narrow and passenger expectations increasingly exacting. Building on this selective reengagement strategy, several airlines have more recently begun deploying capital directly into sustainable aviation fuel (SAF), not merely through purchase agreements but via equity stakes in production plants or SAF technology companies (Martinez-Valencia et al., 2021), with the aim of securing critical energy inputs and influencing the pace at which SAF scales globally.

This renewed interest in vertical integration raises a central question: how can one explain the strategic rationale behind such integration or disintegration choices? « From the individual firm's perspective, vertical strategy choices can be related to three decision domains: boundary setting, boundary bridging, and boundary design » (Wendel et al., 2025). Boundary setting refers to the firm's choices on its vertical scope, that is, the definition of which activities are internalized within the firm's boundaries. Boundary bridging concerns the governance of relationships with upstream and downstream partners beyond those boundaries. Boundary design focuses on the interfaces that connect the firm to its transactional environment. These domains, while subject to managerial discretion, have been shown to exert a significant influence on organizational performance.

The reasons most often advanced to justify vertical integration include the need to ensure access to critical resources, to reduce the risks of opportunistic behavior by partners, and to mitigate the adverse effects of uncertainty. As Grant (2021) highlights, by determining which upstream and downstream segments of the value chain to internalize, firms delineate their organizational boundaries and establish their vertical scope. These decisions are fundamentally strategic, as they shape the firm's identity, determine the allocation of resources, and define the degree of control it exercises over value creation processes. In the context of strategic management, such choices are closely linked to the pursuit of sustainable competitive advantage. Vertical integration can enable a firm to secure the quality and availability of key inputs, to capture a larger share of value created along the chain, and to develop distinctive capabilities that are difficult for rivals to imitate. Yet, this pursuit is never without risk: expanding vertical scope entails considerable capital commitments and managerial complexity, while also potentially reducing organizational flexibility.

These trade-offs are particularly salient in the airline industry, which is widely recognized for operating with persistently thin margins. For decades, airlines have consistently delivered lower shareholder returns than other stakeholders in the aviation ecosystem, including manufacturers, airports, service providers, and intermediaries such as global distribution systems. This process can also be interpreted through the lens of dynamic capabilities theory (Teece, 2007), which emphasizes firms' ability to sense opportunities and threats, seize them through investment and partnerships, and transform their internal structures accordingly. Airlines' renewed engagement in SAF exemplifies this adaptive capacity. They "sense" regulatory and environmental pressures as both risks and opportunities; they "seize" them through long-term offtake contracts and equity investments; and they "transform" their organizational boundaries by internalizing new technological and logistical competences. In this sense, the vertical turn in airlines' SAF strategies reflects not a mere return to ownership, but an exercise in strategic renewal—mobilizing dynamic capabilities to reconfigure resources, partnerships, and routines under conditions of technological uncertainty.

The aviation value chain as a whole may appear financially viable, but airlines themselves—despite being its central component—regularly emerge as the least profitable actors. This paradox has often led analysts to portray airlines as the weakest link in the industry's economic framework, generating profits for nearly every other player in the aviation chain—except themselves. Button (2002) provocatively asked whether the « airline industry has an empty core » invoking the economic concept used to describe markets where competition prevents firms from reaching a stable, sustainable equilibrium. A key reason for this chronic underperformance is the sector's exceptionally intense competition. It stems not only from the number of rivals but also from deep structural interdependencies across the industry. Rather than operating as independent firms, the aviation ecosystem functions as a tightly woven network where decisions by airports, manufacturers, or service providers directly affect airlines' costs, revenues, and viability. Small efficiency gains are rarely enough to secure long-term stability. Instead, airlines must seek innovative strategies not only to differentiate their services but also to reshape their position within this complex web of interdependencies.

It is in this context that recent vertical investments in SAF take on their full significance. Several airlines have moved beyond standard fuel procurement to invest directly in SAF production and technology ventures. United Airlines notably launched the United Airlines Ventures Sustainable Flight Fund, backed by major corporate partners including Air Canada, Boeing, GE Aerospace, JPMorgan Chase and Honeywell, initially committing $100 million in 2023 to support early-stage SAF innovators—now expanded to over $200 million as of early 2025—to scale SAF availability and reduce its cost of production. United had earlier invested $30 million in Fulcrum BioEnergy in 2015, securing access to long-term SAF volumes produced from municipal solid waste, with options to purchase up to 90 million gallons yearly for at least a decad. Qantas and Airbus jointly injected $15 million into Climate Tech Partners in early 2025, part of a broader fund to back SAF start-ups, aligning with Qantas's commitment to reach 10 percent SAF use by 2030 and 60 percent by 2050. Meanwhile Virgin Australia, in partnership with Renewable Developments Australia and Qatar Airways, is developing a sugarcane-derived aviation fuel facility in Charters Towers capable of producing up to 96 million litres annually by 2029. These examples underscore how multiple carriers are investing in feedstock or conversion capacity rather than limiting their role to purchase contracts. It is in this context that recent vertical investments in SAF take on their full significance. On February 28, 2024, Southwest Airlines announced a $30 million equity investment in LanzaJet, a leading producer of SAF, to develop a U.S.-based facility using its patented Alcohol-to-Jet conversion technology. Shortly thereafter, Southwest took a further step upstream by acquiring Saffire Renewables, a company specializing in the production of cellulosic ethanol derived from corn stover, a key feedstock for LanzaJet's SAF production. This dual move exemplifies a deliberate strategy of vertical integration, internalizing both feedstock and conversion processes. From the perspective of Transaction Cost Economics, such integration can be interpreted as a response to uncertainty, asset specificity, and the risks of opportunism in external feedstock markets. By internalizing critical activities, Southwest reduces transaction costs, mitigates exposure to volatile market conditions, and secures a more reliable and cost-effective SAF supply. Other carriers have followed similar paths: United Airlines invested $30 million in Fulcrum BioEnergy in 2021, Alaska Airlines committed $10 million to Red Rock Biofuels in 2022, British Airways invested approximately £15 million in Velocys that same year, JetBlue dedicated $25 million to Green Plains Inc. in 2023, and Air France-KLM contributed USD 4.7 million to DG Fuels in late 2023. Together, these initiatives illustrate a broader industry trend towards re-internalizing critical activities along the SAF value chain. Placed in this wider perspective, such investments highlight the growing emphasis on supply chain resilience. Christopher and Peck (2004) define resilience as the adaptive capability of the supply chain to prepare for unexpected events, respond to disruptions, and recover while maintaining continuity of operations at the desired level. « It has become apparent that many organisations have not fully recognised the nature of systemic supply chain risk and have continued to focus on seeking efficiency improvements through 'lean' solutions. (...) a new priority has emerged for business planning. This priority has to be the search for supply chain strategies that embody a significantly higher degree of resilience ». Airlines' strategic reengagement in SAF thus not only secures a critical input but also strengthens their capacity to withstand shocks in an increasingly volatile regulatory and economic landscape.

Finally, vertical strategy in such contexts must be seen as a continuum rather than a binary choice between full integration and market reliance. As Peterson and Wysocki (1997) observed, industries characterized by transaction complexity, incomplete contracts, and frequent ex post adaptation compel firms to balance efficiency with control. While integration offers safeguards against opportunism and coordination failures, it is not always optimal. Williamson (1985), Oxley (1997), and Ménard (2004) emphasized the importance of hybrid governance forms—alliances, joint-ventures, long-term contracts, or franchising—that combine elements of hierarchy and market. These hybrids allow firms to capture some benefits of control while retaining flexibility, often proving more efficient than relying solely on either pure market transactions or full internalization. In the case of airlines and SAF, the growing variety of equity partnerships, joint-ventures, and long-term offtake agreements exemplifies this spectrum of solutions, each calibrated to balance cost efficiency, resilience, and strategic positioning.

III.e) Emerging Collective Governance Models: Consortia, Co-Investment, and Book-and-Claim Systems

Beyond bilateral deals and vertical integration, the SAF ecosystem is seeing new forms of collective governance that address early-stage market challenges. In the beginning, initiatives were fragmented, and producers, airlines, and investors often waited for others to act first—a situation known as the « coordination trap ». Transition theorists describe this as a common barrier for scaling niche innovations: when no single actor can shoulder the cost of major changes, progress depends on collaborative frameworks that share risk and align incentives across the value chain.

Industrial consortia and multilateral co-investment platforms have become key tools for coordinating the aviation sector. Initiatives such as Clean Skies for Tomorrow (World Economic Forum), the Sustainable Flight Fund (United Airlines Ventures), and the European Alliance for Zero-Emission Aviation show how airlines, aircraft manufacturers, fuel producers, and financial institutions can work together. By pooling resources to co-finance SAF production and support early-stage technologies, these platforms reduce individual risk, make projects more financially viable, and accelerate collective learning. They represent a middle ground between competitive markets and full vertical integration, relying on long-term cooperation to drive innovation.

A parallel innovation is the rise of "book-and-claim" systems, which create a new market mechanism for certifying and claiming SAF use. In these systems, the physical SAF is injected into a local fuel pool, while an airline—possibly operating elsewhere—purchases the corresponding emission reduction certificate. This allows airlines without local SAF access to financially support its production and claim the carbon benefits in their reporting. For example, one tonne of SAF added to an airport in Country A can be "booked" by an airline in Country B, which then accounts for the associated emission reduction. Initiatives such as the Roundtable on Sustainable Biomaterials (RSB) and the World Economic Forum's SAF Certificate Registry implement these systems, making SAF participation possible for a wider range of airlines beyond regions with local supply.

These emerging models suggest that the governance of aviation decarbonization is gradually shifting from bilateral contracting toward network-based coordination. This transition echoes the shift from hierarchical and market-based governance to network forms of coordination described by Powell, Koput, and Smith-Doerr (1996). In their framework, innovation no longer resides within isolated firms but emerges from interorganizational networks where knowledge, capabilities, and trust circulate among diverse actors. The SAF ecosystem—uniting airlines, energy companies, technology startups, and public agencies—perfectly illustrates such "networks of learning." These collective arrangements enable experimentation, mutual adjustment, and shared learning across organizational boundaries, turning cooperation itself into a driver of innovation. By combining consortia, co-investment mechanisms, and digital certification systems, the sector is building a hybrid institutional architecture capable of mobilizing diverse actors around common standards. Such pluralistic governance not only complements traditional offtake agreements but also mitigates the risk of technological and financial lock-in by distributing both costs and benefits more equitably across the ecosystem.

Ultimately, the success of SAF deployment will depend on the coexistence of these governance forms—hierarchical, bilateral, and collective—each fulfilling distinct coordination functions. Long-term contracts and vertical integration secure investment in specific assets; collective platforms and book-and-claim schemes, meanwhile, provide the systemic trust, liquidity, and visibility needed to scale the market. Together, they form a multi-layered governance architecture that reflects the complexity and interdependence of the aviation energy transition.

CONCLUSION

This chapter has shown that the future of aviation's decarbonization rests not only on technological progress but on strategic choices that redefine the industry's relationship to its own value chain. Sustainable Aviation Fuel offers a pathway that is both promising and constrained, demanding bold departures from traditional governance models. Vertical integration, far from being a mere defensive move, emerges as a proactive strategy to overcome structural bottlenecks, mobilize investment, and recalibrate power dynamics within the aviation ecosystem. In this sense, the challenge of SAF is less about chemistry than about vision: who among today's carriers is willing to reimagine their role not just as consumers, but as architects of a sustainable future?

Yet, looming over this transition is the prospect of disruptive technological breakthroughs—hydrogen propulsion, electric aviation, or other radical innovations—that could, if scaled, redefine the very foundations of the sector. For investors, this possibility is never far from mind. However, the risk has thus far been deemed moderate, given the 25- to 30-year lifespan of aircraft fleets and the persistence of strong secondary markets. Many airlines, particularly in regions with limited financial capacity and lower historical responsibility for emissions, will continue to rely on conventional aircraft compatible with drop-in SAF rather than bearing the prohibitive costs of replacing fleets, engines, and infrastructure all at once. This reality ensures a significant and lasting role for SAF in the global decarbonization agenda, not as a perfect solution, but as the most operationally viable bridge between the carbon-intensive present and a radically transformed future.

REFERENCES

Ahire, J. P., Bergman, R., Runge, T., Mousavi-Avval, S. H., Bhattacharyya, D., Brown, T., & Wang, J. (2024). Techno-economic and environmental impacts assessments of sustainable aviation fuel production from forest residues. *Sustainable Energy & Fuels, 8*(19), 4602–4616. DOI: 10.1039/D4SE00749B

Air Transport Action Group. (2021). *Waypoint 2050: Balancing Growth in Connectivity with a Comprehensive Global Air Transport Response to the Climate Emergency*. 2nd ed. Geneva: Air Transport Action Group. Figure: "Mid-range scenario contribution to net-zero by 2050."

Ansel, P. J. (2023). Review of sustainable energy carriers for aviation: Benefits, challenges, and future viability. *Progress in Aerospace Sciences, 141*, 141. DOI: 10.1016/j.paerosci.2023.100919

Braun, M., Grimme, W., & Oesingmann, K. (2024). Pathway to net zero: Reviewing sustainable aviation fuels, environmental impacts and pricing. *Journal of Air Transport Management, 117*, 117. DOI: 10.1016/j.jairtraman.2024.102580

Bullerdiek, N., Neuling, U., & Kaltschmitt, M. (2021). A GHG reduction obligation for sustainable aviation fuels (SAF) in the EU and in Germany. *Journal of Air Transport Management, 92*(May), 102020. DOI: 10.1016/j.jairtraman.2021.102020

Buse, J. (2024). *Sustainable Aviation Fuels, transitioning towards green aviation*. CRC Press.

Button, K. (2002). Empty cores in airlines markets, *5th Hamburg* **Aviation** *Conference*. Hamburg.

Cameron, P. D. (2006). The consumer and the internal market in energy: Who benefits. *European Law Review*.

Carvalho, P., & Marques, R. C. (2014). Computing economies of vertical integration, economies of scope and economies of scale using partial frontier nonparametric methods. *European Journal of Operational Research, 234*(1), 292–307. DOI: 10.1016/j.ejor.2013.09.022

Christopher, M. (1992). *Logistics and Supply Chain Management*. Pitman Publishing.

Christopher, M., & Peck, H. (2004). Building the resilient supply chain. *International Journal of Logistics Management, 15*(2), 1–14. DOI: 10.1108/09574090410700275

Creti, A., & Villeneuve, B. (2004). *Long-term contracts and take-or-pay clauses in natural gas markets*. Energy Studies Review.

Crocker, K. J., & Asten, S. E. M. (1985). Efficient adaptation in long-term contracts: Take-or-pay provisions for natural gas. *The American Economic Review.*

Environmental Protection Agency. (2022). Inventory of U.S. Greenhouse Gas Emissions and Sinks: 1990–2020. Tables A-98 and ES-6, last updated July 13.

European Commission. (2021). Proposal for a Regulation of the European Parliament and of the Council on Ensuring a Level Playing Field for Sustainable Air Transport: COM(2021) 561 Final. (2023). European Commission.

GAO. (2023). Report to Congressional Committees. Sustainable Aviation Fuel, Agencies should track progress toward ambitious federal goals.

Geels, F. W. (2002). Technological transitions as evolutionary reconfiguration processes: A multi-level perspective and a case-study. *Research Policy*, *31*(8/9), 1257–1274. DOI: 10.1016/S0048-7333(02)00062-8

Global Market Insights (GMI). (2024). Sustainable Aviation Fuel Market Size - By Fuel Type, Technology, Blending Capacity, End Use Analysis, Share, Growth Forecast, 2025 – 2034, December.

Gonzalez-Garay, A., Heuberger-Austin, C., Fu, X., Klokkenburg, M., Zhang, D., van der Made, A., & Shah, N. (2022). Unravelling the potential of sustainable aviation fuels to decarbonise the aviation sector. *Energy & Environmental Science*, *15*(8), 15. DOI: 10.1039/D1EE03437E

Grimme, W. (2023). The Introduction of Sustainable Aviation Fuels—A Discussion of Challenges, Options and Alternatives. *Aerospace (Basel, Switzerland)*, *10*(3), 218. DOI: 10.3390/aerospace10030218

Hartley, P. R., & Medlock, K. B.III. (2017). The valley of death for new energy technologies. *The Energy Journal (Cambridge, Mass.)*, *38*(3), 33–61. DOI: 10.5547/01956574.38.3.phar

IATA (2023), *Industry Economic Performance Mid-Year Report*

IATA (2023), *Air Passenger Forecast*

ICAO. (2023).. . *Environment Reporter.*

ICAO (2023), *Long-Term Aspirational Goal Report*

ICAO. (2023). *Long-Term Aspirational Goal (LTAG) Report.* Montréal.

ICCT (International Council Clean Transportation). (2022). [*Aligning Aviation with Paris Agreement Goals.* International Energy Agency] [IEA]. *Vision (Basel)*, •••, 2050.

IEA (2020), World Energy Balances

IEA. (2023). *Tracking Clean Energy Progress Intergovernmental Panel on Climate Change. (2022). Climate Change 2022: Mitigation of Climate Change. Contribution of Working Group III to the Sixth Assessment Report of the Intergovernmental Panel on Climate Change*. Cambridge University Press.

Intergovernmental Panel on Climate Change Sixth Assessment Report. (2022). Impacts, Adaptation and Vulnerability.

Kemp, R., Rip, A., & Schot, J. W. (2001). Constructing transition paths through the management of niches. In Garud, R., & Karnoe, P. (Eds.), *Path Dependence*.

Klimczyk, V., Jasinski, R., Niklas, J., Siedlecki, M., & Ziolkowski, A. (2025). Sustainable Aviation Fuels: A Comprehensive Review of Production Pathways, Environmental Impacts, Lifecycle Assessment, and Certification Frameworks. *Energies*, *18*(14), 3705. DOI: 10.3390/en18143705

Lee, D. S., Fahey, D. W., Skowron, A., Allen, M. R., Burkhardt, U., Chen, Q., Doherty, S. J., Freeman, S., Forster, P. M., Fuglestvedt, J., Gettelman, A., De León, R. R., Lim, L. L., Lund, M. T., Millar, R. J., Owen, B., Penner, J. E., Pitari, G., Prather, M. J., & Wilcox, L. J. (2021). The contribution of global aviation to anthropogenic climate forcing for 2000 to 2018. *Atmospheric Environment*, *244*, 244. DOI: 10.1016/j.atmosenv.2020.117834 PMID: 32895604

McKinsey and Company. (2020). A Fact-Based Study of Hydrogen Technology, Economics, and Climate Impact by 2050 Hydrogen-powered Aviation.

Martinez-Valencia, L., Garcia-Perez, M., & Wolcott, M. P. (2021). Supply chain configuration of sustainable aviation fuel: Review, challenges, and pathways for including environmental and social benefits. *Renewable & Sustainable Energy Reviews*, *152*, 152. DOI: 10.1016/j.rser.2021.111680

Muscio, A., Simonelli, F., & Vu, H. (2023, November). Bridging the valley of death in the EU renewable energy sector: Toward a new energy policy. *Business Strategy and the Environment*, *32*(7), 4620–4635. DOI: 10.1002/bse.3384

Nakano, Y., Sano, F., & Akimoto, K. (2022). Impacts of decarbonization technologies in air transport on the global energy system. *Transportation Research Part D, Transport and Environment*, *110*, 110. DOI: 10.1016/j.trd.2022.103417

Nemmour, A., Inayat, A., Janajreh, I., & Ghenai, C. (2023, September). Green hydrogen-based E-fuels (E-methane, E-methanol, E-ammonia) to support clean energy transition: A literature review. *International Journal of Hydrogen Energy*, *48*(75), 29011–29033. DOI: 10.1016/j.ijhydene.2023.03.240

Padder, S.A., Khan, R., Rather, R. A. (2024). Biofuel generations: New insights into challenges and opportunities in their microbe-derived industrial production. *Biomasse and Bioenergy*, Vol. 185, June.

Powell, W. W., Koput, K. W., & Smith-Doerr, L. (1996). Interorganizational collaboration and the locus of innovation: Networks of learning in biotechnology. *Administrative Science Quarterly*, *41*(1), 116–145. DOI: 10.2307/2393988

Teece, D. J. (2007). Explicating dynamic capabilities: The nature and microfoundations of (sustainable) enterprise performance. *Strategic Management Journal*, *28*(13), 1319–1350. DOI: 10.1002/smj.640

Tretheway, M. W., & Markhvid, K. (2014). The aviation value chain: Economic returns and policy issues. *Journal of Air Transport Management*, *41*, 3–16. DOI: 10.1016/j.jairtraman.2014.06.011

Ubard, G. H., & Weiner, R. (1986). Regulation and long-term contracting in US natural gas markets. *The Journal of Industrial Economics*.

U.S. Department of Energy (DOE), U.S. Department of Transportation (DOT) & U.S. Department of Agriculture (USDA). (2022). SAF Grand Challenge Roadmap: Flight Plan for Sustainable Aviation Fuel. Bio*energy Technologies Office*. September.

U.S. Department of Energy, Federal Agencies Publish. (2025). SAF Grand Challenge Progress Report Highlighting Historic Efforts to Grow America's SAF Industry, *Bioenergy Technologies Office*. January.

Wendel, L. M., Albers, S., & Dewulf, W. (2025). Towards a typology of airline vertical strategies. *Journal of Air Transport Management*, *127*, 127. DOI: 10.1016/j.jairtraman.2025.102810

Williamson, O. (1989). Transaction cost economics, Chapter 3, *Handbook of Industrial Organization.* Elsevier. Vol.1.

World Economic Forum. (2023). Sustainable Aviation Fuel, Offtake Manual, White Paper, June. Nakano, Y., Sano, F., & Akimoto, K. (2022). Impacts of decarbonization technologies in air transport on the global energy system. *Transportation Research Part D, Transport and Environment*, *110*, 103417.

Zhao, X., Taheripour, F., Malina, R., Staples, M. D., & Tyner, W. E. (2021). Estimating induced land use change emissions for sustainable aviation biofuel pathways. *The Science of the Total Environment*, *779*, 146238. DOI: 10.1016/j.scitotenv.2021.146238 PMID: 33744564

ENDNOTES

1 Purchasing new aircraft can reduce CO_2 emissions by around 25%, but doing so requires a substantial financial investment. Take the example of Air France, which has committed to this strategy. In 2024, next-generation aircraft made up 7% of its fleet. By 2030, that share is expected to reach 70%, supported by an annual investment of one billion euros starting in 2025.

2 It is not zero emissions but net-zero emissions when you look at the life cycle. It's still the same combustion process. Burning SAF produces carbon dioxide, as well many of the same pollutants as conventional fuels. It takes energy to refine, process, and transport SAF.

3 Despite this development, the use of SAF in day-to-day flight operations « is still in its infancy, with less than 1% of all European flights being operated with SAF today » (Braun et al., 2024).

4 Life cycle CO2 emissions of SAF depend on the feedstock and conversion pathway. An overview is provided in the article of Grimme, 2023.

5 Some stakeholders in the aviation industry take a broader view of sustainable alternatives to kerosene, including technologies such as pure hydrogen. Thanks to its high specific energy by mass, hydrogen is seen as a promising long-term energy carrier. It can be produced industrially via electrolysis using renewable electricity, without generating direct CO_2 emissions (IEA, 2023). However, under current regulatory definitions, hydrogen is not classified as a SAF. Because of its low density in gaseous form, it must be stored as a liquid at cryogenic temperatures (around −253°C), which requires specially designed tanks, entirely new aircraft architectures, and dedicated airport infrastructure.

6 As Gonzalez-Garay et al. (2022) have studied, the production of sustainable aviation fuels with renewable energy sources, electrolysis, and the Fischer–Tropsch process would have an impact on increasing flight ticket prices, with increases of up to 100%.

7 The U.S. Inflation Reduction Act (IRA), signed into law in August 2022, introduced a two-year blender's tax credit for SAF (2023–2024) ranging from $1.25 to $1.75 per gallon, depending on lifecycle emissions performance. From 2025, this incentive shifts to a production-based credit (Section 45Z). The IRA also includes $290 million in grants to support SAF production, infrastructure, and low-emission aviation technologies. Eligible SAF must achieve at least a 50% reduction in lifecycle GHG emissions compared to conventional jet fuel.

8 As a case in point, in July 2024, Airbus invested $50 million in LanzaJet to accelerate its industrial SAF production capacity based on Alcohol-to-Jet (ATJ) technology. This strategic investment aims to optimize the ethanol-to-SAF conversion process, which is fully compatible with current aircraft engines. Boeing signed a strategic partnership with European provider Norsk e-Fuel (march 2025) to develop one of the continent's first industrial-scale facilities to produce e-SAF.

9 « The network of organizations that are involved, through upstream and downstream linkages, in the different processes and activities that produce value in the form of products and services in the hands of the ultimate consumer » (Christopher, 1992).

10 A SAF production plant using forest residues and operating on the Fischer-Tropsch synthesis process -processing 960 tonnes of biomass per day- can yield approximately 90 tonnes of SAF per day, or about 33,000 tonnes annually. The total investment required for such a facility is estimated at around $422 million, with direct costs (equipment, piping, installation) accounting for more than 70% of that amount. For further details, see Ahire et al. 2024.

11 For a better understanding of take-or-pay clauses (as applied to the natural gas market rather than SAF), refer to the following works of Creti and Villeneuve (2004).

12 Crocker and Asten, 1985; Ubard and Weiner,1986.

13 The former CEO of American Airlines once admitted that, if he ever had to sell an asset, he would struggle to choose between the airline itself and its Sabre GDS—such was the financial power of these systems before the rise of the internet and low-cost carriers.

Chapter 4

Green Technologies and Sustainable Transport Towards Ecological Logistics

Ahlem Hamri
https://orcid.org/0000-0002-2052-021X
Higher School of Business of Tunis, University of Manouba, Tunisia & Laboratory of Accounting, Financial and Economic Modeling (MOCFINE), ISCAE, University of Manouba, Tunisia

Noomen Guirat
Higher Institute of Industrial Managment, Sfax, Tunisia

ABSTRACT

This work delves into the intersection of sustainability and logistics, beginning with a case study on Tunis, which highlights the challenges of congestion and emissions in urban freight transport. The study quantifies road freight emissions in Tunis, providing valuable insights into their environmental impact and the urgent need for sustainable alternatives.The chapter then transitions to exploring innovations in sustainable transportation, including electric vehicles, drones, and emerging technologies like intelligent transportation systems (ITS). These innovations are reshaping urban logistics, offering pathways to reduce emissions and enhance efficiency. Further, the chapter examines strategies for reducing the carbon footprint in logistics, such as route optimization, the use of biofuels and renewable energies, and reverse logistics practices. These approaches demonstrate how companies can minimize their environmental impact while maintaining operational effectiveness.

DOI: 10.4018/979-8-3373-3790-6.ch004

INTRODUCTION

As global concerns over climate change and environmental sustainability intensify, the logistics and supply chain sectors face increasing pressure to reduce their carbon footprint and adopt greener practices. Indeed, traffic congestion is a growing problem in Tunisia. This problem is becoming critical in our major cities, where daily commute times are reaching unprecedented levels that compare unfavorably with those observed in some foreign cities of equivalent size. This problem is also growing in medium-sized cities. As a result, it reduces the quality of life of Tunisians, wastes energy, and produces greenhouse gases and other pollutants that are harmful to Tunisia's environment. But above all, congestion generates significant economic costs. Investment and employment decisions depend on the quality of transportation infrastructure and the free movement of goods and people within and across our cities. Congestion increases current costs and discourages future investment. The effects of congestion are apparent, but its causes are complex. Fundamentally, congestion results from the overuse of a scarce, low-cost resource: road space. In its simplest form, congestion appears to result from insufficient investment in road and transit capacity. But this explanation underestimates the many factors that determine why and how people travel. A multitude of factors, including changes in the economy and employment, demographics, land use, and housing costs, influence congestion levels.

The goal is therefore not to reduce mobility, but to organize it in such a way that it does not cause negative externalities. Indeed, everyone has become aware of the need to seek adequate remedies for these ills and find relevant solutions to develop an efficient, sustainable, and high-quality transport system for their cities and citizens. With an eye on road traffic issues (particularly its automobile component, which remains a major source of controversy), road congestion is a major concern in terms of transport sustainability (Akhtar & Moridpour, 2021 ; Xu et al., 2024). Nowadays, it is primordial to explore innovative solutions and strategies to achieve these goals, focusing on the integration of advanced technologies and renewable energy sources into supply chain operations.

Road freight transport accounts for approximately 30% of total transport emissions (ITF, 2023). In Europe, it contributes approximately 5% of CO_2 emissions (EEA, 2022). Notably urban freight transport accounts for 10 to 20% of CO_2 emissions in cities (Coulombel et al., 2018) and contributes significantly to local pollution (NOx, fine particles). The sustainable supply chain seeks to reduce these impacts through innovative strategies (green logistics, pooling, clean vehicles). The growing demand for last-mile deliveries is increasing urban emissions (Hurlimann et al., 2021).

This work delves into the intersection of sustainability and logistics, beginning with a case study on Tunis, which highlights the challenges of congestion and emissions in urban freight transport. Using the CORINAIR and ATESAME models, the study quantifies road freight emissions in Tunis, providing valuable insights into their environmental impact and the urgent need for sustainable alternatives.

The chapter then transitions to exploring innovations in sustainable transportation, including electric vehicles, drones, and emerging technologies like intelligent transportation systems (ITS). These innovations are reshaping urban logistics, offering pathways to reduce emissions and enhance efficiency. Further, the chapter examines strategies for reducing the carbon footprint in logistics, such as route optimization, the use of biofuels and renewable energies, and reverse logistics practices. These approaches demonstrate how companies can minimize their environmental impact while maintaining operational effectiveness.

Finally, the chapter concludes with a summary of key points, emphasizing the importance of integrating green technologies into sustainable supply chains. It also offers suggestions for future research and best practices, encouraging continued innovation and collaboration in the pursuit of sustainability. By combining theoretical insights with practical case studies, this chapter aims to provide a comprehensive understanding of how technology and renewable energy can drive sustainable supply chains, offering valuable guidance for academics, industry practitioners, and policymakers alike.

1. CONGESTION: FROM CAUSES TO CONSEQUENCES

Congestion can have different definitions. In fact, congestion can be defined as a situation where transport users cannot move as they are accustomed to or as they would like. All vehicles, as well as pedestrians, can be affected. It is a widespread phenomenon when the capacity of an infrastructure is saturated. By definition, this capacity corresponds to the number of users per unit of time who pass through a given location. A traffic flow can be defined as a queue of transport users moving in the same direction. If we compare it to the flow of a river, congestion occurs when the usual or desired quantity of water does not flow in a given direction.

Road congestion is described as results from a mismatch between the flow of vehicles seeking to travel on an infrastructure and the capacity that the latter can offer. By capacity of an infrastructure, we mean the maximum number of vehicles that can travel on it in one hour, i.e. the maximum flow rate. It has an impact on social life and must studied (Shahid et al., 2021). Boiteux report (2001, p.53) defines congestion as follows: "congestion is defined as the inconvenience that vehicles impose on each other due to the relationship that exists between the traffic density on a route or on a network and the speed of traffic flow, taking into account capacity." In practice, this translates into the difference between "the measured speed of vehicles on an infrastructure during a given period (peak hour, in particular) and a certain reference speed"

We can talk about occupancy rates when the space occupied by a vehicle is proportional to its speed: driving fast means occupying more space because safety distances increase. Regardless of the initial traffic situation, the introduction of new vehicles will lead to increased competition in terms of occupancy rates, and maintaining a minimum level of safety will force motorists to slow down. This phenomenon is further amplified when there are significant speed differences between vehicles (as is the case at motorway entrances).

Another way to present this phenomenon is that as speeds increase, the number of vehicles on a given road space decreases: vehicle density is inversely proportional to their speed. These two equivalent relationships can be visualized graphically in a simple way if we do not consider curvature issues. These occupancy rate - speed or density - speed relationships have an impact on the flow rates observed on the road. Indeed, when traffic is relatively fluid, the addition of a few additional vehicles leads to a decrease in occupancy rates more quickly than the reduction in speeds. Thus, even if vehicles hinder each other and are forced to slow down, the flow recorded on the road can still increase. However, if the road continues to be busy, there comes a time when occupancy rates can hardly be compressed any further while speeds continue to drop significantly. At this limit point corresponding to the maximum capacity of the road, the flow will also decrease with speeds. We then obtain a flow-speed curve with a very characteristic shape

1.1. Types of Congestion

When dealing with congestion, it is important to distinguish between the two main types of congestion: recurring congestion and incident congestion.

1.1.1. Recurring Congestion

It is caused by demand, i.e. traffic volume, exceeding the (fixed) supply of the road network; it is mainly observed during the morning and evening peak periods; Recurrent congestion occurs mainly when too many vehicles use the road network at the same time (Gorji et al., 2022). Typically, recurrent congestion occurs on weekdays, in the morning and afternoon, when the vast majority of people travel to and from work or school at the same time. In large urban areas, the peak period is generally observed between 6:00 and 9:30 in the morning, and between 15:30 and 19:00 in the afternoon. In smaller urban areas, the peak period normally extends over a shorter period (one or two hours).

1.1.2. Incident Congestion

It, for its part, is associated with random or special conditions and particular events. It is the one that is caused locally by an accident, a breakdown, a construction site, etc.; it can occur at any time of day, but it is particularly penalizing when it is added to recurring congestion. It is caused by particular and fortuitous events, even minor ones. These include fender benders, mechanical breakdowns, running out of fuel, accidents, construction sites, weather conditions, etc. Slowdowns are often even caused by road users' curiosity about off-road or oncoming traffic incidents. Thus, congestion can temporarily appear on sections of the network that are not usually congested. Unlike recurring congestion, incidental congestion is not necessarily associated with heavy traffic. Incidental congestion is generally attributable to a temporary reduction in the intrinsic (or design) capacity of road links following an incident. It is therefore a completely different type of congestion from recurrent congestion in that, in the case of the latter, fluctuations in travel demand are the explanatory factor, whereas for incident congestion, it is the capacity factor, more precisely its temporary reduction, that explains the phenomenon.

This type of congestion can occur at any time, but also during peak periods, which exacerbates the delays attributable to recurrent congestion. Due to their unpredictable nature, delays caused by incidents are more penalizing for road users than traffic delays because they cannot adapt to them. However, traffic management systems can mitigate the impacts of incidents on users by informing them of disruptions or ensuring rapid intervention by emergency services.

2. THE NUISANCES OF URBAN ROAD TRANSPORT

This section is intended to provide a static image of the issues related to noise and air pollution from urban road transport. And this can be established at three levels: physical, by presenting the phenomena of air and noise pollution in urban areas as well as the growing share of transport in these phenomena; social, by questioning the categories of people most affected as well as the conflicts and stakeholder interactions that develop around these issues; economic, in the narrow sense of the term, by seeking to establish the level of the "external costs" related to these nuisances. A first section focuses on a physical presentation of pollution and its consequences as well as a brief evocation of the social issues they cover. However, it is the last point, economic, which has been favored here to address the issue of the issues related to these pollutions because it gives a measure of the effect to be consented to in order to avoid the nuisances they generate.

In traditional economic logic, it is often assumed that there is a "market" for avoidance or, if it does not exist, an attempt is made to create it in a fictitious manner. This logic crosses the cost of damages with the expenses necessary to avoid them, the objective being to minimize all of these costs. The methods presented can be used through this grid: they attempt to find a monetary expression of the tension that arises between a demand for avoidance (equivalent to the damage felt due to any pollution, translated at the individual or community level) and an offer of avoidance (whether tangible, such as double glazing for soundproofing, or monetary in the case of financial compensation). Depending on the case, the economic evaluation is more concerned with the curve of damages or more with that of the avoidance of these damages.

However, behind this simple reasoning distinguishing between damages and avoidance, practical implementations have led to a very wide variety of evaluation methods and results. To summarize this proliferation, we have adopted the fairly traditional division between microeconomic and macroeconomic approaches:

- There are purely economic approaches that seek to find or artificially create a market in which everyone would reveal their preferences. They are based on the classic microeconomic approach, and the methodological effort essentially focuses on finding ways to monetize these preferences.

It can already be emphasized that between the freely objective description of pollution phenomena and the subjective perception of the nuisances they cause, the space is wide enough for monetary transcriptions to easily vary from 1 to 10. Faced with such divergences in results, the material difficulties of understanding "external effects" and the statistical and technical uncertainties they generate are often mentioned. Obviously, these difficulties cannot be denied. But it is also certain that the problems at work behind the methods do not allow for identical results to be obtained. Here as elsewhere, there is no object entitled "negative external effect", "transport noise" or "air pollution" that can be viewed neutrally, in detail or with hindsight, to be able to fully grasp its reality.

2.1. Road Transport Nuisances: Physical and Social Aspects

2.1.1. Air Pollution Phenomena Due to Transport: Physical Aspects

In addition to the significant noise-related nuisances, road transport emits various pollutants: dust, hydrocarbons H(HC), sulfur oxides (SO2), carbon (CO) and nitrogen oxides (NOx), endangered lead (Pb), carbon dioxide (CO2), not to mention various esters. Their consequences are spread over a very wide temporal and spatial scale. Practically all of these pollutions are sources of immediate or very short-term nuisance. Air pollutants are also partly responsible for the formation and toxicity of urban smog; some (SO2 and NOx in particular) play a significant role in the acidification of the atmosphere (acid rain phenomena, etc.); finally, CO2 emissions contribute to the greenhouse effect.

The problem with this heterogeneity of nuisances is that the ecological levels concerned, and therefore the issues and social actors involved, are diverse. This diversity can lead to contradictory policies with regard to the seemingly simple objective of reducing air pollution. Moreover, the mechanisms at work in global phenomena such as the greenhouse effect are still far from being well understood and explained. This tends to dilute responsibilities and leave free rein to the games of actors who take up environmental problems according to their own interests: decision-making then results more from the balance of power present than from a "scientific and rational" assessment of the ecological, social and economic risks.

This first section aims to describe the effects of air pollution linked directly or indirectly to road transport in urban areas. Its objective is to make more tangible phenomena whose traditional economic approach, which we will revisit later, ultimately reveals only an abstract monetary valuation of their consequences.

2.2. At the Urban Level

At the local level, urban in our case, the nuisances to which road transport contribute are diverse. Their responsibility is also very unequal and is likely to vary given changes in lifestyles and technologies used in different sectors of economic activity. Overall, there has been an improvement in noise levels and urban atmosphere, although road transport continues to lag behind due to the continued increase in its use.

Before presenting the consequences of air pollution and briefly describing the effects of the main pollutants found in cities, we will focus on describing the specific characteristics of the urban climate. Indeed, these contribute to preventing the rapid dispersion of pollutants and reinforcing their effects. Finally, in a third point, we will look at the local nuisance that is noise.

2.2.1 Climatic Particularities of the Urban Environment

From one city to another, the urban atmosphere can vary greatly, depending on the latitude, the configuration of the site, the extent of human activities within the urban environment, the types of energy used, etc. Nevertheless, while not all large cities experience the same problems as Mexico City, Athens, or Los Angeles, it is still commonplace to observe that the air we breathe there is more polluted than in the surrounding countryside.

At a time when environmental issues are becoming a priority for everyone, it seems interesting to highlight the problems that many large cities will face. Indeed, just as with global climate change, taking urban climate into account in city planning is a long-term commitment, which requires immediate consideration. Indeed, if cities, and particularly large ones, want to remain attractive places to live, they will need to integrate new environmental data such as urban climate.

2.2.2. Polluting Gases from Road Transport

The various pollutants found in the urban atmosphere have diverse effects on living things—human health, flora, and fauna. They also affect inert materials, metals, and rocks, which is a concern for buildings in general and certain historic monuments in particular. They are often grouped into two broad categories based on their solid (dust) or gaseous state. Air pollutants from transport are classified into two categories: primary pollutants emitted directly from exhausts, and secondary pollutants resulting from the chemical transformation of the former in the atmosphere.

Carbon monoxide (CO) is caused by the rapid and incomplete combustion of combustibles and fuels. It reacts strongly with oxygen molecules. As a result, it transforms rapidly in the open air and is considered a proximity pollutant. This gas is mainly linked to car traffic and is found near busy streets and/or streets with very narrow spaces between buildings. Hydrocarbons and non-methane volatile organic compounds (NMVOCs), particularly benzene (C6H6), also result from incomplete combustion of fuel and oil during the engine operating cycle. Hydrocarbons play a significant role in the transformation of NOx into ozone (Aminzadegan et al., 2022).

Nitrogen oxides (NO and NO2) result from the combination of nitrogen and oxygen at high temperatures and, again, come mainly from road transport (nearly 60%) and combustion plants. At the exhaust, the mixture is approximately 90% NO (which then transforms into NO2) and 10% NO2. NOx reacts with water vapor to transform into nitric acid (NO3H). Furthermore, under the action of sunlight, through a complex interaction with water and certain volatile organic compounds present particularly in fuels, they produce ozone. Sulfur compounds (SO2) are primarily emitted by diesel vehicles. They are primarily the result of the combustion of sulfur-containing fossil fuels (coal, fuel oil, diesel, etc.).

Micro-particles (PM10)2, emitted overwhelmingly by diesel vehicles, are classified as fine particles, such as those from engine fumes, and large particles from road surfaces. Currently, vehicle emissions are regulated for four of these pollutants: carbon monoxide (CO), hydrocarbons (HC), nitrogen oxides (NOx), and particulate matter (PM10). Lead (Pb) Locally, it is found near factories that prepare this metal in a molten state, but more generally, its presence in cities is due to the introduction of organic compounds into liquid combustion fuels for motor vehicle engines.

Ozone (O3) is a secondary pollutant that results from the photochemical transformation of certain primary pollutants in the atmosphere (particularly NOx and VOCs) under the influence of ultraviolet radiation.

It can be added that transport contributes significantly to total emissions. Since 1994, studies concerning the share of transport in atmospheric pollution in several large cities, show that in terms of CO, hydrocarbons, and NOx, their responsibility systematically exceeds 50% (more than 90% for CO) (Faiz et al, 1996). A study carried out on the Lyon metropolitan area in 1990 confirms this trend by attributing respectively 96, 84 and 87% of CO, non-methane hydrocarbon and NOx emissions to transport.

3. URBAN AND PERI-URBAN TRAVEL TUNIS CITY

3.1. Legal and Institutional Framework

The organization of urban and regional transport is governed by Law No. 2004-33 of April 19, 2004, on the organization of land transport. This law, which replaced the 1985 law, established and developed different principles: the citizen's right to transport and the freedom to choose its mode, the need to prioritize the use of public transport, particularly rail transport, due to the advantages it offers for mass transport and certain specific goods, the recognized possibility for private companies to provide public collective passenger transport etc.

Urban transport is a sector involving several highly diverse stakeholders, both in nature (from the central government to the private sector) and in roles (planning, financing, operation, etc.). Furthermore, the complexity of the institutional framework, which is not specific to Tunis or Tunisia, is due to objective factors, including:

- the type of services and infrastructure characterizing transport systems: technical diversity (multiple modes), diverse infrastructure (roads, rail, etc.);
- the strong links between the different subsystems; in other words, users understand and use the transport offering as a whole independently of the structures ensuring its diversity.

3.2. Evaluation of Freight Urban Transport of the City of Tunis

The National Agency for Environmental Protection (NAEP) publishes an annual report on air quality by making firstly a spatial analysis allowing to draw the particularities and the critical situations and to compare them with the Tunisian standard and secondly to make a temporal analysis by following the evolution of each pollutant through time and finally, in a last phase, the analysis focuses on the critical situations, if they exist, otherwise on the particularities with respect to the current or on the highest measurements with respect to the others and this by making a coupling of this pollutant with the different meteorological parameters.

Given that the transport of goods and notably urban freight transport is an important element of logistics chains, we seek in this section to show the share of this transport in polluting emissions in the city of Tunis.

3.2.1 Methodology

Road traffic is a complex reality. Given the complexity of the process, it is necessary to use tools that facilitate understanding of it. This is where models come in. The justification for such an approach can easily be explained if we think about the difficulty, or even the impossibility, of, for example, testing the effects on traffic of a new infrastructure by building it on a trial basis (and then demolishing it if it does not meet expectations).

When developing the model, the user must then make choices regarding the use he wishes to make of this model, as well as the data available or accessible through observation. However, we do not wish to dwell on the modeling strategies

In order to be able to answer certain questions more easily (calculating the shortest path with or without deviation for example), the GRT has developed the ATESAME software. This computer program is therefore a tool for providing a model of a situation averaged over a period considered on a traffic network. To do this, the user can choose from five allocation methods: user balance, system balance, stochastic user balance, probit allocation and all or nothing allocation (Sheffi, 1985). Another very important facet of this software is to be able to update an origin/destination (O/D) matrix based on counts (the method used is a Furness type method), ATESAME model is used to determine all origin-destination matrices and allocate traffic based on the emissions formulas extracted from the CORINAIR methodology.

Thus, a schematic representation of the territory is first required. This description consists of dividing the studied territory into different homogeneous zones according to the study's motivations, the socioeconomic context of the region, and the available data. Each of these zones is represented by a point, called a centroid. Next, the roads in the network under consideration are represented by arcs whose intersections are also points called intersections. These arcs are associated with a series of characteristics, such as length, speed limit, number of lanes, capacity, sensitivity to congestion, etc. We thus obtain a network of the region studied.

3.2.1.1. Activities Generating the Movement of Freight

We distinguish - Industrial zones - Office and commercial zones - Recreational zones This last category of zone only generates occasional trips while the first two activity zones generate daily trips.

The industrial zones were established at the beginning of the century in the south of the agglomeration in relation to the port of Tunis. The concentration of industrial activities was at the origin of the development of working-class neighborhoods, as well as the first nuclei of gourbivilles which appeared in Tunis from 1942. After independence and more precisely from the 70s, the concentration of industrial zones in the southern zone of Tunis was reinforced by the extension of these industrial zones over several hundred hectares. During the 80s the development of urbanization took place in the north and west of the agglomeration. This spatial fragmentation of industrial functions and residential functions generated significant home-work travel. The rebalancing attempted within the framework of the regional development plan of Tunis did not ensure the rebalancing between residential functions and industrial functions because the western zone only integrated about a hundred hectares of industrial zones and the concentration of large areas (relay points) facilitates delivery to the relay point or delivery from commercial or industrial areas to urban areas

3.2.2. General Issue

A first avenue that we could take to evaluate transport emissions would be to use traffic models that provide the most accurate description possible of traffic conditions. The kinematic characteristics of vehicles appear to be decisive in explaining their emission levels. Each vehicle is assigned one of the following four phases: acceleration, cruising speed, deceleration and stopping. Initial work in this direction has shown that emissions are not uniformly distributed along the same arc. In particular, there are strong pollution peaks at intersections due to traffic lights, frequent stops and left turns. This method appears very effective for establishing a balance sheet of local nuisances from road pollution; it is also interesting for testing the immediate environmental impact of an infrastructure modification to improve traffic flow and thus reduce or increase the journey time of vehicles in circulation. Unfortunately, the complexity of these calculations limits this approach to a network that does not contain too many arcs and intersections.

Therefore, it is more appropriate for the study area to refer to a static aggregated model to obtain a more comprehensive assessment. This model is said to be aggregated in the sense that each trip is not considered in isolation. However, these models have some disadvantages compared to dynamic models. Note, for example, that static models only provide an average speed per arc. However, they allow for larger-scale assessments. We return in more detail to the disadvantages of this type of model by interpreting the results, particularly by examining the influence of using average speed to assess emissions. Using this model, we obtain a static image of the flows along the different arcs, meaning we obtain an average balance of what we would obtain if we made a series of observations over several days over the same period. These results are in no way relevant to a specific moment in time. However, for the purpose of obtaining an assessment of pollutant emissions over long periods, this period proves to be entirely adequate.

Once the allocation phase is complete, we can assess the overall emissions of polluting gases. We perform this final step using the formulas provided by the CORINAIR group (Eggleston et al., 1993). Indeed, these formulas provide average emissions per vehicle for one kilometer traveled at a given speed and for a given temperature. Once these total emissions have been assessed, we will redistribute them according to vehicle type or by study area. The last step is the most complicated to model. In fact, during this step, two types of data mainly come into play: meteorological data and topographical data. This last step consists, from the total emissions per arc, in determining the actual concentration of polluting gases present in the atmosphere by taking into account the different meteorological phenomena (temperature can act as a catalyst during certain chemical reactions and thus reduce the actual concentration of polluting gases in the atmosphere) and topographical data (the topography of certain streets can increase the prevailing wind phenomena and thus reduce the concentrations of pollutants present in the atmosphere).

We call it the diffusion phase and intentionally omit it. Its main purpose is to focus on the dispersion of pollutants in the atmosphere and not on the actual assessment of gas emissions. We can therefore summarize this problem into four distinct phases: the allocation, emissions assessment, emissions distribution, and diffusion phases (the latter part being omitted).

3.3. The CORINAIR Methodology

Developed since the mid-1980s, this methodology aims to obtain inventories that allow the assessment of emissions for a wide range of polluting gases (ranging from CO to alkanes, including zinc, etc.). These inventories must offer essential fundamental qualities: they must be consistent, complete, and comparable. To achieve this result, the researchers based their research on two main sources: manufacturers' technical data and vehicle fleet data. A series of field tests confirmed and sometimes refuted the formulas obtained.

3.3.1. Program Execution and Implementation of Emission Formulas

The ATESAME model is applied to the study area, and we will supplement it by implementing various emission formulas. Our approach during implementation was to continue the work done by the ATESAME software. We wanted to allow ATESAME users to obtain an assessment of gas emissions without excessive manipulation.

4. RESULTS

All calculations made, our network comprises: 70% of lightweight vehicle running on gasoline, 20% of lightweight vehicle running on diesel and 10% of lightweight vehicle on gas,

Results show that Carbon monoxide (CO), carbon dioxide (CO2), sulfur dioxide (SO2) emissions decrease when traffic becomes fluid. Emissions of this gas by means of transport are significant and increase as the vehicle fleet ages. Otherwise, the emissions of carbon dioxide are more important than carbon monoxide. Indeed, to fuel the current debate on how to address the problems related to urban road traffic nuisances, this work raises the issue of congestion and the problems related to urban road traffic nuisances, we raise the issue of congestion and the problems of internalizing its environmental effects. We can touch on the development of diesel within the light vehicle fleet and its consequences in terms of polluting emissions. A third refers to a general reflection on the conditions for sustainable growth in mobility. Transport and congestion have an ambiguous relationship. On the one hand, only a few types of pollutants are very sensitive to variations in traffic conditions: hydrocarbons and carbon monoxide (CO).

Emission levels of other pollutants remain much more stable when speeds vary, and some even appear to decrease, such as NOx. This suggests that combating congestion, in itself, does not improve the environmental performance of road traffic, or only slightly. The main factor explaining emission levels remains the overall level of traffic, i.e. the total distance traveled by all vehicles in the territory and period considered. In addition, a policy aimed at maintaining or improving speeds may not succeed due to criteria of economic efficiency and supply chains. Besides, today, it is necessary for car manufacturers to make even more efforts to develop the ecological performance of their vehicles, but these efforts are not enough because the users of these vehicles must optimize their use.

The growing concerns over environmental degradation, climate change, and urban air quality have placed significant pressure on the transportation sector to adopt more sustainable practices. Road traffic, a major contributor to greenhouse gas emissions and air pollutants, plays a central role in this challenge. Recent studies, such as the one discussed above, highlight the complex relationship between traffic conditions, vehicle emissions, and environmental sustainability. While reducing congestion and optimizing traffic flow can mitigate emissions to some extent, the broader issue lies in the overall volume of traffic, the composition of vehicle fleets, and the efficiency of transportation systems. These findings underscore the urgent need for innovations in sustainable transportation that address both the environmental and logistical challenges of modern mobility.

5. INNOVATIONS IN SUSTAINABLE TRANSPORTATION

The discussion above provides a foundation for understanding the critical areas where innovation can drive sustainability in transportation. For instance:

- Technological Advancements: The study highlights the role of vehicle fleet composition and aging in emissions, emphasizing the need for cleaner technologies. Innovations such as electric vehicles (EVs), hydrogen fuel cells, and hybrid systems can significantly reduce emissions, especially in logistics and urban delivery fleets.
- Traffic Management and Optimization: While congestion reduction alone has limited environmental benefits, integrating smart traffic management systems and AI-driven route optimization can enhance efficiency and reduce emissions across transportation networks.
- Sustainable Mobility Solutions: The study's focus on total distance traveled aligns with innovations like shared mobility (e.g., carpooling, ride-sharing), public transit electrification, and active transportation (e.g., cycling, walking), which can reduce the overall traffic volume.
- Policy and Infrastructure Development: Innovations in urban planning, such as low-emission zones, congestion pricing, and micro-distribution hubs, can create environments that support sustainable logistics and reduce the environmental impact of transportation.
- Behavioral and Operational Changes: The study emphasizes the role of user behavior in emissions, pointing to the potential of eco-driving technologies, driver training programs, and data-driven logistics to optimize vehicle use and minimize environmental harm.

By leveraging these innovations, the transportation sector can address the challenges identified in the study, paving the way for a more sustainable and efficient future. The integration of technology, policy, and behavioral change will be key to achieving this transformation. Indeed, as EVs represent one of the most significant innovations in addressing the challenges of traffic-related emissions and sustainability in transportation. Here's how the two are linked:

The study highlights that emissions like carbon monoxide (CO), carbon dioxide (CO_2), and sulfur dioxide (SO_2) are significant contributors to air pollution, particularly from aging vehicle fleets and diesel engines and Electric vehicles produce zero tailpipe emissions, eliminating the release of CO, CO_2, and other harmful pollutants. This directly addresses the environmental concerns raised in the study, especially in urban areas where traffic congestion exacerbates air quality issues. Then, The study emphasizes that emissions increase as vehicle fleets age, particularly in the case of diesel and gasoline-powered vehicles. Also, EVs offer a cleaner alternative to traditional internal combustion engine (ICE) vehicles. By transitioning to EVs, logistics companies and urban fleets can modernize their operations and significantly reduce their environmental footprint.

While the study notes that congestion reduction alone has limited environmental benefits, EVs can still play a role in improving traffic efficiency. For example, electric delivery vans and trucks are often quieter and more efficient in stop-and-go traffic, making them ideal for urban logistics. Additionally, smart charging infrastructure and vehicle-to-grid (V2G) technologies can optimize energy use and reduce the strain on power grids, further enhancing sustainability.

Sustainable Mobility and Total Distance Traveled

The study identifies the total distance traveled by vehicles as a key factor in emissions. EVs, especially when paired with shared mobility solutions (e.g., electric car-sharing or ride-hailing services), can reduce the number of vehicles on the road and the overall distance traveled.

For logistics, electric last-mile delivery vehicles can replace traditional diesel vans, reducing emissions in densely populated urban areas.

Policy and Infrastructure Development

The study calls for integrated policies to address emissions and congestion. Governments and cities are increasingly incentivizing EV adoption through subsidies, tax breaks, and the development of charging infrastructure.

Innovations like fast-charging networks and battery-swapping stations are critical to supporting the widespread use of EVs in both personal and commercial transportation.

Behavioral and Operational Changes

The study highlights the importance of user behavior in reducing emissions. EVs encourage eco-driving practices due to their regenerative braking systems and energy-efficient designs.

In logistics, companies can optimize EV usage through route planning software and fleet management systems, ensuring that vehicles are used efficiently and sustainably.

Electric vehicles are a key innovation that directly addresses the challenges outlined in the study. By reducing emissions, modernizing vehicle fleets, and supporting sustainable mobility, EVs play a pivotal role in creating a cleaner, more efficient transportation system. Their integration into urban logistics and personal transportation is essential for achieving long-term sustainability goals and mitigating the environmental impact of road. Indeed, this innovation highlights several key challenges in transportation sustainability, such as emissions from aging vehicle fleets, the limitations of congestion reduction, and the need for cleaner technologies. Electric vehicles (EVs) emerge as a critical innovation to address these challenges, offering a direct link to the findings and implications of the study. Here's how EVs connect to the discussion:

Reducing Emissions from Aging Fleets

- o The study emphasizes that emissions increase as vehicle fleets age, particularly for pollutants like CO, CO_2, and SO_2. EVs, which produce zero tailpipe emissions, can significantly reduce these pollutants, especially in urban areas where air quality is a major concern.
- o By replacing older, combustion-engine vehicles with EVs, cities and logistics companies can achieve immediate reductions in harmful emissions.

Addressing the Limitations of Congestion Reduction

- o While reducing congestion has limited environmental benefits, EVs can complement traffic management efforts by ensuring that vehicles on the road are cleaner and more efficient.
- o Even in congested conditions, EVs contribute less to air pollution compared to traditional vehicles, making them a valuable tool for improving urban air quality.

Supporting Sustainable Mobility and Reducing Total Distance Traveled

- o The study identifies the total distance traveled by vehicles as a primary factor in emissions. EVs, particularly when paired with shared mobility models (e.g., electric car-sharing or ride-hailing), can reduce the need for individual car ownership and lower overall traffic volume.

- o In logistics, electric delivery vans and trucks can enable last-mile delivery with minimal environmental impact, supporting sustainable urban freight systems.

Enhancing Energy Efficiency and Renewable Integration

- o EVs are inherently more energy-efficient than internal combustion engine vehicles, converting a higher percentage of energy from the grid to power at the wheels.
- o When charged with renewable energy sources (e.g., solar or wind), EVs can operate with near-zero carbon emissions, aligning with broader sustainability goals.

Driving Innovation in Policy and Infrastructure

- o The adoption of EVs requires supportive policies, such as incentives for EV purchases, investment in charging infrastructure, and low-emission zones. These measures can accelerate the transition to cleaner transportation systems.
- o For logistics, the development of fast-charging networks and battery-swapping technologies can ensure that electric fleets operate efficiently and reliably.

Encouraging Behavioral Change

- o The study highlights the role of user behavior in emissions. EVs, with their lower operating costs and environmental benefits, can incentivize drivers and companies to adopt more sustainable practices.
- o In logistics, training drivers to optimize EV performance and integrating EVs into eco-friendly delivery strategies can further enhance sustainability.

5.1. Electric Vehicle Technologies

5.1.1. Overview of Electric Vehicle Technologies

Electric vehicles are a cornerstone of sustainable transportation innovation, directly addressing the challenges identified in the study. By reducing emissions, improving energy efficiency, and supporting cleaner urban mobility, EVs offer a practical and scalable solution to the environmental and logistical issues posed by traditional road traffic. Their widespread adoption, supported by technological advancements, policy measures, and behavioral changes, can play a transformative role in achieving sustainable transportation systems.

Electric vehicles (EVs) represent a transformative shift in transportation, offering cleaner, more efficient alternatives to traditional internal combustion engine (ICE) vehicles. Below is an overview of the key technologies that define EVs and enable their operation:

Types of Electric Vehicles

- Battery Electric Vehicles (BEVs): Fully electric vehicles powered solely by rechargeable batteries. Examples include the Tesla Model 3 and Nissan Leaf.
- Plug-in Hybrid Electric Vehicles (PHEVs): Combine a battery-powered electric motor with an ICE. They can run on electricity for short distances and switch to gasoline for longer trips. Examples include the Toyota Prius Prime and Chevrolet Volt.
- Hybrid Electric Vehicles (HEVs): Use both an ICE and an electric motor but cannot be plugged in. The battery is charged through regenerative braking and the ICE. Examples include the Toyota Prius and Honda Accord Hybrid.

Core Components of EVs

- Electric Motor: Converts electrical energy from the battery into mechanical energy to drive the wheels. EVs often use AC induction motors or permanent magnet motors for their efficiency and reliability.
- Battery Pack: Stores electrical energy to power the motor. Most modern EVs use lithium-ion batteries due to their high energy density, long lifespan, and declining costs.
- Power Electronics: Includes the inverter, which converts DC power from the battery to AC power for the motor, and the onboard charger, which manages the charging process.
- Regenerative Braking: Captures kinetic energy during braking and converts it into electrical energy to recharge the battery, improving overall efficiency.
- Thermal Management System: Ensures optimal operating temperatures for the battery, motor, and power electronics, enhancing performance and longevity.

Charging Infrastructure

- Level 1 Charging: Uses a standard 120V household outlet. Slow but convenient for overnight charging.
- Level 2 Charging: Requires a 240V outlet and is faster, commonly used in homes, workplaces, and public charging stations.
- DC Fast Charging: Provides rapid charging (up to 80% in 20-30 minutes) using high-power DC stations, ideal for long-distance travel.
- Wireless Charging: Emerging technology that allows EVs to charge without physical cables, using inductive charging pads.

Battery Technology

- Lithium-Ion Batteries: The dominant technology due to their high energy density, efficiency, and declining costs.
- Solid-State Batteries: A promising next-generation technology offering higher energy density, faster charging, and improved safety by replacing liquid electrolytes with solid materials.
- Battery Management Systems (BMS): Monitor and optimize battery performance, ensuring safety, efficiency, and longevity.

Vehicle-to-Grid (V2G) Technology

Allows EVs to feed electricity back into the grid during peak demand, turning them into mobile energy storage units. This supports grid stability and renewable energy integration.

Autonomous and Connected Features

Many EVs are equipped with advanced driver-assistance systems (ADAS) and connectivity features, paving the way for autonomous driving, and integration with smart grids and IoT enables real-time monitoring, route optimization, and energy management. EVs contribute to reducing greenhouse gas emissions, especially when charged with renewable energy. Battery recycling programs are being developed to recover valuable materials like lithium, cobalt, and nickel, minimizing environmental impact

5.1.2. Case Studies on Electric Vehicle Adoption in Urban Logistics

Electric vehicles (EVs) are increasingly being adopted in urban logistics to reduce emissions, lower operating costs, and comply with stricter environmental regulations. Below (Table.1) are real-world case studies highlighting their environmental benefits and challenges.

Table 1. Case studies on electric Vehicule adoption in Urban Logistics

Company	Location	Fleet	Adoption & Implementation	Environmental Benefits	Challenges
Amazon's Electric Delivery Vans (Rivian Partnership)	United States & Europe	100,000 custom electric delivery vans (EDVs) by 2030	Amazon partnered with Rivian to develop electric delivery vans (EDVs) with a range of ~150 miles per charge. Deployed in major U.S. cities (e.g., Los Angeles, Chicago) and Europe (e.g., Munich, London). Uses AI-powered routing to optimize deliveries and charging schedules.	CO_2 Reduction: Expected to eliminate 4 million metric tons of CO_2 annually by 2030. Zero Tailpipe Emissions: Reduces urban air pollution (NOx, PM2.5). Energy Efficiency: Regenerative braking improves efficiency in stop-and-go traffic.	Charging Infrastructure: Requires large-scale depot charging stations. Upfront Costs: High initial investment compared to diesel vans. Cold Weather Performance: Reduced range in extreme temperatures.
DHL's StreetScooter Electric Vans	Germany & Europe	~20,000 StreetScooter EVs deployed	DHL developed its own StreetScooter electric vans for last-mile deliveries. Range: ~80-120 miles, suitable for urban routes. Used in cities like Berlin, Bonn, and Paris.	CO_2 Savings: Each van reduces 5 tons of CO_2 per year compared to diesel equivalents. Noise Reduction: Quieter operations enable nighttime deliveries in noise-sensitive areas. Lower Urban Pollution: Eliminates diesel particulate emissions.	Battery Degradation: High daily usage leads to faster battery wear. Limited Payload Capacity: Smaller cargo space than conventional vans. Production Halted: DHL discontinued manufacturing due to financial issues, highlighting scalability challenges.

continued on following page

Table 1. Continued

Company	Location	Fleet	Adoption & Implementation	Environmental Benefits	Challenges
UPS's Electric & Hybrid Fleet in London	London, UK	Over 300 electric & hybrid vehicles	UPS uses electric trucks, e-bikes, and drones for urban deliveries. Partners with Arrival for custom-built electric vans (~150-mile range). Uses micro-depots to reduce last-mile emissions.	CO_2 Reduction: Cut 12% of CO_2 per package in London since 2018. Congestion Relief: Smaller EVs navigate narrow streets better than large trucks. Renewable Charging: Uses solar-powered charging hubs.	High Initial Costs: EVs are ~50% more expensive than diesel equivalents. Limited Range for Heavy-Duty Routes: Long-haul logistics still rely on diesel. Grid Capacity: High-power charging requires grid upgrades in dense cities.

continued on following page

Table 1. Continued

Company	Location	Fleet	Adoption & Implementation	Environmental Benefits	Challenges
FedEx's Electric Trucks in California (BrightDrop Partnership)	California, USA	Hundreds of BrightDrop Zevo 600 electric vans	FedEx collaborated with BrightDrop, a subsidiary of General Motors, to deploy the Zevo 600 electric vans. Range & Capacity: The Zevo 600 has a range of up to 250 miles and a payload capacity of 2,200 pounds, making it suitable for urban and suburban delivery routes. Deployment: FedEx has integrated these electric vans into its operations in California, a state with stringent emissions regulations and a robust EV charging infrastructure. Charging Infrastructure: FedEx has installed on-site charging stations at its facilities to support the fleet.	CO_2 Reduction: Each Zevo 600 van is estimated to reduce 10 metric tons of CO_2 emissions annually compared to diesel-powered vehicles. Over the lifetime of the fleet, this translates to significant reductions in greenhouse gas emissions. Zero Tailpipe Emissions: Eliminates harmful pollutants like NOx and particulate matter, improving air quality in urban areas. Energy Efficiency: Electric vehicles are more energy-efficient than internal combustion engines, especially in stop-and-go urban traffic.	High Upfront Costs: The initial investment in electric vans and charging infrastructure is significantly higher than traditional diesel vehicles. Charging Infrastructure Limitations: While California has a relatively developed charging network, scaling up to support large fleets requires further investment. Range Constraints for Long Routes: The Zevo 600's range of 250 miles may not be sufficient for longer rural or intercity delivery routes, limiting its applicability. Battery Degradation: Frequent charging and high usage can lead to battery degradation over time, impacting performance and li

Source author's work

5.2. Drones and Drone Delivery

Drones and drone delivery are increasingly important in the realm of sustainable transportation. Their ability to perform quick and efficient deliveries can contribute to reducing carbon emissions and improving logistics efficiency. Here are several case studies of companies using drones for delivery, along with their impacts on emission reduction and logistics efficiency.

5.2.1. Case Studies of Companies Using Drones for Delivery

Amazon Prime Air

Amazon has developed a drone delivery service called Prime Air, which aims to deliver packages in less than 30 minutes. This system has the potential to significantly reduce CO2 emissions compared to traditional delivery vehicles, especially in urban areas. By replacing delivery trucks with electric drones, Amazon hopes to minimize the carbon footprint of its operations while enhancing service speed.

Zipline

Zipline specializes in delivering medical supplies via drone, primarily operating in Africa. They deliver blood products, vaccines, and other essential supplies to remote areas. By using drones, Zipline has reduced the need for ground transportation, which decreases greenhouse gas emissions and improves access to healthcare in underserved regions.

Wing (a subsidiary of Alphabet)

Wing has launched a drone delivery service in Australia and the United States. Their technology allows for the delivery of small packages, such as medicines and food, directly to consumers. Wing has demonstrated that drones can reduce the need for motor vehicles, contributing to emission reductions in urban areas while providing a faster and more efficient delivery solution.

Flirtey

Flirtey is a drone delivery company that has conducted deliveries in the United States and Australia. Their service has been used to deliver essential items like medical supplies and food. Flirtey has shown that drone delivery can cut down transport times and associated emissions, particularly in areas where road infrastructure is lacking.

5.2.2. Impacts on Emission Reduction and Logistics Efficiency

Reduction of Carbon Emissions: Electric drones emit significantly less CO2 than delivery trucks, contributing to efforts against climate change.

Logistics Efficiency: Drones can bypass road congestion and make deliveries more quickly, enhancing the overall efficiency of the supply chain.

Accessibility: Drones enable delivery in hard-to-reach areas, reducing reliance on more polluting vehicles.

In conclusion, the use of drones for delivery represents a significant advancement toward more sustainable transportation. Companies adopting this technology can not only improve their logistics efficiency but also contribute to reducing greenhouse gas emissions, which is essential in the current context of transitioning to more sustainable practices.

5.3. Sustainable Public Transportation Technologies

5.3.1. Electric Buses

Example: BYD Electric Buses (China, Worldwide): BYD manufactures electric buses used in cities like Los Angeles, London, and Shenzhen. Shenzhen became the first city in the world to fully electrify its public bus fleet. These buses help cut down noise pollution and eliminate tailpipe emissions.

5.3.2. Hydrogen Fuel Cell Buses

Example: Toyota Sora (Japan): Toyota's hydrogen-powered Sora buses are used in Tokyo’s public transport network. They emit only water vapor and are designed to support Japan's goal of a hydrogen-based economy.

5.3.3. Modern Trams and Light Rail Systems

Example: Alstom Citadis (Europe, Australia): These energy-efficient trams are used in cities like Sydney, Paris, and Barcelona. They often run on renewable electricity and are designed with regenerative braking systems to improve energy efficiency.

5.4. Intelligent Transportation Systems (ITS)

5.4.1. Smart Traffic Management Systems

Example: SCOOT (Split Cycle Offset Optimization Technique) – United Kingdom: Used in London and other cities, SCOOT adjusts traffic light timings in real-time based on traffic flow data. This reduces congestion and improves fuel efficiency.

5.4.2. Route Optimization for Public Transit

Example: Google's Transit Data with AI Integration: Some cities use AI-powered route optimization tools integrated with real-time data (e.g., traffic, weather) to reroute buses and reduce delays, ultimately cutting emissions.

5.4.3. Connected Vehicle Infrastructure

Example: V2X (Vehicle-to-Everything) in Ann Arbor, Michigan, USA: The University of Michigan's MCity project tests how connected vehicles can communicate with traffic signals and road infrastructure to improve safety and reduce idling time.

6. REDUCING THE CARBON FOOTPRINT IN LOGISTICS

The integration of emerging transportation technologies—such as drone delivery systems, electric public transit, and intelligent transportation systems (ITS)—marks a crucial shift toward more sustainable logistics. These innovations not only improve operational efficiency but also align with broader strategies to reduce the carbon footprint across the logistics and transportation sectors. This includes advancements in route optimization, the adoption of renewable energies and biofuels, and the implementation of reverse logistics strategies.

6.1. Route Optimization

Efficient route planning is one of the most immediate and impactful ways to reduce carbon emissions in logistics. By shortening travel distances, reducing idle time, and avoiding traffic congestion, companies can significantly lower fuel consumption and CO_2 output. Technologies such as GPS-based fleet management software, AI-powered route optimization tools, and real-time traffic data (enabled by ITS) are transforming how logistics providers plan and execute deliveries. For example, companies like Amazon and FedEx use sophisticated algorithms to dynamically adjust delivery routes based on weather, traffic, and delivery priority—leading to measurable emissions reductions.

6.2. Use of Biofuels and Renewable Energies

As fossil fuels remain a major source of emissions in logistics, the adoption of biofuels and renewable energy is increasingly essential. Biofuels, such as biodiesel and ethanol, offer a cleaner-burning alternative to diesel and gasoline, and are already being used in hybrid and heavy-duty delivery fleets. Meanwhile, renewable electricity is powering everything from electric delivery vans to drone charging stations. For instance, UPS has invested in fleets that run on renewable natural gas (RNG), while DHL is expanding the use of electric vehicles and powering logistics hubs with solar energy. These approaches not only reduce greenhouse gas emissions but also support long-term energy resilience.

6.3. Reverse Logistics Strategies

Reverse logistics refers to the process of moving goods from their final destination back to the manufacturer or distributor for return, reuse, recycling, or disposal. This strategy plays a vital role in building circular supply chains, which reduce waste, lower emissions from manufacturing, and extend product life cycles. Companies like IKEA have implemented reverse logistics systems to collect and refurbish old furniture, while HP offers cartridge return programs and recycles used electronics. These initiatives not only reduce landfill waste but also minimize the need for energy- and resource-intensive new production.

Strategies for reverse logistics are essential for controlling the flow of returned goods, cutting expenses, and improving sustainability. Establishing unambiguous return rules that make the process easier for consumers—such as simple instructions and pre-paid return labels—is the first step in effective return management. Recurring problems with particular items can be found by analyzing return data, which enables quality enhancements. Furthermore, recycling and refurbishment initiatives might offer chances to salvage value from products nearing the end of their useful lives. The return process can be streamlined by working with outside logistics companies, and technology and automation can improve return tracking and management. Employee education on these procedures guarantees effective handling, and a smooth system for handling returned items can be achieved by incorporating reverse logistics into the supply chain as a whole.

Reverse logistics techniques are used successfully by a number of sectors to improve customer satisfaction and operations. Here are few instances:

Retail and e-commerce: Businesses with strong return policies and procedures in place include Amazon and Zappos. Customers can return things with no difficulty because to their simple return policies and pre-paid shipping labels. These businesses examine return data in order to spot patterns and enhance the caliber of their products.

Consumer electronics: Reverse logistics is used by companies like Apple and Best Buy to handle returns of faulty goods. In order to reduce waste and maximize resource recovery, they frequently recycle parts or fix returned gadgets for resale.

Automobile: To handle returned parts and components, the automobile sector uses reverse logistics. Businesses such as Ford and General Motors have initiatives for reusing materials or repairing parts.

Pharmaceuticals: In this sector, handling unused or expired drugs requires the use of reverse logistics. Systems for the safe return and disposal of medications are in place at businesses like Walgreens and CVS, guaranteeing regulatory compliance and reducing environmental hazards.

Fashion and Apparel: To manage returns from online purchases, retailers such as ASOS and Zara employ reverse logistics. In addition to putting recycling processes for unsold or returned clothing into place, they examine return trends to enhance the fit and quality of their products.

Companies in the food and beverage industry, like Coca-Cola, handle reverse logistics for returns of unsold or expired goods. To reduce their influence on the environment, they frequently recycle packaging materials and make sure food waste is disposed of properly.

These sectors show how efficient reverse logistics techniques can improve operational

6.3.1. Connected Framework for Sustainable Logistics

To encourage environmental, social, and economic sustainability in supply chains, A Connected Framework for Sustainable Logistics combines cutting-edge technologies, cooperative collaborations, and the concepts of the circular economy. Organizations may improve visibility and streamline processes by utilizing advances like IoT and AI, which will ultimately lower expenses and emissions. Stakeholder collaboration promotes pooled resources and coordinated efforts, which reduce waste and boost productivity. Environmental effects are further reduced by using eco-friendly packaging and sustainable shipping techniques.

Tracking sustainability indicators is made possible by supply chain transparency, and encouraging consumers to adopt ethical behavior strengthens brand loyalty. By taking this all-encompassing strategy, businesses not only meet legal obligations but also establish themselves as industry leaders in sustainability, encouraging creativity and adaptability in their logistical processes.

These practices—optimized routing, renewable energy use, and reverse logistics—complement emerging transportation technologies like drones, electric buses, and ITS. They form an interconnected framework for reducing emissions, improving efficiency, and advancing the transition to sustainable logistics. As global supply chains grow increasingly complex, integrating these strategies will be key to meeting environmental targets and building more resilient, eco-friendly transportation systems.

CONCLUSION

This chapter has explored the intersection of sustainable transportation and logistics through the lens of the Tunis case study, offering a detailed analysis of the challenges, innovations, and strategies necessary for reducing environmental impacts in urban supply chains.

The case of Tunis highlights the urgency of addressing road congestion and freight-related emissions, underscoring the environmental pressures faced by many growing urban centers. Using the CORINAIR methodology and the ATESAME model, the study provided a quantitative assessment of the emissions generated by road freight transport in the city. The results revealed significant levels of CO_2 and other pollutants, reinforcing the need for more sustainable freight strategies and greener infrastructure.

Building on these findings, the chapter examined a range of innovations in sustainable transportation, including electric vehicles, drone delivery systems, and other emerging technologies such as intelligent transportation systems (ITS) and electric public transit. Case studies from various global contexts demonstrated the environmental and operational benefits of these technologies, though challenges in scalability and infrastructure development remain.

The chapter also discussed critical strategies for reducing the carbon footprint in logistics, including route optimization, the use of biofuels and renewable energy, and reverse logistics. These measures not only support emissions reduction but also enhance overall supply chain efficiency and resilience. Companies implementing advanced fleet management tools, transitioning to cleaner energy sources, and adopting circular logistics practices are setting valuable precedents for future efforts.

- The integration of green technologies, such as electric vehicles, drones, and ITS, is essential to building sustainable urban logistics systems.
- The Tunis case study offers a concrete example of how emissions modeling can inform policy and infrastructure planning in developing cities.

- Optimizing transport routes, adopting renewable energy, and embracing reverse logistics are actionable pathways for reducing environmental impact.
- Collaborative efforts between governments, private sectors, and academia are necessary to scale innovations and ensure that sustainability becomes a core element of logistics planning.

Recommendations for Future Research and Practice

Future research should focus on the long-term environmental and economic impacts of green logistics technologies, especially in developing urban areas where data is often limited. Comparative studies between cities can also help identify context-specific best practices. Furthermore, there is a need to explore policy mechanisms that can accelerate the adoption of sustainable transportation, such as subsidies, carbon pricing, and public-private partnerships.

Practically, logistics operators and city planners should prioritize data-driven decision-making, invest in smart mobility infrastructure, and adopt life-cycle approaches to assess the true sustainability of their operations. By doing so, the path toward greener, more efficient, and resilient supply chains becomes not only achievable but essential.

REFERENCES

Akhtar, M., & Moridpour, S. (2021). A review of traffic congestion prediction using artificial intelligence. *Journal of Advanced Transportation*, *2021*(1), 8878011. DOI: 10.1155/2021/8878011

Aminzadegan, S., Shahriari, M., Mehranfar, F., & Abramović, B. (2022). Factors affecting the emission of pollutants in different types of transportation: A literature review. *Energy Reports*, *8*, 2508–2529. DOI: 10.1016/j.egyr.2022.01.161

Boiteux, M., & Baumstark, L. (2001). France. Commissariat général du plan. Transports: choix des investissements et coût des nuisances (p. 325). La documentation française.

Coulombel, N., Dablanc, L., Gardrat, M., & Koning, M. (2018). The environmental social cost of urban road freight: Evidence from the Paris region. *Transportation Research Part D, Transport and Environment*, *63*, 514–532. DOI: 10.1016/j.trd.2018.06.002

Eggleston, H. S., Gaudioso, D., Gorissen, N., Joumard, R., Rijkeboer, R. C., Samaras, Z., & Zierock, K. H. (1993). *CORINAIR working group on emission factors for calculating 1990 emissions from road traffic*. Office for Official Publications of the European Communities.

Faiz, A., Weaver, C. S., & Walsh, M. P. (1996). *Air pollution from motor vehicles: standards and technologies for controlling emissions*. World Bank Publications.

Gorji, M. A., Akbarzadeh, M., & Shetab-Boushehri, S. N. (2022). Evaluation and improvement of the urban transportation networks resilience in short-term non-recurring traffic congestion: a novel graph connectivity-based criteria. *Transportation engineering, 10*, 100152.

Hurlimann, A., Moosavi, S., & Browne, G. R. (2021). Urban planning policy must do more to integrate climate change adaptation and mitigation actions. *Land Use Policy*, *101*, 105188. DOI: 10.1016/j.landusepol.2020.105188

ITF. (2023). ITF Transport Outlook 2023. OECD. Available at: https://oecd.org/en/publications/itftransport-outlook-2023_b6cc9ad5-en.html

Shahid, N., Shah, M. A., Khan, A., Maple, C., & Jeon, G. (2021). Towards greener smart cities and road traffic forecasting using air pollution data. *Sustainable Cities and Society*, *72*, 103062. DOI: 10.1016/j.scs.2021.103062

Sheffi, Y. (1985). *Urban transportation networks equilibrium analysis with mathematical programming methods*. Prentice-Hall.

Xu, S., Sun, C., & Liu, N. (2024). Road congestion and air pollution-analysis of spatial and temporal congestion effects. *The Science of the Total Environment, 945*, 173896. DOI: 10.1016/j.scitotenv.2024.173896 PMID: 38880138

Chapter 5
Impact of Digital Infrastructure on Renewable Energy Consumption Across Countries

Ajay Kumar Singh
https://orcid.org/0000-0003-0429-0925
Department of Humanities and Social Sciences, Graphic Era University, Dehradun, India

S. Srinivasan
https://orcid.org/0009-0002-0179-9849
Department of Humanities and Social Sciences, Graphic Era University, Dehradun, India

Arakhita Behera
https://orcid.org/0000-0002-6761-9313
Regional Institute of Education, Ajmer, India

ABSTRACT

This chapter develop empirical model to observe the implication of digital infrastructure (DI) on renewable energy consumption (REC) as compiling a country-wise imbalanced panel data during 2004 – 2022. Three functional forms of empirical models are formulated to achieve its objective. REC is considered as a core explained variable. Mobile cellular subscriptions and individuals using the internet promote DI, thus both the variables are considered as core explanatory variables. Moreover, education expenditure, inflation, per capita GDP, energy use per capita, popula-

DOI: 10.4018/979-8-3373-3790-6.ch005

tion growth and technological development are applied as CVs in the regression analysis. DTs are found beneficial to increase REC. Population growth and carbon intensity have a negative impact on REC. It suggests crucial policy implications to increase the renewable energy management and energy security as using DTs. Finally, this chapter discovers the scope of further research in different dimensions is explicated in this study.

1. INTRODUCTION

High economic growth is caused to increase the demand for non-renewable energy consumption (NREC) in global countries. NREC is highly harmful for environmental quality and human health (Pan et al., 2024), and it also enhance ecological footprint (EF) (Anser et al., 2021), and CO_2 emissions (CO_2E) (Hu et al., 2021). NREC is caused to increase CO_2E and resource depletion (Sharma et al., 2021). NREC, thus, is creating an obstacle for sustainable development and SDGs. While renewable energy consumption (REC) may be an essential driver to increase economic growth in a sustainable manner in the long-run. However, increasing REC is a crucial challenge for global countries to sustain energy security (Zhou et al., 2023). Use of clean fuels, green innovation and technologies would be effective to reduce CO_2E (Hu et al., 2021). Practices of renewable energy would be imperative to increase sustainability of growth and its determinants (Venkatraja, 2020).

Creation and development of energy is an important option to increase sustainable development. Thus, renewable energy creates a clean production path and reduces occurrence of climate change (Su et al., 2022; Akan, 2024; UNEP, 2025). REC can work as climate resilience. The users can increase the energy intensity (EI) using DTs. While renewable energy is crucial for environmental development (Hu et al., 2021; Ashfaq et al., 2024; UNEP, 2025). Climate change and global warming are crucial challenges for sustainable development and SDGs (Adebayo et al., 2023). Therefore, REC would be a key driver for mitigating climate change impacts as renewable energy reduces emissions (Zarrad et al., 2025; Liu et al., 2025). Raza et al. (2020) noted a significant impact of residential energy consumption on environmental degradation in BRICS nations.

Application of DTs can avoid the additional wastages and enhance the multiple uses of energy. DTs provide the best alternatives to increase energy efficiency. Mobile communication can be effective to increase energy efficiency as it reduces its costs (Chen et al., 2024). Energy efficiency would be positive to attain SDG 7.3 (Wu et al., 2023), and sustainable development in firms (Guan et al., 2025). Around 40% of greenhouse gas can be minimized as improving energy efficiency (Chen et al., 2024). DTs can be used as an innovative tool to increase the sources of renewable energy and its intensity. Renewable energy is significant to reduce environmental damage (Zarrad et al., 2025). Low value of energy intensity of a country implies its high resource efficiency. Many emerging countries have low environmental performance due to high energy intensity and efficiency. While renewable energy development (RED) may be supportive to reduce energy intensity (Yu et al., 2022). Therefore, renewable energy helps to reduce CO_2E (Zafar et al., 2019). Moreover, reduction of CO_2E plays an important role for improving renewable energy (Raza et al., 2020). Use of renewable energy resolves multiple economic and environmental problems. Many emerging countries could make a significant difference in the economic and social sectors due to an increase in energy consumption. Source of renewable energy would also assist the global countries to achieve carbon neutrality goals in future (Deka et al., 2023). While high energy efficiency would be most crucial for it (Shen et al., 2024). China is expected to achieve neutrality goals in future (Guan et al., 2025). Carbon neutrality can be achieved by implementing more renewable energy projects. NREC leads to increased pressure on ecosystem services and environmental degradation, while renewable energy reduces environmental stress. Application of renewable energy would increase the transformation of world economies towards green growth and green economy (Ashfaq et al., 2024). Green economy is highly essential to increase green growth progress (Huang & He, 2023). Green economic growth also helps to increase REC (Ashfaq et al., 2024). Energy efficiency also reduces waste materials (Chen et al., 2024).

Past studies highlighted the role of DTs and technological innovation in different sectors like economic, social, agricultural, financial, environmental and energy sectors (Adebayo et al., 2023; Zheng & Wong, 2024). The findings of most studies found that DTs are positive to increase the transformation of the above-mentioned sectors (Xu et al., 2022). However, limited researchers paid little attention to observe the implications of DTs in the energy sector. Energy sector is an emerging area in which more research is required for increasing its sustainability. While non-renewable energy sources are limited and scarce in nature. While energy consumption is increasing due to several reasons like population growth, industrialization, urbanization, infrastructural development and digitalization. Extensive energy consumption would increase another form of energy crisis in the world in near future. Therefore, the global countries should move their intention to search for scientific and innovative platforms to reduce the energy crisis. It can be avoided through production of renewable energy from natural sources. Thereupon, *REC* would be valuable to increase environmental sustainability as it minimizes CO_2 emissions (CO_2E) (Zafar et al., 2019; Namahoro et al., 2021; Chen et al., 2024). Thus, existing scholar estimated the impact of REC, renewable energy development, and other dimensions of energy systems on CO_2E, environmental quality, carbon neutrality, economic development, green growth, sustainable development, sustainable economic growth, and SDGs

(Ummalla & Samal, 2019; Namahoro et al., 2021; Adebayo et al., 2023; Behera et al., 2023; Huang & He, 2023; Murshed, 2024; Ashfaq et al., 2024).

Environmental quality also improves as REC increases (Nathaniel et al., 2020; Mahalik et al., 2021; Chen et al., 2024). Past studies also examined the impact of digital technologies, digital infrastructure, digitalization, digital economy and digital devices on various components of energy sectors like energy production, energy innovation, sustainable energy management, etc. (Sultanova et a., 2022; Lei & Tong, 2022; Xu et al., 2022; Alofaysan et al., 2024; Dong et al., 2024; Zheng & Wong, 2024; Guan et al., 2025; Liu et al., 2025; Guan et al., 2025;). Scholars also investigated the role of digitalization for increasing energy security (Thanh et al., 2023). Few scholars also examined the implication of DTs and ICTs on energy efficiency and intensity and their mechanism (Lei & Tong, 2022; Wu et al., 2023). Zein and Gebresentbert (2024) reviewed the existing literature for exploring the association of digitalization with the energy sector. Bergman and Foxon (2023) explored the role of digitalization in energy demand. Despite that, limited research examines the association of digital infrastructure (DI) with REC in the global countries. Moreover, limited studies could assess the impact of DI and other macro level variables on REC in global countries using advanced empirical models. Hence, this chapter is designed to access the answers on following research questions:

- What is the role of REC in environmental sustainability?
- How are DTs helpful to increase the renewable energy sources?
- How can DTs be useful to increase energy intensity?
- What is the impact of DTs on renewable energy consumption?
- What may be policy actions to increase energy efficiency such as using DTs in global countries?

This chapter is achieved following objectives:

- To explore the environmental implication of renewable energy consumption as per the existing literature.
- To provide the theoretical and empirical literature on the impact of DTs on renewable energy consumption.
- To examine the impact of DTs on renewable energy consumption.

2. REVIEW OF LITERATURE

2.1. Implications on *REC*

REC is primarily essential to increase the sustainable use of available resources. Consequently, Thereupon, REC reduces enormous pressure of population growth on ecosystem services. Thereupon, it may be positive to meet the rising energy demand of the growing population (Zafar et al., 2019). Fossil fuels, petrol and oil are a main source of non-renewable energy. Therefore, extensive application of non-renewable energy consumption (NREC) would produce adverse impact on environment and human health. Most scholars have indemnified the impact of REC on multiple activities like economic growth, environmental quality, environmental performance, CO_2E in a specific and across countries. It is unanimously accepted that REC is supportive for economic growth, and reduces CO_2E. Peng et al. (2023) found that REC can be effective to reduce CO_2E. Similar results are also reported by Namahoro et al. (2021); Zhou et al. (2024) in E-7 countries and African countries, respectively. Thus, REC is an environmentally friendly energy which helps to enhance environmental sustainability (Yüksel & Dinçer, 2021; Chen et al., 2024).

Chen et al. (2024) investigated the impact of energy efficiency on environmental sustainability in selected 10 countries. Ummalla and Samal (2019) examined the impact of REC on CO_2E and economic growth in India and China. It noted that REC has a bi-directional causal relationship with economic growth in India. Zafar et al. (2019) also investigated the impact of REC on CO_2E in emerging countries. It reported that CO_2E declined as REC increased.

Nathaniel et al. (2020) explored the association between renewable energy and ecological footprint (EF) in CIVET nations. REC also reduces ecological footprint (Anser et al., 2021). Renewable energy is found as a useful driver to improve environmental quality. Further, renewable energy improves as CO_2E decreases. Hu et al. (2021) detected a negative impact of renewable energy consumption on CO_2E in India. Sharma et al. (2021) claimed that trade expansion and NREC are caused to increase CO_2E in South and Southeast Asia.

Education can be supportive to search for renewable energy sources (Zhou et al., 2023). Green economic growth also signifies the improvement in REC (Ashfaq et al., 2024). In BRICS nations, Mahalik et al. (2021) found a positive contribution of education in REC. Adebayo et al. (2023) also reported the decreasing trend in CO_2E as implementation of REC in BRICS countries. While energy intensity (EI) is leading to increase CO_2E (Namahoro et al., 2021). Murshed et al. (2023) investigated the impact of energy productivity gain on CO_2E in 11 countries. It noted that energy productivity gain and REC help to reduce CO_2E. Zhou et al. (2024) also clarified that energy productivity is positive to reduce CO_2E. Furthermore, renewable energy helps to mitigate CO_2E, while non-energy consumption, energy-import and energy intensity enhance CO_2E. Murshed (2024) explored the impact of renewable energy transitions on green growth in N11 countries. Table 1 provides the concise summary of previous research which investigated the impact of REC on CO_2E and other variables. This table also includes the findings of previous scholars who have estimated the implication of DI on REC in different countries.

Table 1. Summary of review of literature

Author(s)	Study Area	*DVs*	*Main IVs*	Empirical Model	Results
Ummalla and Samal (2019)	India and China	CO_2E and economic growth	REC, natural gas,	Autoregressive distributed lag model	REC and economic growth have a causal association in India.
Zafar et al. (2019)	Emerging countries	CO_2E	REC, non-REC, trade openness	VECM model	REC is caused to reduce CO_2E. While non-REC increase it.
Nathaniel et al. (2020)	CIVETS	Environmental quality	Urbanization, trade, economic growth and ecological footprint	AMGE model	Renewable energy has a positive impact on environmental quality.
Venkatraja (2020)	BRICS countries	GDP per capita	REC, renewable electricity output	Panel regression model	Economic growth is significantly associated with renewable energy.
Raza et al. (2020)	BRICS countries	CO_2E	Residential energy, financial development, economic growth	Panel regression model	Existence EKC hypothesis between energy consumption and environmental degradation.
Hu et al. (2021)	India	CO_2E	Energy consumption, economic growth, technological innovation	Time series data analysis	CO_2E are positively influenced by energy consumption, and REC reduces CO_2E
Sharma et al. (2021)	South and Southeast Asia	CO_2E	NRE, trade expansion and financial development	Westerlund cointegration test	NREC and trade expansion are caused to increase CO_2E in the long-run.

continued on following page

Table 1. Continued

Author(s)	Study Area	*DVs*	*Main IVs*	Empirical Model	Results
Mahalik et al. (2021)	BRICS nations	CO_2E	REC, education, urbanization, and globalization	Random effect and GMM models	Environmental quality improves as *REC* and secondary education increase.
Namahoro et al. (2021)	African countries	CO_2E	REC, energy intensity (EI), economic growth	Variance decomposition regression analysis	REC helps to reduce CO_2E and EI enhance CO_2E.
Khan et al. (2021)	USA	CO_2E	Natural resources, renewable energy, population growth	Comprehensive empirical analysis	Non-energy consumption and population growth are caused to increase CO_2E.
Yu et al. (2022)	82 countries	Energy intensity	Renewable energy	Dynamic panel regression analysis	Renewable energy is positive to reduce energy intensity.
Yu et al. (2022)	Developing countries	CO_2E	REC, GDP and trade freedom	Panel quantile regression model	REC helps to achieve SDG7 and minimize CO_2E.
Murshed et al. (2023)	11 countries	CO_2E	Energy productivity gain, renewable energy use, economic growth and urbanization	Advanced econometric model	REC is useful to reduce CO_2E.
Adebayo et al. (2023)	BRICS countries	CO_2E	Technological innovation, natural resources and REC	ARDL model	REC is supportive to reduce CO_2E.
Behera et al. (2023)	18 Emerging countries	CO_2E	Green technologies, urbanization, energy-import and energy intensity	Autoregressive distributed lag model	Renewable energy is vital to mitigate CO_2E. While CO_2E increases due to increase in GDP, urbanization and energy imports.
Yüksel and Dinçer (2021)	G7 countries	Renewable energy use	Technological development	Pedroni panel cointegration analysis	Renewable energy use improves as technological development increases.
Sultanova et a. (2022)	European countries	REC	Digital economy	Panel data analysis	DTs is important to increase clean energy.

continued on following page

Table 1. Continued

Author(s)	**Study Area**	*DVs*	*Main IVs*	**Empirical Model**	**Results**
Xu et al. (2022)	International level	Energy and its mechanisms	Digitalization	GMM model	Digitalization increases the minimum energy consumption.
Wu et al. (2023)	China	Energy efficiency	Digital technology	Two-way fixed effect model	Use of DTs help to increase energy efficiency.
Ashfaq et al. (2024)	G20 countries	REC	Green economic growth and globalization	ARDL model	Green economic growth enhances REC.
Thanh et al. (2023)	European countries	Energy security	Digital transformation Energy sustainability may be conducive to SDGs.	PCSEs and FGLS model	Digital transformation is positive to increase acceptability and sustainability of energy security.
Yi et al. (2024)	65 cross countries	REC	Digital economy	Panel data regression analysis	Digital economy helps to enhance renewable energy innovation.
Zheng and Wong (2024)	China	Renewable energy	Digital economy	Fixed effect model	Digital economy helps to increase renewable energy development.
Alofaysan et al. (2024)	European countries	Energy efficiency	Green technology innovation, population, digitalization and GDP	PCSEs model	Digitalization and renewable source of energy promote energy efficiency.
Herbert and Tchawa (2024)	Africa	REC	ICT	Augmented mean group estimator model	Internet and ICT services produce positive impact on REC.
Liu et al. (2025)	68 countries	Renewable energy	Digital economy	Panel data regression analysis	Renewable energy improves as digital economy increases.

continued on following page

Table 1. Continued

Author(s)	Study Area	*DVs*	*Main IVs*	Empirical Model	Results
Guan et al. (2025)	China	Energy intensity	DTs		Adoption of DTs is helpful to reduce energy intensity.
Khan et al. (2025)	All income group countries	Energy efficiency	DTs and trade openness	GMM model	Digitalization and trade openness are reported positive to increase energy efficiency.

Source: Author's compilation based on previous literature.

2.2. Factors Affecting *REC*

REC depends on many indicators in social, economic, trade and environmental dimensions (Su et al., 2022). High economic growth is creating a significant challenge to ensure energy sustainability. Economic factors determine the use of renewable energy (Yüksel & Dinçer, 2021). Economic stability has a significant influence on CO_2E and it enhances renewable energy (Akan, 2024). A nation requires an additional quantity of energy to maintain the dynamic economic growth and development. Thus, dynamic economic growth diminishes energy and its non-renewable sources. Political factors like government initiatives and institutional support play a vital role in enhancing the practices of renewable energy (Yüksel & Dinçer, 2021). Political stability may be effective to increase renewable energy (Dong et al., 2024).

Production of goods and services are highly dependent on availability of non-renewable energy sources. There is significant diversity in REC and NREC in developed and developing countries. For instance, many developing countries have a higher dependence on NREC. While extensive use of NREC is increasing CO_2E in many developing countries (Khan et al., 2021). Hence, the population is increasing more energy (Mombekova et al., 2024). Moreover, economic development is not possible without industrial development. The industrial sector has a high dependency on the energy sector (Yu et al., 2022). Thus, extensive industrialization is harmful for energy resources (Mombekova et al., 2024). Thereupon, international trade may be negative to increase the sustainable use of energy. Most countries could improve their economic growth and development by increasing foreign trade which is significantly improved after trade liberalization and globalization. However, energy demand has greatly increased due to globalization (Bingxin et al., 2025). Conversely, highly advanced and industrial countries are capable of producing energy from natural sources like water, sus and wind. Moreover, most developed countries are doing innovative practices in the energy sector through technological development (TD). Therefore, TD is helping to improve the innovation in the energy sector.

2.3. Technological Development (TD), Digitalization and Energy Sector

TD is a core originator point for creating digital infrastructure. Accordingly, (TD) brings multiple opportunities to increase the scientific practices in energy sector. Yüksel and Dinçer (2021) investigated the association of TD with renewable energy use in G7 countries. It noted that TD is highly effective to increase renewable energy use. TD is also positive to increase the transparency in the energy sector. TD can be conducive to improve energy efficiency. Khan and Su (2023) highlighted that technological innovation has a significant impact on renewable energy. Green technology (GT) is an important driver to reduce CO_2E (Behera et al., 2023). GT may be positive to increase production and saving of renewable energy. Moreover, green energy products and renewable energy will reduce burden on coil and oil to mee the increasing demand of energy (Chen et al., 2023).

Digitization is helpful to increase and manage the network to increase the appropriate distribution of energy (Zarrad et al., 2025). Digitalization is found as an important driver to increase energy security in European countries (Thanh et al., 2023). Alofaysan et al. (2024) observed the impact of green technology innovation on energy efficiency in European countries. Xu et al. (2022) detected that digitalization is helping to increase energy transitions. Digitalization can provide the smart solution to reduce scarcity of energy and environmental issues (Zhou et al., 2023). Digitalization is an engine of clean energy (Bingxin et al., 2025). Digital innovation creates a conducive platform to increase renewable energy innovation (Yi et al., 2024).

2.4. Digital Economy and Energy Sector

Digital economy promotes efficient use of resources and reduces information asymmetry (Yi et al., 2024). Transformation of the digital economy is highly imperative to increase sustainable use of energy. Digital economy enhances sustainability and efficiency of renewable energy (Yi et al., 2024). Digital economy is a significant driver to energy transition (Zheng & Wong, 2024). Digital economy is found positive to increase RED and renewable energy innovation (Yi et al., 2024). Bingxin et al. (2025) also found that digitalization is an effective driver of renewable energy development. Digital economy is also increasing the transformation of energy consumption (Wu & Zhang, 2025). In 52 countries, Dong et al. (2024) found that the digital economy could be helpful to increase renewable energy development. Sultanova et a. (2022) examined the impact of the digital economy on *REC* in European countries.

2.5. DTs, Energy Efficiency and Intensity

Use of DTs in the energy sector enhances productivity and reduces resource consumption (Khan et al., 2025). Application of ICTs in the energy sector is significant for increasing transparency in the energy market (Yi et al., 2024). Herbert and Tchawa (2024) examined the implication of ICT in REC in Africa. DTs are increasing the diffusing and adoption of renewable energy in different sectors (Zein & Gebresentbert, 2024). Digital tools and DTs are also helping to increase energy efficiency, productivity, efficiency and sustainability (Shen et al., 2024; Zarrad et al., 2025; Wu & Zhang, 2025). Wu et al. (2023) explored the impact of digital technology on energy efficiency in China. It noted the energy efficiency significantly improved as application of DTs. Guan et al. (2025) also recognized the impact of DTs on energy efficiency in China's firms. It also found that DTs is found helpful to reduce energy intensity by 52.69%. Shen et al. (2024) also found negative impact of DTs on energy intensity in China. Lei and Tong (2022) also noted a positive contribution of ICT for improving energy efficiency. Energy efficiency can be improved through extensive application of technological innovation (Wu & Zhang, 2025). In the context of China, Wu et al. (2023) found that DTs is helping to increase energy efficiency. However, DTs is increasing more energy efficiency in high income countries due to advanced infrastructure and appropriate accessibility of energy efficient technologies (Khan et al., 2025). DTs can be effective to increase the distribution of energy at the lowest price to the low-income communities (UNEP, 2025).

2.6. DTs and Renewable Energy Consumption (REC)

DTs and digital infrastructure (DI) are worthy to increase REC. For instance, affordability and accessibility of energy depend on use of DTs. DTs also enhance the equitable distribution of energy among the communities. DTs are also maintaining the balance between energy demand and supply effectively. DTs are minimizing the energy price. Accordingly, the economic power of people to acquire energy automatically improves. DTs can also increase the progress of all components of the energy sector. However, low-income group countries could not get significant benefits from the energy sector due their low digital technological adoption capacity and ineffective digital infrastructure. High digital illiteracy and poor DI are creating barriers for digital transformation of the energy sector in developing countries. While high income and industrialized countries have the capacity to adopt DTs in the energy sector. Xu et al. (2022) detected that digitalization is found effective to reduce energy consumption at international level. DTs are providing incentive to people to use energy in a sustainable manner (Zarrad et al., 2025).

Figure 1. Conceptual framework of this chapter

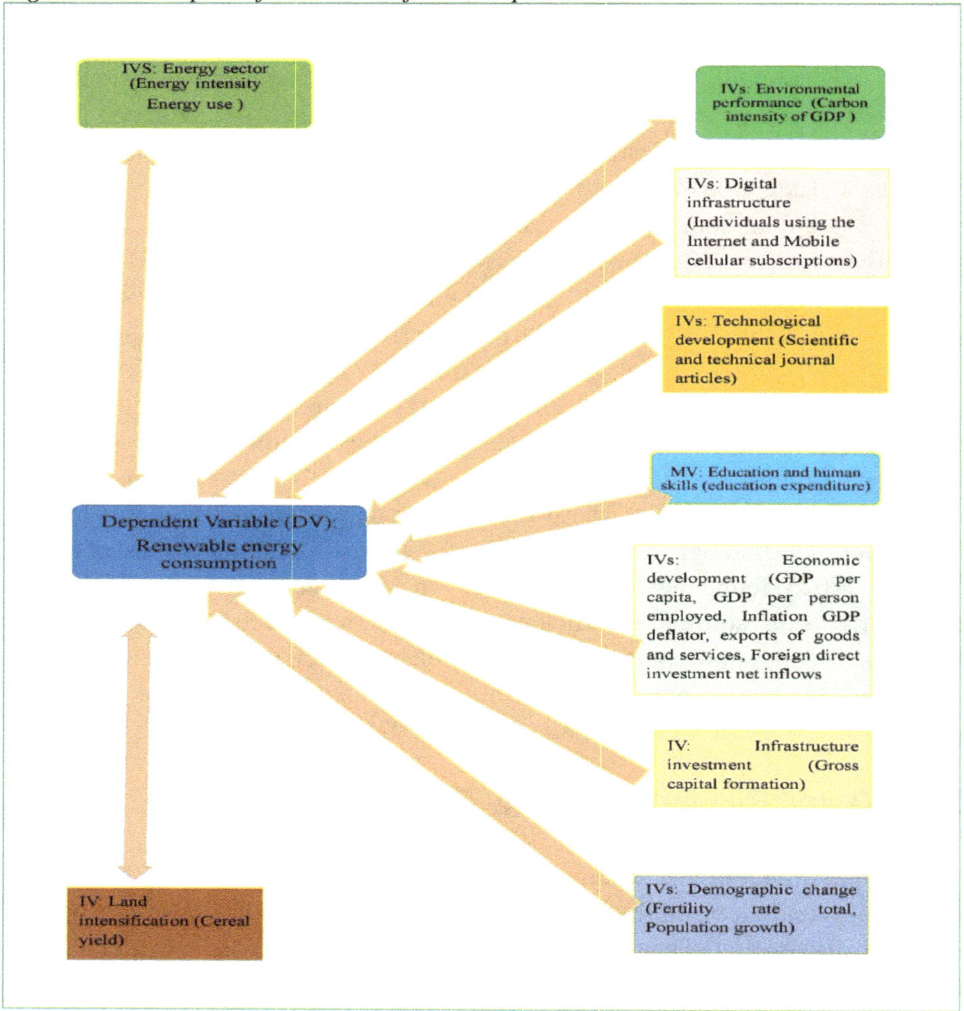

Source: *Developed by authors.*

2.7. Advanced DTs and Energy Sector

Advanced DTs like artificial intelligence (AI), blockchain, big data analysis (BDA), IoT, etc. are helping to increase the progress of the energy sector (Zarrad et al., 2025; Khan et al., 2025). For instance, AI increases the transparency in institutions which distribute energy among the people. AI may be effective to increase the multiple usages of energies in different sectors. IoT helps institutions to produce more renewable energies from various sources. IoT is helping to detect energy inefficiencies in this sector (Zarrad et al., 2025; Khan et al., 2025). AI and IoT are positive to increase the fair distribution of energies. Thereupon, these DTs are increasing renewable energy development and innovation in this sector. Advanced DTs are used to make the future prediction of energy output (Zarrad et al., 2025). BDA is an appropriate analytical tool to examine the future prediction of energy demand and supply. Hence, DTs can be useful for institutions to make further plans to meet the energy demand of different sectors.

2.8. Negative Effects of Digitalization in Energy Sector

Development of DI may require more energy like electricity, and resources. DI also needs more raw materials energy (UNCTAD, 2024). The production of digital devices also consumes more energy. Thus, digitalization may also bring a negative impact in the energy sector in the short-run. Moreover, multiple studies also noticed that extensive digitalization may be caused to increase digital wastage in a country. UNCTAD (2024) claimed that application of ICT consumes around 6% to 12% of electricity globally. Digital wastages are harmful for the environment and energy sectors (UNCTAD, 2024). Moreover, digital devices would cause more demand for minerals. Therefore, Bingxin et al. (2025) argued that initially digitalization is caused to increase the requirement of non-renewable energy. Initial stage of DI hampers renewable energy.

2.9. DTs, REC and SDGs

DTs can be proactive to increase the equitable distribution of energy among the various groups of communities. Thus, DTs enhance the accessibility of energy and are supportive to increase SDGs. For instance, Sultanova et a. (2022) highlighted that DTs could be helpful to increase clean energy. DTs is found positive to increase energy security and sustainability in European countries (Thanh et al., 2023). Improving energy efficiency would be a key tool to reduce global warming (Khan et al., 2025). Hence, it can be used as climate action for global countries to achieve SDGs.

2.10. Reliable Research Gap and Conceptual Framework of the Study

The above-mentioned literature emphasizes that DI is significant to improve efficiency and intensity of energy sector in a broader perspective. However, previous literature could not develop an empirical association of DI with REC. Therefore, this chapter explores the interlinkages of REC with DI creating indicators and other macro level variables which are divided in 9 categories. Figure 1 divides the undertaken variables in 9 specific categories i.e., energy sector, environmental performance, digital infrastructure, technological development, education and human skills, economic development, infrastructure investment, demographic change, and land intensification. Past literature did not estimate the impact of the mentioned variables on REC in global countries. Therefore, this research is significantly different from previous studies.

3. RESEARCH METHODOLOGY

3.1. Study Area and Sources of Data

This chapter includes the cross countries which are chosen as per the availability of associated data. This chapter could include only 95 countries which have the statistical data of explained, explanatory and control variables. While 32, 5, 26 and 32 countries belong to high, low, lower-middle and upper-middle-income groups, respectively are included in this chapter (Table 2). The time period and income group of undertaken countries is given in Table 2. Hence, this study develops a country-wise imbalanced panel data during 2004 – 2022 to achieve the prescribed objectives. The required data is taken from the website of world development indicators (The World Bank) and OECD.

Table 2. Time span and income groups of selected countries

Countries	Time Period
High income group countries	
Ireland, Belgium, Israel, France, Italy, Korea Rep., Qatar, Japan, Austria, Latvia, Canada, Lithuania, Chile, Luxembourg, Slovak Republic, Netherlands, Denmark, Switzerland, Hungary, New Zealand, Norway, Portugal, Spain, United Arab Emirates, Uruguay, Poland, Australia	2004-2020
Croatia, Greece, United Kingdom	2005-2021
Finland	2006-2021
Germany	2007-2020
Low-income group countries	
Chad, Madagascar, Mozambique, Rwanda, Uganda	2004-2021
Lower middle income group countries	
Egypt Arab Rep., Cameroon, El Salvador, Algeria, Honduras, Angola, Bangladesh, Kenya, Congo Rep., Kyrgyz Republic, Ukraine, Nicaragua, Pakistan, Philippines, Tunisia, Morocco	2004-2020
	2004-2021
Tanzania	2005-2014
Iran, Islamic Rep.	2005-2018
Cambodia, India, Tajikistan	2005-2021
Zimbabwe	2006-2021
Sri Lanka	2007-2015
Bolivia	2007-2021
Ghana, Mongolia	2008-2021
Upper middle income group countries	
Libya	2004-2018
Russian Federation	2004-2020
Albania, Argentina, Armenia, Belarus, Botswana, Brazil, Bulgaria, Costa Rica, Georgia, Indonesia, Kazakhstan, Lebanon, Mauritius, Mexico, Moldova, Peru, Romania, South Africa, Thailand, Turkiye	2004-2021
Gabon	2005-2013
Malaysia	2005-2021
Azerbaijan, China, Namibia	2006-2021
Colombia, Jordan, Panama	2007-2021
Iraq	2008-2021
Paraguay	2009-2021

Source: Author's compilation.

3.2. Finalization of Dependent Variable (IV) and Independent Variables (IVs)

Renewable Energy Consumption (REC): *REC* is used as an explained variable in this study. Previous scholars also applied *REC* as dependent variable (*DV*) and independent variable (*IV*) for evaluating its implications on CO_2E, energy intensity, etc. (Ummalla & Samal, 2019; Venkatraja, 2020; Raza et al., 2020; Hu et al., 2021; Sharma et al., 2021; Su et al., 2022; P Esposito, 2023; eng et al., 2023; Ashfaq et al., 2024; Zarrad et al., 2025). Accordingly, previous studies examined the causal association of *REC* with the mentioned variables.

Digital Infrastructure: DI is a scientific and technological ecosystem which promotes online activities in a country. A precise variable cannot be included to capture the appropriate progress of a country in digital infrastructure. DI improves renewable energy development (RED) and its distribution (Dong et al., 2024). Mobile cellular subscriptions (*MCS*) and internet users help to enhance the progress of DI (Herbert & Tchawa, 2024). MCS and internet users are useful indicators of DTs. DTs play a positive to increase energy efficiency (Wu et al., 2023). Thus, *MCS* and individuals using the internet (*IUI*) are included as proxy variables for DI.

Technological Development (TD): TD is the originator for new digital devices. Hence, TD may be crucial to create DI and nurture an efficient path to meet the energy requirement of the population. Technological innovation is crucial to reduce CO_2E (Adebayo et al., 2023). Subsequently, renewable energy consumption and its use may be progressive as TD increases (Yüksel & Dinçer, 2021). Scientific and technical journal articles are used as a representative variable for TD in this chapter. These are the important sources to increase dissemination of new knowledge and ideas among the scientists. Accordingly, scientists develop and discover new innovation and technologies in scientific field. Hence, it can be applied as proxy indicator for TD.

Energy Sector: Energy intensity (*EI*) is caused to increase CO_2E and it has a negative impact on the environmental sector (Namahoro et al., 2021). *EI* may be negative for *REC*. However, renewable energy development (RED) may be crucial to reduce energy intensity (Yu et al., 2022). Energy use (*EUPC*) is increasing due to population growth, urbanization and industrialization. Thus, it is proven that per capita energy use may be negative for *REC*. Also, abatement of CO_2E assists to improve REC (Raza et al., 2020). Carbon intensity (*CIGDP*) can be used as a representative variable for environmental performance.

Education and Human Skills: Education level is a vital indicator for increasing the attention of people towards sustainable use of energy (Zhou et al., 2023). High investment in the education sector helps to increase human skills and bring innovation in a nation. High education levels of people may be proactive to increase renewable energy (Mahalik et al., 2021).

Land Intensification: The global countries could improve cereal yield by applying extensive quantities of fertilizer and pesticides in the cultivation. Land intensification and land management practices could be found effective to increase cereal yield. However, the mentioned practices are bringing negative implications in the environmental and health sector. Over practices of land intensification may increase more demand of energy in the agricultural sector.

Economic Development: Renewable energy consumption may increase as per capita GDP (*GDPPC*) increases. Thereupon, GDP has a negative impact on *REC*. GDP per person employed (*GDPPE*) is also included as an explanatory variable in the model. Inflation (*IGDPD*) may lead to increased cost of energy (Akan, 2024). Moreover, exports of goods and services (*EGS*) is an important indicator that explores the progress of trade openness. Further, it shows the progress of trade after economic globalization (Mahalik et al., 2021). Trade openness also increases the consumption of per capita renewable energy (Su et al., 2022). Trade openness is found to increase energy efficiency in all income groups of countries (Khan et al., 2025). Extensive international trade also enhances CO_2E (Murshed et al., 2023). Demand of energy is also determined by *FDINI* (Esposito, 2023). *FDINI* influences energy projects (Bingxin et al., 2025).

Infrastructure Investment: Capital formation signifies the infrastructure investment which may be caused to increase more natural resource extraction (Chen et al., 2023).

Table 3. Summary of the variables

Variables	**Code**	**Unit**	**Category of the variable**
Renewable energy consumption (% of total final energy consumption)	*REC*	%	Dependent variable
Energy intensity level of primary energy (MJ/$2017 PPP GDP)	*EI*	MJ/$	Energy sector
Energy use (kg of oil equivalent per capita)	*EUPC*	Kg	
Carbon intensity of GDP (kg CO2e per constant 2015 US$ of GDP)	*CIGDP*	Kg	Environmental performance
Individuals using the Internet (% of population)	*IUI*	%	Digital infrastructure
Mobile cellular subscriptions (per 100 people)	*MCS*	Number	
Scientific and technical journal articles	*TD*	Number	Technological development
Adjusted savings: education expenditure (% of GNI)	*ASEE*	%	Education and human skills

continued on following page

Table 3. Continued

Variables	Code	Unit	Category of the variable
Cereal yield (kg per hectare)	*CYPH*	Kg./Ha.	Land intensification
GDP per capita (constant 2015 US$)	*GDPPC*	US$	Economic development
GDP per person employed (constant 2021 PPP $)	*GDPPE*	US$	
Inflation, GDP deflator (annual %)	*IGDPD*	%	
Exports of goods and services (% of GDP)	*EGS*	%	
Foreign direct investment, net inflows (% of GDP)	*FDINI*	%	
Gross capital formation (% of GDP)	*GCF*	%	Infrastructure investment
Fertility rate, total (births per woman)	*FRT*	Number	Demographic change
Population growth (annual %)	*PGA*	%	

Source: Author's compilation.

Demographic Change: Annual population growth (*PGA*) is caused to increase land use patterns and emits more CO_2E (Ou et al., 2024). Thus, over population growth also increases environmental pollution (Liu et al., 2025). Furthermore, fertility rate (*FRT*) is a sole determinant of population growth. Thus, *PGA* and *FRT* boosts CO_2E (Khan et al., 2021; Peng et al., 2023; Murshed et al., 2023). Therefore, *PGA* and *FRT* may be accountable for energy demand. Therefore, *PGA* and *FRT* have a negative impact on *REC*. Table 3 describes the brief summary of the variables and their specific categories.

3.3. Statistical Analysis

Simple statistical analysis (i.e., mean, standard deviation and correlation coefficient analysis) is used to explore the statistical data. Linear interpolation method is applied to examine the missing data for few variables (Liu et al., 2025).

3.4. Formulation of Empirical Models

Existing studies have been formulated various forms of empirical models to examine the impact of *REC* on macro level indicators and vice-versa (Zafar et al., 2019; Sharma et al., 2021; Namahoro et al., 2021; Yu et al., 2022; Peng et al., 2023). Ummalla and Samal (2019) observed the impact of *REC* on CO_2 emissions (CO_2E) and economic growth using an autoregressive distributed lag model. Mahalik et al. (2021) introduced a random effect model to explore the impact of *REC* on environmental quality. Yu et al. (2022) applied a panel quantile regression model to detect the impact of *REC*, trade freedom and GDP of CO_2E. Su et al. (2022) assessed the impact of trade, industrialisation, and social and economic change on *REC* using panel regression models. Behera et al. (2023); Zarrad et al. (2025) examined the impact of REC and other variables on CO_2E in emerging and European countries, respectively using the ARDL model. A similar model is applied by Ashfaq et al., (2024) to recognize the impact of green economic growth on *REC*. Liu et al. (2025) applied a fixed effect model to expose the impact of the digital economy on renewable energy. Recently, Guan et al. (2025) preferred a DID model to examine the impact of DTs on energy intensity in Chineses firms.

This chapter develops a country-wise imbalanced data of the mentioned variables which are explained in Table 3. Linear and non-linear models are adopted to examine the linear and non-linear association of *REC* with included IVs. A log-linear regression model is adopted to observe the elasticities of *REC* with *IV*s. This chapter assumes that *REC* is a function of *EI, EUPC, CIGDP, IUI, MCS, TD, ASEE, CYPH, GDPPC, GDPPE, IGDPD, EGS, FDINI, GCF, FRT*, and *PGA*. The functional association of *REC* with these variables are described as:

$$REC = f\,\{EI, EUPC, CIGDP, IUI, MCS, TD, ASEE, CYPH, GDPPC, GDPPE, IGDPD, EGS, FDINI, GCF, FRT, PGA\} \tag{1}$$

Equation (1) explores the functional association of *REC* with core, mediator and control variables. Following forms of linear, log-linear and non-linear regression equations are adopted for estimating the impact of DTs and explanatory variables on *REC*:

$$(REC)_{ct} = \alpha_0 + \alpha_1 (EI)_{ct} + \alpha_2 (EUPC)_{ct} + \alpha_3 (CIGDP)_{ct} + \alpha_4 (IUI)_{ct} + \alpha_5 (MCS)_{ct} + \alpha_6 (TD)_{ct} + \alpha_7 (ASEE)_{ct} + \alpha_8 (CYPH)_{ct} + \alpha_9 (GDPPC)_{ct} + \alpha_{10} (GDPPE)_{ct} + \alpha_{11} (IGDPD)_{ct} + \alpha_{12} (EGS)_{ct} + \alpha_{13} (FDINI)_{ct} + \alpha_{14} (GCF)_{ct} + \alpha_{15} (FRT)_{ct} + \alpha_{16} (PGA)_{ct} + \phi_{ct} \tag{2}$$

$$(REC)_{ct} = \beta_0 + \alpha_1 \ln(EI)_{ct} + \beta_2 \ln(EUPC)_{ct} + \beta_3 \ln(CIGDP)_{ct} + \beta_4 \ln(IUI)_{ct} + \beta_5 \ln(MCS)_{ct} + \beta_6 \ln(TD)_{ct} + \beta_7 \ln(ASEE)_{ct} + \beta_8 \ln(CYPH)_{ct} + \beta_9 \ln(GDPPC)_{ct} + \beta_{10} \ln(GDPPE)_{ct} + \beta_{11} \ln(IGDPD)_{ct} + \beta_{12} \ln(EGS)_{ct} + \beta_{13} \ln(FDINI)_{ct} + \beta_{14} \ln(GCF)_{ct} + \beta_{15} \ln(FRT)_{ct} + \beta_{16} \ln(PGA)_{ct} + \lambda_{ct} \tag{3}$$

$$(REC)_{ct} = \xi_0 + \xi_1 (EI)_{ct} + \xi_2 (EI)^2_{ct} + \xi_3 (EUPC)_{ct} + \xi_4 (EUPC)^2_{ct} + \xi_5 (CIGDP)_{ct} + \xi_6 (CIGDP)^2_{ct} + \xi_7 (IUI)_{ct} + \xi_8 (IUI)^2_{ct} + \xi_9 (MCS)_{ct} + \xi_{10} (MCS)^2_{ct} + \xi_{11} (TD)_{ct} + \xi_{12} (TD)^2_{ct} + \xi_{13} (ASEE)_{ct} + \xi_{14} + \xi_{15} (CYPH)_{ct} + \xi_{16} + \xi_{17} (GDPPC)_{ct} + \xi_{18} + \xi_{19} (GDPPE)_{ct} + \xi_{20} + \xi_{21} (IGDPD)_{ct} + \xi_{22} + \xi_{23} (EGS)_{ct} + \xi_{24} + \xi_{25} (FDINI)_{ct} + \xi_{26} + \xi_{27} (GCF)_{ct} + \xi_{28} + \xi_{29} (FRT)_{ct} + \xi_{30} + \xi_{31} (PGA)_{ct} + \xi_{32} + \mu_{ct} \tag{4}$$

Here, α_0, β_0 and ξ_0 are constant coefficients; α_1, ... α_{16}; β_1, ... β_{16}; ξ_0, ... ξ_{32} are coefficients of corresponding explanatory variables; ϕ_{ct}, λ_{ct} and μ_{ct} are the error-terms; *c* is cross-sectional country; and *t* is time period in equation (2), equation (3) and equation (4). While explanation of the variables is given in Table 3.

3.5. Process for Finalizing an Appropriate Model and Application of Statistical Software

This chapter is applied country-wise to imbalanced panel data during 2004 – 2022. Thus, it applied a scientific process to reduce all statistical remedies. For instance, the presence of multi-correlation among the regressors are tested through statistical mean value of variance inflation factors (*VIF*) (Khan et al., 2025). This chapter applied three regression equations. Hence, it is required to check whether functional forms of these equations are correctly specified. Hence, *Ramsey RESET* tests are used to recognize the suitability of functional forms of proposed regression equations. While AIC and BIC values are determined to select a most consistent model which produces the best coefficients of *IV*s. The existence of a panel root test in a time series of a specific variable is examined using the Im-Pesaran-Shin test. Random and fixed models are also applied (Mahalik et al., 2021). The presence of serial correlation and heteroskedasticity and autocorrelation is also tested. Finally, this validity of coefficients is checked as estimating the correlation coefficients among the various lags of error-terms in the proposed regression equations. The estimates are estimated using *Feasible Generalized Least Square Estimation* model (Thanh et al., 2023). Descriptive analysis is done using SPSS statistical software. The regression analysis is completed through STATA statistical software.

4. DESCRIPTIVE RESULTS

Table 4 explores the statistical summary of the variables. The statistical values of overall mean and standard deviation of respective variables infer a high variation across 95 countries. The statistical values of skewness also imply that all variables (except *IUI*, *MCS* and *ASEE*) have high and significant variation in these countries.

Therefore, natural logarithm of all variables is included in the regression model. The correlation among the variables is presented in Table 5. The estimates showed that *REC* is positively associated with *EI*, *IGDPD*, *FRT* and *PGA*. While *REC* is negatively correlated with *EUPC*, *CIGDP*, *IUI*, *TD*, *ASEE*, *CYPH*, *GDPPC*, *GDPPE*, *EGS*, *FDINI* and *GCF*.

Table 4. Statistical summary of the variables

Variable		*Mean*	*Std. Dev*	*Min*	*Max*	*Obs.*	*Skewness*
REC	Overall	27.8439	24.8684	0.1000	95.3000	N =1619	1.0258
	Between		25.2549	0.2333	92.7944	n =95	
	Within		3.7708	11.903	49.6939	T-bar =17.0421	
EI	Overall	4.6191	2.4760	1.0900	22.4400	N =1619	2.5607
	Between		2.4121	1.5567	16.2944	n =95	
	Within		0.7284	1.7568	12.6968	T-bar =17.0421	
EUPC	Overall	2257.656	2464.08	154.24	21557.47	N =1619	3.3958
	Between		2414.24	214.23	17588.38	n =95	
	Within		334.42	-461.47	6226.75	T-bar =17.0421	
CIGDP	Overall	0.5227	0.4355	0.0488	3.7533	N =1619	2.4788
	Between		0.4270	0.0649	2.7221	n =95	
	Within		0.0967	-0.2875	1.5539	T-bar =17.0421	
IUI	Overall	46.6295	30.4692	0.1990	100.0000	N =1619	0.0111
	Between		25.4477	2.9750	92.4167	n =95	
	Within		17.1563	-3.6205	99.1366	T-bar =17.0421	
MCS	Overall	98.9991	38.7764	1.2795	221.3088	N =1619	-0.4901
	Between		27.5328	29.769	165.2539	n =95	
	Within		27.3877	-9.6665	171.9960	T-bar =17.0421	
TD	Overall	17293.97	48899.92	3.84	766200.4	N =1619	8.1834
	Between		47181.26	10.22	408970.9	n =95	
	Within		17172.70	-198795	374523.5	T-bar =17.0421	
ASEE	Overall	4.1395	1.5496	1.0100	9.4800	N =1619	0.6131
	Between		1.4893	1.5762	9.1750	n =95	
	Within		0.4509	1.9307	7.2706	T-bar =17.0421	
CYPH	Overall	3876.109	2985.22	217.60	36761.90	N =1619	4.2527
	Between		2827.53	467.49	23229.13	n =95	
	Within		887.78	-7240	17408.87	T-bar =17.0421	
GDPPC	Overall	16226.45	21090.66	377.07	112417	N =1619	1.9724
	Between		20774.10	451.84	106177	n =95	
	Within		2207.28	3386.5	51056.07	T-bar =17.0421	
GDPPE	Overall	59691.22	48164.32	2057.8	316742.4	N =1619	1.6439
	Between		47367.07	3086	287361	n =95	
	Within		6684.52	22624	144550	T-bar =17.0421	

continued on following page

Table 4. Continued

Variable		*Mean*	*Std. Dev*	*Min*	*Max*	*Obs.*	*Skewness*
IGDPD	Overall	7.1426	18.5712	0.0002	604.9459	N =1619	22.7408
	Between		9.0410	0.7679	79.1804	n =95	
	Within		16.3484	-71.670	532.9081	T-bar =17.0421	
EGS	Overall	41.1688	24.8131	6.7685	213.2227	N =1619	2.4744
	Between		23.4509	10.930	177.9339	n =95	
	Within		7.3542	3.9382	76.4575	T-bar =17.0421	
FDINI	Overall	5.4281	10.2752	0.0017	234.2487	N =1619	10.2183
	Between		6.9194	0.3777	50.8485	n =95	
	Within		7.5102	-40.696	188.8283	T-bar =17.0421	
GCF	Overall	24.8507	7.5022	1.5712	76.7823	N =1619	1.3894
	Between		6.1486	10.704	43.8821	n =95	
	Within		4.4452	4.7721	59.5775	T-bar =17.0421	
FRT	Overall	2.4400	1.2091	0.8370	7.1600	N =1619	1.5640
	Between		1.1988	1.1374	6.7997	n =95	
	Within		0.2179	1.3436	3.6001	T-bar =17.0421	
PGA	Overall	1.4080	1.4494	0.0002	21.7003	N =1619	5.0954
	Between		1.1778	0.1017	7.9533	n =95	
	Within		0.8437	-6.056	15.1551	T-bar =17.0421	

Source: Author's estimation.

Table 5. Correlation among the variables

Variables	*REC*	*EI*	*EUPC*	*CIGDP*	*IUI*	*MCS*	*TD*	*ASEE*	
REC	1	0.269**	-0.408**	-0.359**	-0.472**	-0.457**	-0.181**	-0.186**	
EI	0.269**	1	0.084**	0.451**	-0.313**	-0.292**	0.046*	-0.115**	
EUPC	-0.408**	0.084**	1	0.01	0.525**	0.347**	0.112**	0.078**	
CIGDP	-0.359**	0.451**	0.01	1	-0.224**	0.001	0.041	-0.002	
IUI	-0.472**	-0.313**	0.525**	-0.224**	1	0.674**	0.207**	0.289**	
MCS	-0.457**	-0.292**	0.347**	0.001	0.674**	1	0.073**	0.233**	
TD	-0.181**	0.046*	0.112**	0.041	0.207**	0.073**	1	-0.093**	
ASEE	-0.186**	-0.115**	0.078**	-0.002	0.289**	0.233**	-0.093**	1	
CYPH	-0.329**	-0.204**	0.464**	-0.167**	0.488**	0.361**	0.164**	0.019	
GDPPC	-0.295**	-0.223**	0.722**	-0.336**	0.657**	0.351**	0.126**	0.234**	
GDPPE	-0.447**	-0.274**	0.730**	-0.254**	0.695**	0.444**	0.102**	0.248**	

continued on following page

Table 5. Continued

Variables	*REC*	*EI*	*EUPC*	*CIGDP*	*IUI*	*MCS*	*TD*	*ASEE*	
IGDPD	0.099**	0.238**	-0.046*	0.122**	-0.143**	-0.101**	-0.067**	-0.083**	
EGS	-0.263**	-0.121**	0.379**	-0.016	0.306**	0.248**	-0.146**	0.120**	
FDINI	-0.052*	-0.034	0.104**	-0.049*	0.091**	0.033	-0.074**	0.044*	
GCF	-0.038	0.093**	0.057*	0.178**	-0.154**	-0.068**	0.166**	-0.120**	
FRT	0.674**	0.311**	-0.396**	-0.101**	-0.633**	-0.558**	-0.220**	-0.202**	
PGA	0.210**	0.179**	0.230**	-0.015	-0.268**	-0.208**	-0.179**	-0.180**	
Variables	*CYPH*	*GDPPC*	*GDPPE*	*IGDPD*	*EGS*	*FDINI*	*GCF*	*FRT*	*PGA*
REC	-0.329**	-0.295**	-0.447**	0.099**	-0.263**	-0.052*	-0.038	0.674**	0.210**
EI	-0.204**	-0.223**	-0.274**	0.238**	-0.121**	-0.034	0.093**	0.311**	0.179**
EUPC	0.464**	0.722**	0.730**	-0.046*	0.379**	0.104**	0.057*	-0.396**	0.230**
CIGDP	-0.167**	-0.336**	-0.254**	0.122**	-0.016	-0.049*	0.178**	-0.101**	-0.015
IUI	0.488**	0.657**	0.695**	-0.143**	0.306**	0.091**	-0.154**	-0.633**	-0.268**
MCS	0.361**	0.351**	0.444**	-0.101**	0.248**	0.033	-0.068**	-0.558**	-0.208**
TD	0.164**	0.126**	0.102**	-0.067**	-0.146**	-0.074**	0.166**	-0.220**	-0.179**
ASEE	0.019	0.234**	0.248**	-0.083**	0.120**	0.044*	-0.120**	-0.202**	-0.180**
CYPH	1	0.481**	0.478**	-0.093**	0.299**	0.078**	-0.110**	-0.435**	-0.032
GDPPC	0.481**	1	0.932**	-0.118**	0.500**	0.279**	-0.061**	-0.404**	-0.002
GDPPE	0.478**	0.932**	1	-0.118**	0.576**	0.291**	-0.094**	-0.503**	-0.058*
IGDPD	-0.093**	-0.118**	-0.118**	1	-0.050*	-0.03	-0.039	0.134**	0.087**
EGS	0.299**	0.500**	0.576**	-0.050*	1	0.498**	0.021	-0.229**	0.042*
FDINI	0.078**	0.279**	0.291**	-0.03	0.498**	1	0.091**	-0.036	0.005
GCF	-0.110**	-0.061**	-0.094**	-0.039	0.021	0.091**	1	0.059**	0.170**
FRT	-0.435**	-0.404**	-0.503**	0.134**	-0.229**	-0.036	0.059**	1	0.508**
PGA	-0.032	-0.002	-0.058*	0.087**	0.042*	0.005	0.170**	0.508**	1

Source: Author's estimation. **Note:** **. Correlation is significant at the 0.01 level and *. Correlation is significant at the 0.05 level.

5. EMPIRICAL RESULTS AND DISCUSSION

The regression results which estimate the impact of explanatory variables on *REC* is given in Table 6. The coefficients of explanatory variables with *REC* are estimated using three different models. The R^2 values are noticed 0.61, 0.70 and 0.84 under model 1, model 2 and model 3, respectively. However, the log-linear regression model provides the lowest values of *AIC* and *BIC* as compared to another model. Hence, it produces most rational coefficients of *IVs* with *REC*. Thereupon, statistical values of *mean VIF* are found less than 10 in model 1 and model 2. Hence, it suggested that explanatory variables are free from the existence of multicollinearity. Furthermore, the statistical values of *Wald* Chi^2 are appeared statistically significant at 1% significance level. Thus, the estimates indicate that undertaken variables have an independent impact on *REC*. Functional forms of the all models are well specified as per the statistical *F*- values under *Ramsey RESET tests*.

The coefficients of *IUI* and *MCS* with *REC* are appeared positive and statistically significant. Hence, *IUI* and *MCS* help to increase renewable energy consumption. The estimates proposed that *REC* can be improved by 0.0039% and 0.0960% as 1% increase in *IUI* and *MCS*, respectively. *IUI* and *MCS* are the most supportive indicators of DI. Thus, the estimates clarify that DI is useful to increase *REC*. The association also suggested global countries should focus on DI promoting activities to increase REC. Thanh et al. (2023); Dong et al. (2024); Liu et al. (2025); Bingxin et al. (2025) which already detected the positive implications of digital infrastructure, digitalization and digital economy on *REC*, energy security and energy efficiency across countries. Findings of Sultanova et al. (2022) also claimed a positive impact of digital economy on clean and REC in European countries. Herbert and Tchawa (2024) detected a positive impact of ICT and internet on REC in African countries. Recent literature also found that DTs are positive to increase energy efficiency (Khan et al., 2025).

TD is an important driver to discover advanced DTs. TD is also creating an appropriate innovative ecosystem to develop new mechanisms in the production sector. TD, thus, infers that *REC* is expected to be improved by 0.0157% due 1% increase in technological development. In G20 countries, Khan and Su (2023) reported that technological innovation is the most significant driver for promoting renewable energy. *EI* measures the requirement of energy for increasing one unit of GDP (Huang & He, 2023). Low intensity is proactive to increase environmental sustainability (Huang & He, 2023). Hence, increase in per capita GDP requires an additional quantity of energy. Accordingly, *EI* produces a negative impact on *REC* and environment (Namahoro et al., 2021). However, the result of this study found that energy intensity is found positive to improve *REC*.

Per capita energy use has a negative impact on *REC*. However, previous studies found a complex association between per capita energy use and REC. The coefficient of *CI* with *REC* is negative. *REC* is expected to decline by 1.0618% due to 1% increase in carbon intensity. Carbon intensity measures the environmental performance. Improvement in CI may be effective to reduce natural resource depletion and increase their productivity. The coefficient of *ASEE* with *REC* is detected positive. The result infers that education expenditure is positive to increase *REC*. It is noted that *REC* is projected to be increased by 0.3086% as 1% increase in education expenditure. Thus, the estimate signifies that public investment in the education sector is also supportive to increase *REC*.

Table 6. Regression results

Models	*LLRM* **[Model 1]**		*LRM* **[Model 2]**		*NLRM* **[Model 3]**	
R-squared	0.61		0.7045		0.8007	
Wald Chi²	88353.70*		154115.09*		648340.62*	
Mean VIF	9.97		3.26		23.66	
Ramsey RESET test using powers of the fitted values of REC [*F* – Value]	162.34*		132.06*		112.40*	
Ramsey RESET test using powers of the *IVs* [*F* – Value]	17.14*		37.01*		28.42*	
AIC	-2795.472		-7499.578		-7499.578	
BIC	-2033.331		-6512.699		-6193.709	
REC =[*DV*]	*Reg. Coef.*	*Std. Err.*	*Reg. Coef.*	*Std. Err.*	*Reg. Coef.*	*Std. Err.*
EI	0.6584*	0.1126	3.4377*	0.1100	8.1818*	0.4484
(EI)²	-	-	-	-	-0.2514*	0.0233
EUPC	-0.9642*	0.1209	-0.0015*	0.0002	-0.0023*	0.0006
(EUPC)²	-	-	-	-	0.0000**	0.0000
CIGDP	-1.0618*	0.0267	-28.8777*	0.8255	-70.2001*	1.2789
(CIGDP)²	-	-	-	-	14.5342*	0.4244
IUI	0.0039	0.0240	-0.1115*	0.0369	-0.4126*	0.0434
(IUI)²	-	-	-	-	0.0040*	0.0003
MCS	0.0960*	0.0225	0.0319*	0.0061	0.0623*	0.0211
(MCS)²	-	-	-	-	0.0000	0.0001
TD	0.0157*	0.0046	0.0001*	0.0000	-0.0001*	0.0000
(TD)²	-	-	-	-	0.0001*	0.0000
ASEE	0.3086*	0.0447	-0.5967*	0.0798	-0.3558	0.5607
(ASEE)²	-	-	-	-	-0.0098	0.0576
CYPH	-0.2645*	0.0187	-0.0005*	0.0001	0.0001	0.0002
(CYPH)²	-	-	-	-	0.0000	0.0000
GDPPC	0.0765	0.0553	0.0005*	0.0000	-0.0006*	0.0001
(GDPPC)²	-	-	-	-	0.0001*	0.0000
GDPPE	-0.3177*	0.0870	-0.0003*	0.0000	-0.0003*	0.0000
(GDPPE)²	-	-	-	-	0.0001*	0.0000
IGDPD	0.0080	0.0186	-0.0019	0.0190	0.0305	0.0205

continued on following page

Table 6. Continued

Models	*LLRM* [Model 1]		*LRM* [Model 2]		*NLRM* [Model 3]	
$(IGDPD)^2$	-	-	-	-	0.0000	0.0000
EGS	0.0221	0.0409	0.0329	0.0150	0.2352*	0.0498
$(EGS)^2$	-	-	-	-	-0.0025*	0.0005
FDINI	0.0560***	0.0218	-0.0138	0.0235	-0.1806*	0.0462
$(FDINI)^2$	-	-	-	-	0.0011**	0.0005
GCF	-0.2665*	0.0390	-0.0752*	0.0271	0.0813	0.1289
$(GCF)^2$	-	-	-	-	-0.0008	0.0022
FRT	-0.8852*	0.0667	5.8856*	0.4725	2.7610*	0.8405
$(FRT)^2$	-	-	-	-	0.0343	0.0898
PGA	-0.1007*	0.0226	-0.7870**	0.3358	-0.9428**	0.4065
$(PGA)^2$	-	-	-	-	0.0515*	0.0173
Con. Coef.	13.0779*	0.5887	33.2237*	2.4413	40.2437*	2.2055

Source: Author's estimation. **Note:** *, ** and *** indicate that coefficient is statistically significant at 1%, 5% and 10% significance level, respectively.

The production and growth of agriculture and its allied activities like dairy farming, meat production and livestock rearing occupation require huge quantities of energy. Thus, agricultural production activities are caused to reduce renewable energy. Cereal yield is a significant variable for land intensification. Hence, land intensification is negative to increase *REC*. Economic development has a negative impact on energy intensity (Yu et al., 2022). Hence, *GDPPC* has a significant impact on *REC*. The estimate implies that *REC* is expected to decline by 0.0765% due to 1% increases in *GDPPC*.

Table 7. Validity of regression results

Log-linear regression model								
No. of lags	*Lag* **(1)**	*Lag* **(2)**	*Lag* **(3)**	*Lag* **(4)**	*Lag* **(5)**	*Lag* **(6)**	*Lag* **(7)**	*Lag* **(8)**
Lag (1)	1.0000							
Lag (2)	0.9663*	1.0000						
Lag (3)	0.9465*	0.9682*	1.0000					
Lag (4)	0.9269*	0.9456*	0.9704*	1.0000				
Lag (5)	0.9022*	0.9270*	0.9437*	0.9722*	1.0000			
Lag (6)	0.8892*	0.9006*	0.9240*	0.9437*	0.9707*	1.0000		
Lag (7)	0.8760*	0.8846*	0.9005*	0.9192*	0.9445*	0.9729*	1.0000	
Lag (8)	0.8623*	0.8709*	0.8851*	0.8947*	0.9201*	0.9462*	0.9713*	1.0000

continued on following page

Table 7. Continued

Log-linear regression model								
No. of lags	*Lag* **(1)**	*Lag* **(2)**	*Lag* **(3)**	*Lag* **(4)**	*Lag* **(5)**	*Lag* **(6)**	*Lag* **(7)**	*Lag* **(8)**
Linear regression model								
Lag (1)	1.0000							
Lag (2)	0.9880*	1.0000						
Lag (3)	0.9758*	0.9880*	1.0000					
Lag (4)	0.9625*	0.9753*	0.9875*	1.0000				
Lag (5)	0.9499*	0.9616*	0.9743*	0.9871*	1.0000			
Lag (6)	0.9364*	0.9487*	0.9600*	0.9732*	0.9868*	1.0000		
Lag (7)	0.9213*	0.9345*	0.9462*	0.9582*	0.9721*	0.9860*	1.0000	
Lag (8)	0.9045*	0.9184*	0.9309*	0.9431*	0.9569*	0.9709*	0.9859*	1.0000
Non-linear regression model								
Lag (1)	1.0000							
Lag (2)	0.9760*	1.0000						
Lag (3)	0.9567*	0.9771*	1.0000					
Lag (4)	0.9354*	0.9548*	0.9771*	1.0000				
Lag (5)	0.9091*	0.9324*	0.9529*	0.9762*	1.0000			
Lag (6)	0.8851*	0.9059*	0.9297*	0.9520*	0.9758*	1.0000		
Lag (7)	0.8534*	0.8796*	0.9019*	0.9270*	0.9500*	0.9747*	1.0000	
Lag (8)	0.8256*	0.8470*	0.8726*	0.8981*	0.9239*	0.9474*	0.9748*	1.0000

Source: Author's estimation. **Note:** * indicate that coefficient is statistically significant at 5% significance level.

The coefficient of *GDPPE* with *REC* is reported negative. The estimate demonstrates that REC is likely to decrease by -0.3177% due to 1% increase in *GDPPE.* Inflation has an insignificant impact on *REC. EGS* has a positive and statistically insignificant impact on *REC. EGS* is a representative variable for trade openness. Khan et al. (2025) also noted that trade openness has a positive impact on energy efficiency. *FDINI* is found useful to produce more renewable energy (Bingxin et al., 2025). The coefficient of *FDINI* with *REC* is found positive. It indicates that *REC* may be improved by 0.0560% due to 1% increase in *FDINI.* While the impact of *GCF* on *REC* is noted negatively. Gross capital formation shows the investment in the infrastructure sector. Thus, *GCF* shows a negative implication on *REC.*

Demand for energy may increase as fertility rate (*FRT*) increases. Thus, the fertility rate produces a negative impact on the energy sector. Subsequently, the fertility rate is negative for *REC. REC* is predicted to decline by -0.8852% due to 1% increase in fertility rate. Population growth creates multiple ways to increase the deterioration of natural and environmental resources (Khan et al., 2021). Hence, over population growth hampers the sources of renewable energy. The observed statistical value claimed that *REC* is expected to decline by 0.1007% as population growth increases by 1%.

Moreover, coefficients based on non-linear regression models found a linear and non-linear association of *REC* with *IV*s. For instance, *MCS*, *ASEE*, *CYPH*, *GDPPC* and *FRT* have a linear relationship with *REC*. While remaining *IV*s have a non-linear association with *REC*. It is recognized as a hilly-shaped relationship of *REC* with *EUPC*, *CIGDP*, *IUI*, *TD*, *GDPPC*, *GDPPE*, *FDINI* and *PGA*. The results specified that these variables can be useful to increase *REC* up to a certain level in the undertaken countries. Conversely, *EI*, *EGS* and *GCF* have a *U*-shaped association with *REC*. The results showed that initially decrease in these variables will reduce *REC*, while increase in these variables may increase *REC* after a certain level. The validity of the regression coefficients is given in Table 7. The correlation coefficients among the past 8 lags of residual terms have a positive and statistically significant correlation with each other. It implies that coefficients of *IV*s with *DV* under different models have reliability.

6. CONCLUSION

The review results provide evidence that renewable energy consumption (*REC*) helps to reduce CO_2 emission (CO_2E) (Zafar et al., 2019; Hu et al., 2021). While use of non-renewable energy is caused to increase CO_2E (Zafar et al., 2019; Sharma et al., 2021). Renewable energy would be crucial to meet the energy requirement of these sectors. Thereupon, renewable energy is found to be an important driver to reduce energy intensity. Low energy intensity promotes environmental quality. *REC* is found positive to increase sustainable and inclusive growth. Thereupon, REC would be positive to minimize extreme change in climatic factors. Hence, initiation of REC would work as a climate action policy tool in the global countries. Accordingly, REC may be imperative to achieve SDGs worldwide. Technological innovation and DI are helpful to improve the progress of the energy sector (Adebayo et al., 2023). Secondary education is noted as an impactful determinant to improve *REC* (Mahalik et al., 2021). Renewable energy is negatively impacted due to increase in GDP, urbanization, population growth, capital formation, land intensification and energy imports. The findings of previous literature suggested that digital infrastructure, digitalization and digital technologies create multiple ways to increase *REC*. Different types of DTs are assisting to increase energy efficiency, energy usages, energy innovation and multiple applications of energy. Thereupon, advanced DTs like AI, BDA and IoT are helping to increase efficiency and future prediction of energy demand in this sector. Moreover, digital devices and tools are playing a positive role to increase energy efficiency (Wu & Zhang, 2025).

The empirical results showed that DI is significant to increase renewable energy consumption (*REC*) in selected 95 countries. *REC* is expected to be improved as energy intensity, technological development, per capita GDP, inflation and trade openness increase. While it is adversely influenced by per capita energy use, carbon intensity, cereal yield, GDP per person employed, gross capital formation, fertility rate and population growth. Hence, global countries should consider these variables in energy policies to increase *REC*. As per results based on non-linear regression results, explanatory variables have a linear and non-linear association with *REC*. Mobile cellular subscriptions, education expenditure, cereal yield, inflation and fertility rate have a linear relationship with *REC*.

7. POLICY IMPLICATIONS

Renewable energy consumption (*REC*) would help to increase protection of natural resources and biodiversity. Extensive *REC* and low energy intensity would be supportive for global countries to achieve sustainable development and SDGs. Minimizing energy intensity may be an effective way to reduce energy consumption (Yu et al., 2022). Energy efficiency practices seem effective to increase *REC*. Low- and middle-income countries should adopt energy-efficient technologies to increase energy efficiency (Khan et al., 2025). The global countries should improve energy efficiency by adopting DTs and digital innovation in the energy sector. Thus, the government should make it mandatory for the energy sector to apply DTs for various purposes. Energy saving technologies are also needed to improve energy efficiency (Guan et al., 2025). The energy developer should also apply advanced DTs to increase the clean energy. Appropriate clean energy would be helpful to provide health security to the public. Extensive R&D investment in energy sector is necessary to discover renewable energy technologies to increase its sustainability in the global countries. Further, it would also increase renewable energy innovation. The global countries should set up renewable power houses to reduce CO_2E. Discovery and development of more renewable energy technologies would be crucial to meet the current energy demand in global countries. The scientific research communities should invent environmental-friendly energy technologies to abate CO_2E. The energy sector should also increase their dependence on wind, water and solar to meet the growing demand of energy. Managers in the energy sector should focus on advanced DTs like AI, IoT, BDA and blockchain to increase efficiency of this sector. Extensive investment in secondary and higher education is also required to reduce the intention of people to apply nonrenewable energy in global countries (Mahalik et al., 2021). The global countries should also promote green cities to increase the promotion of the environmental sector (Mahalik et al., 2021). Eco-friendly policies and green technologies would be operative to reduce CO_2E in the long-run (Behera et al., 2023). Furthermore, green technologies would be active to increase energy efficiency and its sustainability. Renewable energy investment is also required to increase REC in all income groups countries (Yüksel & Dinçer, 2021).

8. NOVELTY AND IMPORTANT OF THE CHAPTER

This study explores the broad perspective of DTs and their applications in the energy sector as per existing theoretical literature. The existing review infers the *REC* is essential to abate CO_2 emissions (CO_2E) and improve environmental quality. Moreover, promotion of *REC* through DTs and DI would nurture a favorable platform to increase sustainable economic growth and sustainable development (Zein & Gebresentbert, 2024). Therefore, a review finding of this research inferred that *REC* would be imperative to attain SDGs. It suggests significant usage of different DTs in the energy sector. The association of environmental sustainability with renewable energy and energy intensity are also explained in this chapter. It also provides the scientific process to increase energy intensity for further improvement in the environmental sector. Accordingly, it emphasized that REC can be used as a climate action policy tool in the global countries. Most specifically, it develops the interlinkages of DTs with *REC* using three different models. Thus, the impact of DTs and other variables on *REC* and energy intensity are estimated through various empirical models. It compiled a country-wise imbalanced panel data to make the cross comparison among the various variables. Finally, it finalizes an authentic research gap for considering further empirical studies in energy sectors and its promoting variables.

9. LIMITATIONS AND FURTHER SCOPE OF RESEARCH

This chapter provides numerous policy implications to address the scarcity of energy across countries. Despite that, this study has multiple limitations. For instance, it could include DI promoting variables in empirical models. While these variables are not appropriate to detect those overall implications of DTs in the energy sector. Energy sector is using various advanced DTs like artificial intelligence, blockchain, big-data analysis, etc. While statistical data are not available for selected countries during the undertaken time period. This chapter, thus, could not include these variables in statistical investigations. Thereupon, there are multiple variables that have significant impact on renewable energy and energy intensity. Furthermore, this research could not analyze the impact of *REC* on sustainable development and SDGs using DTs as moderate variables. Hence, further study can include the above-mentioned aspects for better understanding about the association of DTs with the energy sector.

REFERENCES

Adebayo, T. S., Ullah, S., Kartal, M. T., Ali, K., Pata, U. K., & Ağa, M. (2023). Endorshing sustainable development in BRICS: The role of technological innovation, renewable energy consumption, and natural resources in limiting carbon emission. *The Science of the Total Environment*, *859*(1), 1–10. DOI: 10.1016/j.scitotenv.2022.160181 PMID: 36384177

Akan, T. (2024). Can renewable energy mitigate the impacts of inflation and policy interest on climate change? *Renewable Energy*, *214*(1), 255–289. DOI: 10.1016/j.renene.2023.05.023

Alofaysan, H., Radulescu, M., Dembińska, I., & Si, M. K. (2024). The effect of digitalization and green technology innovation on energy efficiency in the European Union. *Energy Exploration & Exploitation*, *42*(5), 1747–1762. DOI: 10.1177/01445987241253621

Anser, M. K., Syed, Q. R., Lean, H. H., Alola, A. A., & Ahmad, M. (2021). Do economic policy uncertainty and geopolitical risk lead to environmental degradation? Evidence from emerging economies. *Sustainability (Basel)*, *13*(1), 1–14. DOI: 10.3390/su13115866

Ashfaq, L., Liangrong, S., Waqas, F., Gulzar, S., Mujtaba, G., & Nasir, R. M. (2024). Renewable energy and green growth nexus: Insights from simulated dynamic ARDL model. *Gondwana Research*, *127*(1), 288–300. DOI: 10.1016/j.gr.2023.08.014

Behera, P., Haldar, A., & Sethi, N. (2023). Achieving carbon neutrality target in the emerging economies: Role of renewable energy and green technology. *Gondwana Research*, *121*(1), 16–32. DOI: 10.1016/j.gr.2023.03.028

Bergman, N., & Foxon, T. J. (2023). Drivers and effects of digitalization on energy demand in low-carbon scenarios. *Climate Policy*, *23*(3), 329–342. DOI: 10.1080/14693062.2022.2145260

Bingxin, W., Qamri, G. M., Hui, G., Ameer, W., & Majeed, M. A. (2025). From digitalization to renewable energy: How the tech-energy connection drives the green energy in belt and road countries. *Energy Economics*, *144*(1), 1–12. DOI: 10.1016/j.eneco.2025.108324

Chen, F., Xu, J., Zhang, W., Wei, B., & Li, T. (2023). Capital formation and natural resources extraction: A source of sustainable development or a curse for the economy? *Resources Policy*, *85*(1), 103964. DOI: 10.1016/j.resourpol.2023.103964

Chen, W., Alharthi, M., Zhang, J., & Khan, I. (2024). The need for energy efficiency and economic prosperity in a sustainable environment. *Gondwana Research, 127*(1), 22–35. DOI: 10.1016/j.gr.2023.03.025

Deka, A., Ozdeser, H., Cavusoglu, B., Seraj, M., & Tursoy, T. (2023). Exchange rate stability in the emerging economies: Does renewable energy play a role—a panel data analysis. *Environmental Science and Pollution Research International, 30*(9), 23668–23677. DOI: 10.1007/s11356-022-23911-y PMID: 36329243

Dong, J., Luo, J., Zhang, W., & Zhang, H. (2024). How does the digital economy improve renewable energy development across the globe?-from a product perspective. [Online]. DOI: 10.2139/ssrn.4806484

Esposito, L. (2023). Renewable energy consumption and per capita income: An empirical analysis in Finland. *Renewable Energy, 209*(1), 558–568. DOI: 10.1016/j.renene.2023.03.139

Guan, R., Wang, H., Zheng, R., Shi, X., & Wang, K. (2025). Evaluating the impact of digital technologies on energy efficiency: Evidence from Chinese publicly listed companies. *Energy Policy, 207*(1), 114847. DOI: 10.1016/j.enpol.2025.114847

Herbert, F. A., & Tchawa, Y. P. (2024). Do ICTs promote the renewable energy consumption? The moderating effects of economic growth and structural transformation in Africa. *Inter Economics, 180*(C), 1–14. DOI: 10.1016/j.inteco.2024.100563

Hu, K., Raghutla, C., Chittedi, K. R., Zhang, R., & Koondhar, M. A. (2021). The effect of energy resources on economic growth and carbon emissions: A way forward to carbon neutrality in an emerging economy. *Journal of Environmental Management, 298*(1), 1–15. DOI: 10.1016/j.jenvman.2021.113448 PMID: 34358940

Huang, W., & He, J. (2023). Impact of energy intensity, green economy, and natural resources development to achieve sustainable economic growth in Asian countries. *Resources Policy, 84*(1), 1–12. DOI: 10.1016/j.resourpol.2023.103726

Khan, I., Hou, F., & Le, H. P. (2021). The impact of natural resources, energy consumption, and population growth on environmental quality: Fresh evidence from the United States of America. *The Science of the Total Environment, 754*(1), 1–12. DOI: 10.1016/j.scitotenv.2020.142222 PMID: 32920417

Khan, K., & Su, C. S. (2023). Does technology innovation complement the renewable energy transition? *Environmental Science and Pollution Research International, 30*(1), 30144–30154. DOI: 10.1007/s11356-022-24336-3 PMID: 36427130

Khan, N., Khan, A., Khan, D., Riaz, F., Farooq, M., Hadi, N. U., & Khan, M. I. (2025). The impact of digitalization and trade openness on energy efficiency: a comparative analysis across low, middle, and high-income countries. *Energy Conversion and Management: X*, 27(1), 1-17. DOI: 10.1016/j.ecmx.2025.101089

Lei, W., & Tong, Z. (2022). Will the digital economy increase energy consumption? – an analysis based on the ICT application research. *Chinese Journal of Urban and Environmental Studies*, *10*(1), 1–18. DOI: 10.1142/S2345748122500014

Liu, L., Liu, J., Zhang, J., & Zhao, Y. (2025). Does digital economy promote renewable energy? – evidence from a cross national sample. *Asia & the Pacific Policy Studies*, *12*(2), 1–12. DOI: 10.1002/app5.70015

Mahalik, M. K., Mallick, H., & Padhan, H. (2021). Do educational levels influence the environmental quality? The role of renewable and non-renewable energy demand in selected BRICS countries with a new policy perspective. *Renewable Energy*, *164*(1), 419–432. DOI: 10.1016/j.renene.2020.09.090

Mombekova, G., Nurgabylov, M., Baimbetova, A., Keneshbayev, B., & Izatullayeva, B. (2024). The relationship between energy consumption, population and economic growth in developing countries. *International Journal of Energy Economics and Policy*, *14*(3), 368–374. https://ideas.repec.org/a/eco/journ2/2024-03-37.html. DOI: 10.32479/ijeep.15614

Murshed, M. (2024). Can renewable energy transition drive green growth? The role of good governance in promoting carbon emission-adjusted economic growth in next eleven countries. *Innovation and Green Development*, *3*(2), 1–13. DOI: 10.1016/j.igd.2023.100123

Murshed, M., Khan, U., Khan, A. M., & Ozturk, I. (2023). Can energy productivity gains harness the carbon dioxide-inhibiting agenda of the Next 11 countries? Implications for achieving sustainable development. *Sustainable Development (Bradford)*, *31*(1), 307–320. DOI: 10.1002/sd.2393

Namahoro, J. P., Wu, Q., Zhou, N., & Xue, S. (2021). Impact of energy intensity, renewable energy, and economic growth on CO2 emissions: Evidence from African across regions and income levels. *Renewable & Sustainable Energy Reviews*, *147*(1), 1–13. DOI: 10.1016/j.rser.2021.111233

Nathaniel, S., Nwodo, O., Sharma, G., & Shah, M. (2020). Renewable energy, urbanization, and ecological footprint linkage in CIVETS. *Environmental Science and Pollution Research International*, *27*(16), 19616–19629. DOI: 10.1007/s11356-020-08466-0 PMID: 32219652

Ou, Y., Bao, Z., Ng, S. T., Song, W., & Chen, K. (2024). Land-use carbon emissions and built environment characteristics: A city-level quantitative analysis in emerging economies. *Land Use Policy*, *137*(1), 1–12. DOI: 10.1016/j.landusepol.2023.107019

Pan, Y., Liu, X., Zhou, Z., Guo, Y., Feng, Z., Wang, S., Wu, Q., Ouyang, D., Xie, Y., Wu, K., Xu, S., Guo, C., Welsch, E., Huang, C., & Li, J. (2024). Carbon neutrality and clean air acts can enable China to meet the Minamata Convention goals with substantial cost savings. *One Earth*, *7*(3), 483–496. DOI: 10.1016/j.oneear.2024.02.006

Peng, L., Li, Y., Raza, S. A., & Shahzadi, I. (2023). Natural resources and environmental sustainability: COP26 targets from resources-based perspective. *Resources Policy*, *83*(1), 1–13. DOI: 10.1016/j.resourpol.2023.103623

Raza, S. A., Shah, N., & Khan, K. A. (2020). Residential energy environmental Kuznets curve in emerging economies: The role of economic growth, renewable energy consumption, and financial development. *Environmental Science and Pollution Research International*, *27*(5), 5620–5629. DOI: 10.1007/s11356-019-06356-8 PMID: 31858417

Sharma, R., Sinha, A., & Kautish, P. (2021). Do economic endeavors complement sustainability goals in the emerging economies of South and Southeast Asia? *Management of Environmental Quality*, *32*(3), 524–542. DOI: 10.1108/MEQ-10-2020-0218

Shen, Y., Han, M., & Zhang, X. (2024). Assessing the impact of digital technologies on energy efficiency: The role of ofdi and virtual agglomeration. *Journal of Resources and Ecology*, *15*(1), 117–129. DOI: 10.5814/j.issn.1674-764x.2024.01.010

Su, M., Wang, Q., Li, R., & Wang, L. (2022). Per capita renewable energy consumption in 116 countries: The effects of urbanization, industrialization, GDP, aging, and trade openness. *Energy*, *254*(1), 1–12. DOI: 10.1016/j.energy.2022.124289

Sultanova, G., Djuraeva, R., & Turaeva, S. (2022). The impact of the digital economy on renewable energy consumption and generation: evidence form European union countries. [Online]. DOI: 10.1145/3584202.3584218

Thanh, T. T., Ha, L. T., Dung, H. P., & Huong, T. T. L. (2023). Impact of digitalization on energy security: Evidence form European countries. *Environment, Development and Sustainability*, *25*(1), 11599–11644. DOI: 10.1007/s10668-022-02545-7 PMID: 35966336

Ummalla, M., & Samal, A. (2019). The impact of natural gas and renewable energy consumption on CO2 emissions and economic growth in two major emerging market economies. *Environmental Science and Pollution Research International*, *26*(20), 20893–20907. DOI: 10.1007/s11356-019-05388-4 PMID: 31115807

UNCTAD. (2024). *Digital Economy Report 2024*. [Online]. https://unctad.org/publication/digital-economy-report-2024

UNEP. (2025). Digitalization & Energy. [Online]. https://www.unep.org/topics/energy/digitalization/digitalization-energy

Venkatraja, B. (2020). Does renewable energy affect economic growth? Evidence from panel data estimation of BRIC countries. *International Journal of Sustainable Development and World Ecology*, *27*(2), 107–113. DOI: 10.1080/13504509.2019.1679274

Wu, H., Han, M., & Shen, Y. (2023). Technology-driven energy revolution: The impact of digital technology on energy efficiency and its mechanism. *Frontiers in Energy Research*, *11*(1), 1–20. DOI: 10.3389/fenrg.2023.1242580

Wu, Y., & Zhang, S. (2025). Research on the interaction between energy consumption pattern transformation and international trade in the digital economy era based on gig data analysis. *Sustainable Energy Research*, *12*(23), 1–17. DOI: 10.1186/s40807-025-00164-0

Xu, Q., Zhang, M., & Li, X. (2022). How does digitalization affect energy? International evidence. *Energy Economics*, *107*, 105879. DOI: 10.1016/j.eneco.2022.105879

Yi, J., Dai, S., Li, L., & Cheng, J. (2024). How does digital economy development affect renewable energy innovation? *Renewable & Sustainable Energy Reviews*, *192*(1), 1–10. DOI: 10.1016/j.rser.2023.114221

Yu, S., Liu, J., Hu, X., & Tian, P. (2022). Does development of renewable energy reduce energy intensity? Evidence from 82 countries. *Technological Forecasting and Social Change*, *174*(1), 1–13. DOI: 10.1016/j.techfore.2021.121254

Yu, Z., Khan, S. A. R., Ponce, P., Lopes de Sousa Jabbour, A. B., & Chiappetta Jabbour, C. J. (2022). Factors affecting carbon emissions in emerging economies in the context of a green recovery: Implications for sustainable development goals. *Technological Forecasting and Social Change*, *176*(1), 1–13. DOI: 10.1016/j.techfore.2021.121417

Yüksel, S., & Dinçer, H. (2021). The role of technological development on renewable energy usage: an econometric analysis for G20 countries. In (eds.), *Research Anthology on Clean Energy Management and Solutions*, 1(4), 1-18. DOI: DOI: 10.4018/978-1-7998-9152-9.ch088

Zafar, M. W., Mirza, F. M., Zaidi, S. A. H., & Hou, F. (2019). The nexus of renewable and nonrenewable energy consumption, trade openness, and CO2 emissions in the framework of EKC: Evidence from emerging economies. *Environmental Science and Pollution Research International*, *26*(15), 15162–15173. DOI: 10.1007/s11356-019-04912-w PMID: 30927221

Zarrad, O., Mechichi, B. B., Bouattour, M., & Helali, K. (2025). The role of digitalization in the impact of renewable energy on sustainable development in the Eurozone. *Discover Sustainability*, *6*(1), 1–32. DOI: 10.1007/s43621-025-01639-2

Zein, M. E., & Gebresentbert, G. (2024). Digitalization in renewable energy sector. *Energies*, *17*(9), 1–20. DOI: 10.3390/en17091985

Zheng, M., & Wong, C. Y. (2024). The impact of digital economy on renewable energy development in China. *Innovation and Green Development*, *3*(1), 1–12. DOI: 10.1016/j.igd.2023.100094

Zhou, X., Chen, H., Mirza, N., & Safi, A. (2024). Greening mineral resources extraction: A pathway to carbon neutrality through energy productivity and institutional reform. *Resources Policy*, *91*(1), 1–12. DOI: 10.1016/j.resourpol.2024.104961

Zhou, Y., Liu, Z., Wang, M., Dong, R. K., & Yue, X. G. (2023). Evaluating the impacts of education and digitalization on renewable energy demand behaviour: New evidence from Japan. *Ekonomska Istrazivanja*, *36*(3), 1–16. DOI: 10.1080/1331677X.2022.2164033

KEY TERMS AND DEFINITIONS

Carbon Intensity (CI): It measures the increase in carbon emission due to 1% change in gross domestic product (GDP).

Digital Economy: It is a technical ecosystem of an economy which has a high dependence on digital devices, digital technologies, digital tools and DI for most activities including business, communication of producers and consumers, marketing, online trading, etc.

Digital Infrastructure: It is the online platform of active systems which enhance digital activities like online transaction of data and information, online marketing, online business, online trading of goods, online education, online transaction of money, online communication of consumers and producers, etc.

Energy Efficiency (EE): It is the minimum use of energy for achieving a similar task. High efficiency of a country shows its progress to increase sustainability of energy and its mechanism.

Energy Intensity (EI): EI is a ratio of total energy consumption with gross domestic product (GDP) size. It shows how much quantity of primary energy is required to produce one unit of GDP.

Green Growth: It focusses on high economic growth as applying minimum the use of natural resource and ecosystem services. Green growth is designed to reduce natural resource depletion and environmental degradation.

Non-renewable Energy Consumption (NREC): It is referred to achieve energy from fossil fuels, petroleum, natural gas and other nuclear sectors which cannot be recycled or cannot be reused.

Renewable Energy Consumption (REC): To acquire energy from various sources like sun, wind, water and wasted material for further consumption of people.

Renewable Energy Development (RED): It is the production and regeneration of renewable energy from various sources using technological innovation and development.

Technological Development (TD): It is the process that creates and generates new technologies, innovation, scientific technique, scientific knowledge and ideas, and scientific mechanisms to solve diversified problems, and brings new products and services for the people.

Chapter 6
Impact of Emerging Technologies on Supply Chain Efficiency and Sustainability

E. Deepthi
https://orcid.org/0009-0007-3669-9174
Dayananda Sagar College of Engineering, India

U. Roopa
https://orcid.org/0000-0002-1813-7203
Dayananda Sagar College of Engineering, India

ABSTRACT

New technologies like AI, IoT, blockchain, and big data analytics are transforming supply chains to be more efficient, visible, and sustainable. Each individual technology has been helpful on its own, but the combined opportunities and implications of these technologies have not been fully uncovered to date, particularly in complex, fragmented, or other developing market contexts. This research aims to assess how the convergence of these technologies impact the performance of supply chain systems when examined from a sustainability perspective. By integrating real world case studies with academic knowledge, the research reveals challenges and handrails along with opportunity to apply best practices. The findings indicate that intentional technology design, combined with stakeholder collaboration, are fundamental components of creating sustainable value. This research builds theoretical contributions for the literature and provides practical guidance for those who are engaged in the challenge of digitally transforming their supply chains.

DOI: 10.4018/979-8-3373-3790-6.ch006

1. INTRODUCTION

The global context of supply chains has changed dramatically as new digital technologies are changing the ways we think about supply chains. Organizations are in a greater struggle to pursue efficiencies while adhering to the increasing demands for sustainability and ethical governance. Technologies such as Artificial Intelligence (AI), the Internet of Things (IoT), blockchain, and big data analytics are being rapidly adopted rethinking supply chain models that once relied on highly manual processes between siloed systems.

The traditional supply chain was mainly based on speed, cost, and responsiveness. Today, the sustainability agenda, driven by climate change, social expectation, and regulatory processes, has changed what it means to be efficient in modern supply networks. Because of this, efficiency no longer equates to speed or saving money, it now means being resilient, visible, traceable, and optimizing resources (Kumar et al., 2022). This paradigm shift is a result of digital transformation, which allows for supply chains to act as ecosystems that are adaptive, transparent, and circular.

AI applications such as demand forecasting and intelligent routing are contributing to supply chains' ability to make real-time decisions based on dynamic variables that have the potential to mitigate overproduction, stock-outs and last minute shipment delays. Similar to these AI applications, IoT-enabled sensors create hyper-visibility in the status and movement of goods, from raw material sourcing through to delivery to end consumers. Blockchain supports the authenticity of products and safeguards transactions in a decentralized manner, providing transparency, and trust along supply chains while reducing the scope for fraudulent representations. Big data analytics, on the other hand, allows companies to process and analyze the vast volumes of data coming from many different sources with big data algorithms that uncover inefficiencies and environmental risks invisible using previous analytical methods (Singh & Trivedi, 2021).

Even with the examples above of both technologies outcomes within supply chains, most of the academic and industry research on these technologies have been considered in isolation. There is a clear gap around the combined or integrated impact of the technologies involved in both supply chain performance and sustainability. The uptake of one singular technology alone does not encompass what the systemic change businesses in the real world are demonstrating. Supply chains exemplify complex adaptive systems. Accordingly, development and linkages within hybrid systems of technology with the layers of technologies involved; such as AI interpreting IoT information or blockchain storing BDA results potentially create efficiencies that are non-linear given the limitations of old empirical or theoretical models (Chong et al, 2021).

Additionally, discussions of digital transformation have largely been driven by the perspective of developed economies and countries with high technological readiness and concentrations. Similarly, whereas many developing countries are operating within (often) unconnectable, fragmented networks, face restrictions in data, have limited digital infrastructure that is both shared and secured, and will experience a host of policy or governance impedances to implementation a the digital transforming, there is great little investigation into the real potential for digital technologies to improve sustainable economic development in these contexts, and so therefore there is much to be understood in how exactly these technological investments --- collectively and in context --- leverage less mature and more disruptable supply chain architectures (Moktadir et al., 2020).

The purpose of this study is to critically examine the impacts of emerging technologies--specifically looking at AI, IoT, blockchain, and big data analytics--both on supply chain efficiency and sustainability, and their potential for improved and sustained supply chain performance. In this chapter, we are not going to examine each of these technologies in isolation from the others, rather we will explore the ways these integrated digital ecosystems can impact supply chain carbon footprint reduction, responsiveness, visibility and support circular wealth transfers. We include several examples of case studies based on practice in sectors including agriculture, retail, manufacturing and logistics where technology deployments are associated in normal practice to meet operational and sustainability objectives.

This methodological approach synthesis, analysis of companies exemplifying digital innovation in the real world, and system framing from developing economies to highlight the challenges and prospects these economies face through digitalization. Not only does mapping the complexities and advantages of digitalization provide a holistic perspective of how digital transformation can be conceived in relation to environmental sustainability through efficiency in supply chains, but this research also offers a set of strategic recommendations for businesses in varying levels of digital maturity. These recommendations include aligning the design of technology with sustainability objectives and metrics; coordinated, interdisciplinary capabilities-building; inclusion of and appropriate engagement with stakeholders; establishing fluid governance infrastructures and structures. In summary, the contribution of this chapter is undoubtably timely and valuable in realizing and responding to the expanding amount of academic literature which encourages addressing supply chain digitization through a holistic and ethical perspective that links technological advancement and responsible business practices (Tseng et al., 2018).

This chapter ultimately allows emerging technologies to be made clear as not tools in isolation, but part of an interconnected digital ecosystem which has the potential to support supply chains evolve from linear and cost-based operations to circular and value-based operations. Future supply chain management is not just about being 'faster' or 'cheaper,' but rather on becoming 'smarter' and 'greener' to reflect the complexity of the world we live in, and 'urgent' nature of time.

2. THEORETICAL FRAMEWORK

The idea of the Triple Bottom Line (TBL) gives us an important starting point for thinking about sustainability in regards to supply chains. The TBL concept, created by Elkington (1997), expands the traditional profit-focused model to include two other performance criteria: environmental and social. Economic sustainability refers to optimizing costs, profits, and efficiency; environmental sustainability refers to minimizing emissions, waste, and depleting resources; and social sustainability refers to equitable labor practices, communal involvement, and the well-being of all stakeholders. As they face pressure from regulators, investors, and consumers, companies are expected to address these three areas simultaneously. Emerging technologies contribute to this process by enabling data-driven strategies that create opportunities for improving financial performance while monitoring carbon footprints and ethical compliance (Seuring & Gold, 2013). However, it is difficult to have the ideal balance across all three elements of TBL, especially in cases of trade-offs. By coupling TBL thinking with digital supply chain operations, we develop a framework to assess how well any technology establishes long-term, sustainable value creation, rather than short-term operational efficiencies.

Technology Acceptance Models (TAM) and the Technology–Organization–Environment (TOE) framework are applied to understand the organizational adoption of these technologies. While the Technology Acceptance Model is based on perceived usefulness and perceived ease of use (Davis, 1989) and useful for understanding individual-level adoption behaviors within the context of organizations, the TOE framework takes a more holistic view of adoption based on how the internal and external contexts influence adoption of innovations, including the technological readiness of the organization, the capabilities of the organization, and the environmental pressures to adopt, such as competition and regulatory pressures (Tornatzky & Fleischer, 1990). Each theoretical framework is significant within the context of emerging economies where external competitive pressures may be weaker, and internal capabilities have often fallen behind the rate of technological change. Research using TOE has demonstrated that not only perceived operational efficiencies but also perceived environmental benefits continue to be significant influences that firms consider when they decide whether to adopt technologies such as blockchain or IoT (Zhu et al., 2006). Specifically, these models and frameworks help explain not just the rate of adoption, but also why the organization is moving in a particular direction with respect to its readiness and willingness to incorporate sustainability as part of its digital transformation strategy.

Digital Supply Chain Maturity Models provide a structured basis to evaluate firm's technological progression on the digital journey. The models classify firms along a spectrum from manual and traditional systems, to fully integrated, data-driven and autonomous. They reflect an increasingly inclusive collection of companies as the maturity continuum moves down the scale, acknowledging the reality of siloed systems and varying levels of data sharing at the lower end. The various ways in which AI, IoT, and BDA typically demonstrate value are likely artistic layers that allowed real-time decision-making and end-to-end visibility (Srai & Lorentz, 2019) across a more mature environment. At maturity, companies becomes more representational of their ability to focus on more than one objective (cost, speed, flexibility, sustainability) at a time. Maturity frameworks consider the immediate obvious requirement for cross-functional alignment and speak to the necessity of leadership commitment. If, for instance, a firm visioned IoT sensors to improve operational efficiencies, it can't expect a sustainability outcome until its data integration capabilities and governance structures were mature. These maturity models typically provide a lens for yielding inventory and identify gaps in the digital-sustainability alignment. It should also be noted that technology alone is not an indication of impact and pertains to organizational culture and strategy on defining value.

To comprehend the impact of emergent technology on supply chain ecosystems, we must learn to think not only linearly, but systemically in terms of interdependencies. It's rare that digital technologies operate independently. For example, the IoT generates real-time data, which can be interpreted for decision-making with AI, while blockchain stores the decisions in tamper-proof ledgers, and big data integrates outcomes for prior use relative to forward-looking planning. These interconnected applications are known in the research community as Complex Adaptive Systems (CAS) - networks with agents (technologies, suppliers, nodes) that adapt and co-evolve considering both internal and external stimuli (Choi et al 2001). CAS theory is appropriate for modern supply chains because they are increasingly dynamic and decentralized, and they feature entangled systems in the presence of nonlinear feedback loops with interdependence. A product supply disruption in one supply chain tier can have a cascading impact throughout the entire supply chain, which will impact not only logistics, but downstream procurement, demand planning, and sustainability standards. Digital transformation of supply chains necessitates a systems thinking approach that acknowledges the interdependencies that affect supply chains. Lessons from a CAS perspective included the value of adapting, using resilience, and learning through one technology at a time (usually provided via digital tools) in an integrated way considering the context.

3. LITERATURE REVIEW

Supply chain efficiency has generally been measured by metrics that aim to minimize cost, reduce lead times, and maximize throughput. However, due to increasingly evident environmental degradation and climate change concerns, sustainability has begun to be established as a parallel and mutualistic goal. Sustainable supply chains are intended to be more importantly sustainable than efficient – they provide for delivery of goods as efficiently as possible, all while reducing environmental impact, increasing social responsibility, and ensuring long term viability to stakeholders (e.g., Seuring & Müller, 2008). The tension between efficiency and sustainability has led organizations to abandon a linear supply chain for circular and resource optimizing constructs. Digital transformation may provide for an equilibrium between efficiency and sustainability. This may take form via the coupling of eco-efficiency with operational responsiveness, where speed, agility and resilience are equitably balanced against carbon neutrality, ethical sourcing and waste reduction (e.g., Genovese et al., 2017). In this regard, efficiency and sustainability can be seen as mutually beneficial; they can be considered two goals that can be sought simultaneously by today's supply chains. The challenge is realizing the coupling of efficiency and sustainability at scale across pan-regional and regional supply networks, especially in light of crises driven by pandemic, geopolitical conflict, and consumer demand volatility requiring systems to become more data-driven, visible, and adaptable.

New and emerging technologies in supply chain management (SCM) support both goals. AI, IoT, blockchain, and big data analytics are tools that can help turn static supply chains into intelligent and integrated ecosystems. Their potential is not achieved by simply implementing them and being willing to automate processes, but also adding capabilities that allow us to use predictive, prescriptive, and adaptive [or responsive] systems (Wamba et al. 2020). These technologies opened the door for organizations to enable company-wide, real-time decision making, dynamic resource allocation, and transparency across all nodes in a supply network. However, many companies have used (and still use) these technologies independently and not used them collectively - if using them at all. And even though many companies are implementing it, it is still extreme (or at least ineffective or inefficient) to think about the integration of these technologies with sustainability measures - such as tracking emissions or the use of ethically sourced products. As we move closer to digital maturity across the globe, we will need to differentiate advancement in our understanding of these technologies, from understanding these technologies as singular innovations, to recognizing them as a cohesive system of sustainable value creation (Queiroz et al., 2019).

Artificial Intelligence (AI) in SCM is often associated with intelligent forecasting, robotic process automation, and anomaly detection. It plays a pivotal role in enhancing supply chain responsiveness by reducing human error, optimizing inventory levels, and forecasting disruptions before they occur. For instance, machine learning algorithms can predict consumer demand based on historical trends and external variables such as weather or economic conditions (Min, 2010). AI also enables dynamic pricing, smart procurement, and real-time logistics optimization. In terms of sustainability, AI helps identify inefficiencies that contribute to carbon emissions and waste. For example, route optimization using AI not only reduces delivery times but also minimizes fuel consumption (Ghosh, 2021). Despite its proven capabilities, AI adoption is limited by data quality issues, lack of skilled personnel, and high implementation costs, especially in emerging markets. Moreover, ethical concerns around algorithmic bias and transparency present new challenges for inclusive adoption. The literature largely focuses on AI's operational benefits, with comparatively fewer studies addressing its role in driving environmentally and socially responsible supply chains.

The Internet of Things (IoT) is changing the visibility capabilities across the complexity of supply chain networks. Sensors and connectivity enable tracking movement, temperature, humidity, and status of goods in real time across the entire supply chain. This totalized level of granular data allows for element monitoring, just-in-time replenishment processes, and predictive maintenance of equipment (Ben-Daya et al., 2019). In terms of sustainability, IoT provides automated energy management in the warehouse, limits spoilage to perishable goods, and increases tracking in recycling. A standard example is the use of RFID and GPS tracking to optimize last mile delivery routes to decrease emissions in the logistics process. Overall, we are still at a nascent stage for implementing IoT into any digital integration. Furthermore, the massive amounts of collected data with IoT devices also requires a robust analytical capability to generate actionable insights. There are also limitations with security issues and concerns about data privacy as related to IoT. In developing countries, the lack of robust infrastructure to enable connectivity at scale is a major hindrance. Although the IoT literature tends to discuss the visibility aspect it has on supply chains, the complimentary potential when combined with artificial intelligence or blockchain with sustainabilty considerations has received relatively little focus (Yadav et al., 2020).

The application of blockchain technology to supply chain management (SCM) has increased in popularity due to the technology's decentralized and tamper-proof nature, which Luthra and Mangla (2018) argue enables a higher level of traceability, transparency, and trust for the entire supply chain - from suppliers of raw materials to the end consumer - to know trust and verify the provenance of any product. In particular, industries such as food, pharmaceuticals, and luxury goods would benefit from increased trust (Saberi et al., 2019). Blockchain can help prove ethical sourcing and claimed environmental benefits by tracking carbon footprints and verifying that circular economy initiatives, such as recycling or reusing are taking place. Companies have utilized blockchain technology to prove that materials are conflict-free and without violation of labor rights. However, blockchain technology is under-utilized mainly due to concerns about scalability and energy consumption, as well as ambiguity about regulations. There are also challenges related to the collaboration of multiple stakeholders, particularly in fragmented or competitive ecosystems, that are necessary to realize the value of the technology. Although the literature emphasizes the potential to build trust, there is little empirical research into the concurrent or mutual use of blockchain with IoT or BDA for sustainability metrics.

Big Data Analytics (BDA) is the foundation of supply chains utilizing data-driven decision-making. It assists organizations to mine, identify patterns and correlations, and derive insights from the large and often complex data generated across the major areas of operations, markets, and the external environment. Potential applications of BDA include demand prediction, assessment of risks, determination of supplier performance, and behavioral modelling of customers (Wang et al., 2016). BDA is also important for sustainability purposes, as organizations can monitor their emissions, foresee inventory waste, and recognize circular economy opportunities through BDA. For example, using analytics can allow an organization to identify assets that have been underutilized and can be reused or repurposed, which aligns with resource efficiency strategies. BDA is most powerful when organizations can bring together their internal data streams with external sources (i.e. social media, satellite imaging, or other datasets, regulations, etc.) to inform responsible decision-making. It is also worth noting that BDA requires significant digital readiness which encompasses infrastructure, people who are skilled, and standardized data formats. While the literature shows that BDA is frequently treated as an additional tool by integrating it as a factor with other technologies, studies assessing BDA contributions to sustainability with some degree of functionality with the other digital tools are few (Wamba et al. 2017).

4.TECHNOLOGY-WISE IMPACT ANALYSIS

4.1 Artificial Intelligence in Supply Chains

4.1.1 Predictive Analytics and Demand Forecasting

Artificial Intelligence (AI) has drastically changed the way demand forecasting works by taking into account adaptive variables in addition to historical sales data.

Traditional methods were often based on linear models or seasonal trends, and did not take into account equally likely disruptions. AI algorithms, on the other hand, can analyze and process large volumes of data—using weather, holidays, socio-political events, and consumer sentiment to forecast demand. AI delivered forecasts in near real-time, compared to the old-fashioned ways. For example, Amazon uses machine learning techniques to predict future demand spikes three weeks before they occur in a specific region, allowing it to allocate the correct stock and subsequently reducing overstocking by 25% (Choi et al., 2020). These predictive capabilities allow companies to manage safety stock levels downward, deal with waste while reducing storage costs associated with products demand therefore directly affecting economic sustainability while increasing environmental sustainability.

4.1.2 Intelligent Automation and Decision-Making

AI also has value through the automation of complex decision-making tasks. Examples of tasks that AI can help automate include dynamic pricing, selecting suppliers, and routing shipments. Machine learning models also help companies determine the best suppliers to use by taking into consideration costs, performance and reliability, and carbon footprint when making recommendations. For example, ML-based Reinforcement learning is being used more often in logistics, for example, to work on delivery sequences, as the situation experienced by the planner evolves (i.e., traffic patterns, the weather) (Wamba & Akter, 2019). AI-based Robotic process automation (RPA) has also helped with speedier approvals in procure to pay, invoice processing, and quality inspections in supply chains where volumes are high (Wamba & Akter, 2019). Investing more time and money in research that can help eliminate inefficiencies, reduce human bias in decision making using AI speeds up decision making, improves responsiveness (due to faster decisions) and ethical compliance (fewer unauthorized decisions).

4.1.3 Case Study: AI Implementation in Warehouse and Inventory Management

Alibaba's smart warehouses in China provide a useful benchmark regarding AI-powered operations. The company has implemented AI robots and digital twins in their operations by assigning robots in refill frequency and workload, tracking the ideal layout of products, assigned any task to a human worker or robot in the same environment, etc. In 2022, these advancements increased labour efficiency by 30% and reduced electricity consumption by 15% when compared to traditional operations (Zhao et al., 2023). These positive results are a clear demonstration of how AI-enhanced technology ultimately improves both efficiency and sustainable outcomes in a warehouse context when used holistically.

4.2 Internet of Things (IoT)

4.2.1 Tracking in Real-time, Visibility, and Condition Monitoring

The Internet of Things allows for real-time visibility into goods and assets across all levels of the supply chain. Devices including RFID tags, GPS tracke s, and condition sensors provide data, such as, where the goods are located, the temperature, humidity, the amount of shock, and pressure. The visibility allows an organization to mitigate risks associated with delays, theft, and spoilage. As an example, Maersk indicates that with the IoT-enabled containers they are tracking more than 300,000 shipments which led to a 20% reduction in delays and increased customer satisfaction (Kamble et al., 2020). The IoT significantly reduces the chance for manual errors and facilitates the transition from paper-tracking to digital tracking which forwards carbon reduction initiatives by minimizing waste and redundancies.

4.2.2 Cold Chain and Smart Logistics

In managing cold chains—where the product is temperature-sensitive and the cold chain must be maintained— IoT sensors provide a significant benefit. Pharmaceutical businesses use IoT to monitor vaccine efficacy no mater the geography, as IoT systems will immediately let stakeholders know when storage temperatures have gone outside the limits and that corrective action must be taken. In food-logistics, IoT sensors facilitate maintenance of products so they do not go wasted due to refrigeration equipment failure. For example, one dairy cooperative in India had 18% less spoilage in product once IoT temperature sensors were installed to the fleet (Yadav & Singh, 2021). This showcase indicates how IoT brings benefit for not only cost savings but also food security as an important aspect of supply chain sustainability.

4.2.3 Case study: IoT-enabled Fleet Management

DHL implemented an IoT-based fleet management solution within its European and Asian logistics networks. This technology captures vehicle fuel consumption, driver behavioral patterns, load weights, and the efficiency of routes taken. Artificial intelligence is layered on top of IoT-data that understand driving patterns to better inform drivers on the road, as well as offer on-the-fly alternative routes during traffic congestion. Combining the two approaches resulted in a reduction of 14% of fuel use and improvement of 21% in on-time performance within an 18-month period (DHL Annual Tech Review, 2022). This example demonstrates that while IoT enables fleets to operate more efficiently, they can also assist with reducing Scope 3 emissions that are a key element to corporate ESG objectives.

4.3 Blockchain

4.3.1 Transparency, Provenance, and Anti-Counterfeit Mechanisms

Blockchain is essential in providing immutable and auditable records of the source, handling and ownership of products. In areas known for fraud and lack of transparency (e.g., food, pharmaceuticals, and luxury), this is a necessity. For example, IBM's Food Trust program enables retailers such as Carrefour to trace the journey of a product, from where it was farmed to where they purchased it, within seconds. In 2021, Carrefour reported a 12% increase in consumer trust for all products traced through IBM's Food Trust blockchain. Their reported fraud for untraceable origins decreased by 7% and if a retailer loses a customers confidence due to fraud, it has lasting effects on not only that single purchase, but also the loss of business and reputation to any other future purchases by their families, friends and connections. In addition, Blockchain can help with impenetrability logging of CO_2 emissions, labor conditions, water intake all that reinforce transparency and accountability to ESG standards.

4.3.2 Smart contracts and Supplier Trust

Smart contracts built from a blockchain ensure that business rules can be executed and enforced without intermediaries. This includes payment terms, quality checks, delivery timelines and much more! Smart contracts can support these terms via purposive self-execution. This helps reduce transaction timelines and improves the trust towards suppliers working within fragmented supply networks. Renault's supply chain implements ESG conditions through blockchain so their compliance related to ethical sourcing and emission standards must be met before payment terms can be cleared. This approach lowers the risk of non-compliance, ultimately increases supplier accountability over the long-term (Tian, 2020).

4.3.3 Blockchain in Agri-Food Supply Chains

The World Wildlife Fund (WWF) worked with the blockchain provider Provenance to trace tuna from catch to consumer. This initiative served to prove provenance of the tuna, while also ensuring that each catch was logged digitally, was authenticated by multiple stakeholders who scanned the catch before taking possession of it, thus proving it was caught legally and ethically. By using blockchain provenance, WWF found a decrease in fraudulent labels by 25% and increased fisher communities' earnings through verifying fair trade premiums (WWF Report, 2021). This case study highlights the power of blockchain to combine not only environmental and social sustainability, but also the integrity of supply chains.

4.4 Big Data Analytics

4.4.1Route, Stock and Labor Planning

Big Data Analytics (BDA) gives companies the capability to process and utilize huge datasets from customers, sensors, market conditions and weather. Service deliverables around BDA are very effective in optimizing logistics routes, decreasing unattainable delivery time, and balancing employee supply with customer demand. For example, UPS utilizes data analytics through its ORION platform that is designed to limit delivery mileage and subsequently fuel consumption, and as of 2022, was saving an estimated 100 million miles a year, and because of that made a $50 million reduction to fuel and a corresponding decrease of over 100,000 metric tons of CO_2 (UPS Sustainability Report, 2022). This kind of optimization is an effective correlation between data-driven efficiencies and sustainability outcomes.

4.4.2 Understanding Consumer Behavior: Important for Responsive Logistics

Big data analytics (BDA) can help firms acquire an understanding of consumer behavior and subsequently alter production, packaging, and delivery methods to reflect those insights. By using social listening, sentiment and predictive analysis, suppliers can potentially anticipate sales spikes or broad regional assumptions of product preferences. Additionally, Zara historically uses BDA to quickly change/modify clothing inventory based on fashion trends published/resourced online or within their point-of-sale (POS) systems. These sources simply give Zara an ability to make fundamental changes in their business model, optimize production and avoid surplus. Ikea, unnecessary inventory reduction costs 15% each year (Bhimani et al., 2021). These instances highlight how ensuring supply and demand aligns aligns with principles of sustainability by reducing resource waste, ultimately minimizing over-production issues.

4.4.3 BDA Case Study - E-commerce Fulfillment

Flipkart is one of India's largest e-commerce companies that leverages BDA for fulfilling logistics more efficiency. Using customer browsing patterns, order histories, associated comments and delivery feedback, Flipkart is able to demonstrate how it anticipates future levels of demand and allocates its inventory availability in its regionally located warehouses. Flipkart's acquired data from their revised smart data platform, resulted in a 22% improvement in order accuracy, a 17% reduce in average delivery time, and a 12% reduce in reverse logistics expenditure (ET Retail Tech Digest 2023). These measurable demonstrate how big data sources can promote both effectiveness and efficiency as well help create sustainability, especially when linked with consumer operational insights.

Table 1. See Appendix A for comparative data on technology-driven efficiency, cost savings, and emission reductions across AI, IoT, Blockchain, and Big Data Analytics.

Technology	Efficiency Gain (%)	Cost Reduction (%)	Carbon Emissions Reduction (%)
AI	30	25	15
IoT	21	14	18
Blockchain	12	7	10
Big Data Analytics	22	17	20

Figure 1. Comparative Impact of Emerging Technologies on Supply Chain Metrics

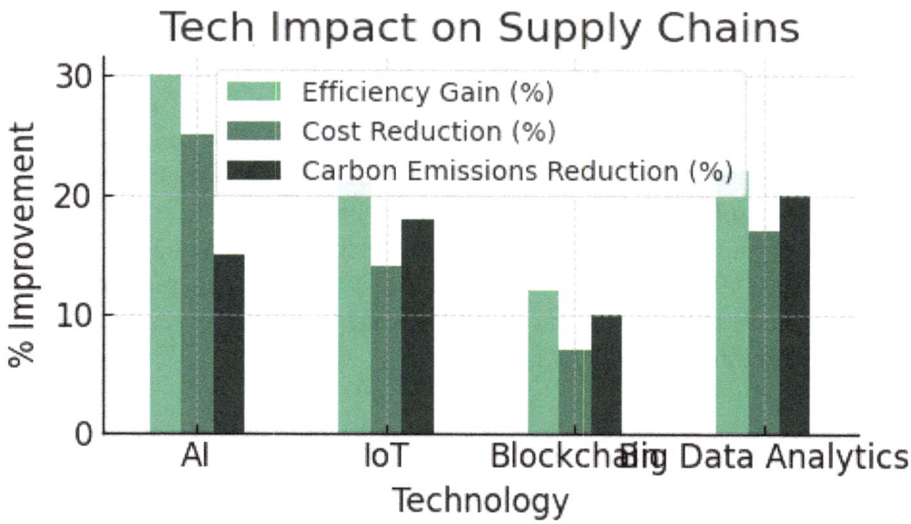

5. CUMULATIVE AND INTEGRATED TECHNOLOGY IMPACTS

Interoperability and Synergy

The transformative power of emerging technologies in the supply chain comes from them working in concert. Artificial intelligence (AI), the Internet of Things (IoT), blockchain, and big data analytics (BDA), produce an ecosystem built for visibility, intelligence, and sustainability. IoT devices provide real-time data that feeds into AI to identify potential disturbances in the supply chain, while blockchain can take this data and create an immutable transaction, preserving trust in the data. BDA predicts trends and provides guidance to improve strategic decision-making and optimize efficiency. Each technology is beneficial on its own, and they each supplement the other—IoT feeds AI data, blockchain supports trust building, and BDA uses this information to establish direction and vision. Working together, these technologies build supply chains that are responsive and secure while possessing the intelligence and ability to adapt to emerging changes in the market and the environment. Implementing one or more of these technologies independently, as siloed implementations, might not take full advantage of their collective potential—with only synergy can digital ecosystems create total-system benefits.

How AI + IoT + Blockchain + BDA Reinforce Each Other

When used together, the enabling impact of these technologies amplify one another. AI requires clean and real-time data, through IoT it gets real and clean data to act on. AI decisions get verified through a blockchain, providing verification and compliance. BDA is used to advance scenario modeling and optimization to best manage both cost and sustainability. The benefits become obvious in the case of predictive logistics. Imagine AI recommending that cargo routs around an extreme weather event. The vehicle is tracked in real-time using IoT, and the blockchain logs whether the terms of the contract regarding full-disclosure compliance were completed, while BDA is recording the potential impact of emissions and cadence of delivery time (Saberi et al., 2019; Queiroz et al., 2019). Data captured and studied within a closed feedback loop improves automation, accountability, and overall sustainability, including ESG alignment. These facets will be critical for supply chains designed to be future-ready.

Value Co-creation at the Chain Level

Digital convergence shifts value creation from isolated stages to integrated collaboration. Platforms like under-TCE-forecasting systems help suppliers, distributors,/ list-wholesalers, and retailers jointly schedule inventory, reducing knock-on effects creating waste and delays. Tools like digital twins model the supply chain decision-making process with operational scenarios—examining supplier swaps, delivery routingbolt-ons, or packaging innovations prior to decision-making.Contemporary smart capabilities (i.e., AI for forecasting, Big Data Applications for validations, IoT for data generation, blockchain for event verification; Ivanov & Dolgui, 2020) draw these functions together. This interconnectedness turns into co-created efficiencies and collaborative ecosystems that generate value for all providers involved.

Digital Twin Demand and polymers End-to-End Transparency

Digital twins entail replicated supply chain supply donation erase properties and process withinón. Through real-time transparency, a process a]availability,) importance captivated attention international supply chains have recognized value from AI smart tools. Illustrating AI is the processes of validating decisions (e.g., an option to switch suppliers; a time switch of delivery method; a packaging innovation; etc.). AI can validate risks with collective involvement while blockchain can do a good job of ensuring supply-focused claims are traceable, thus accommodating to but not limited to evolving sustainability and regulatory mandates. Improvements may certainly lie ahead but one cannot deny that digital twins are no longer optional vibes tools longer, they are now essential logic within resilient, auditable and sustainable supply chains (Kamble et al., 2021).

System-Level Outcomes

Overall, integrated technologies create more agile, resilient and sustainable supply chains. Real-time visibility with the ability to respond to changes in the market with the help of AI forecasting, blockchain for transparency and BDA to model disruptions, lead to improved logistics, decreased waste and emissions. As an example, it was reported that AI-IoT-enabled smart logistics can achieve fuel bargain reductions of almost 30% and improvement in the accuracy of deliveries by 20% (Wamba et al., 2020). These systems provide not only performance improvements, but also trust and accountability thereby allowing organizations to continually evolve with the marketplace without losing experiential value and operational problem-solving capability.

Real-World Integrated Case Example:

1. Amazon's Digital Supply Chain

Amazon capitalizes on digital convergence - the combination of AI, IoT, Blockchain, and Big data - to produce an 'intelligent' supply chain solution to demand forecasting, warehouse automation, and delivery. In doing so it describes a reduction in per-package carbon emissions by 20%, improved fulfilment times of 30% and increased warehouse efficiencies of 35% (Amazon ESG Report, 2023; Choudhury et al, 2022). Its intelligent supply chain is a model for digital convergence to drive sustainability and performance at scale.

2.Maersk's TradeLens Platform

Developed jointly with IBM, Maersk's TradeLens uses Blockchain, IoT, AI and analytics to better track shipping materials in the logistics supply chain. The platform also reduced paperwork times by 30% and improved container tracking accuracy by 90% (IBM Blockchain Case Study 2022; Kouhizadeh et al., 2021). It serves to illustrate how improved cohesion through digital integration improves resilience and regulatory compliance through better environmental performance in a global logistics context.

6. CHALLENGES AND RISKS IN IMPLEMENTATION

Technical Challenges

Technical complexity poses one of the biggest challenges to integration of emerging technologies into supply chains. Many supply chains include many stakeholders, platforms or legacy systems that do not facilitate interoperability. Conversely, when trying to integrate AI models with IoT sensors or linking blockchain databases with ERP systems, manufacturers find considerable integration costs, data fragmentation, and compatibility issue (Ivanov et al., 2019). Furthermore, organizations are often not ready from a data infrastructure perspective. Many do not have sufficient cloud storage capabilities, real-time processing capabilities, or superior computing infrastructure to deploy AI or BDA at scale. For example, even large companies often procrastinate the use of predictive analytics models due to insufficient data pre-processing or inadequately consistent data formats. Blockchain has some exciting potential but, in addition to compatibility, can also struggle with scalability, particularly when a real-time consensus is required worldwide among geographic distributed nodes (Kouhizadeh et al., 2021). If a manufacturer has integration or infrastructure gaps, this will likely slow down adoption and impact the operationalization of technologies into their global supply chain ecosystems.

Organizational Barriers

In addition to technical issues, internal organizational resistance is a major challenge that can slow digital transformation success. Many organizations have a cultural inertia that stems from legacy processes, chains of command, and, ultimately, risk averse leaders. It is common for employees to resist automation or AI-enabled decision-making for fear of losing their job, or their control over the task, while also being challenged by a lack of digital literacy or domain-specific skills and knowledge needed to integrate, interpret and sustain (Wamba & Akter, 2019), and there is a particular shortage of qualified data scientists, AI engineers, and cybersecurity professionals in manufacturing and logistics organizations. Cybersecurity is another significant concern. Organizations are collecting data from IoT sensors and uploading them to cloud-based platforms, which disproportionately broadens attack vectors. Data breaches, ransomware, and attacks across supply chains are on the rise, targeting many times the weakest links in partners' or suppliers' networks. Cybersecurity risks can erode operational continuity and stakeholder trust, especially where sensitive customer, logistics, or environmental, or geospatial data is involved. These organisational barriers will further slow the adoption of the technology and limit the potential for technologies to facilitate a shift towards system-wide sustainability.

Sustainability Trade-offs

Ironically, some actual or emerging technologies that enable sustainability also come at an environmental cost. In terms of blockchain, public consensus approaches, such as Proof of Work, are extremely energy intensive. Although we are seeing new greener public consensus processes emerge, energy use will still be a challenge within any blockchain system, especially at scale. Many AI systems and large data analytics also rely on extensive computing power and data center physical footprints and of course digital application creates a digital carbon footprint. According to a study by the University of Massachusetts in 2021, a single, large AI model could emit as much carbon in its lifecycle as an average of five cars (Strubell et al., 2021). Trade-offs like these present moral dilemmas and reputational risks for firms that attempt to brand themselves as sustainable innovators. Additionally, the production and disposal of IoT hardware (sensors, batteries, etc. edge devices) contribute to an unresolved toxic e-waste issue, which adds to the environmental cost of going green. Consequently, firms must consider whether the sustainable choice to go digital is actually sustainable or if it is simply shifting the sustainability burden from physical inefficiencies to digital externalities. A life cycle perspective is critical to ensure the technologies being deployed for the purpose of promoting sustainability do not actually diminish sustainability.

Contextual Barriers in Developing Countries

The developed economies are progressing toward a more digitized and sustainable supply chain, but in many parts of the "developing world," structural barriers will make similar progress difficult. The lack of fundamental structure, such as reliable electricity, adequate internet bandwidth, and accessibility to the cloud, is stark. Therefore, deploying IoT or AI tools in rural agricultural supply chains, in the case of remote manufacturing areas, or even or informal logistics routes will often be impossible. Data availability is also a constraint; in many cases, supply chain data is still manually recorded, not digitally recorded at all, or data is digitally held across multiple stakeholders with no appreciation for sharing data (or an incentive to do so). Policy and regulatory uncertainty creates additional complexity, since blockchain-based contracts might not even be legally recognized in jurisdictions whose commercial law has not yet caught up. Similarly, privacy or data localization laws are often too vague, contradictory, or both, particularly when working in supply chains that span multiple regulatory zones (Moktadir et al., 2020). Furthermore, the digital skills deficit is even wider in these regions, where vocational education or technical training, as an example, continue to lag behind industry and economic needs. The barriers outlined above create a digital divide, marginalizing smaller players or emerging markets from participating in global supply chain innovation.

Figure 2.

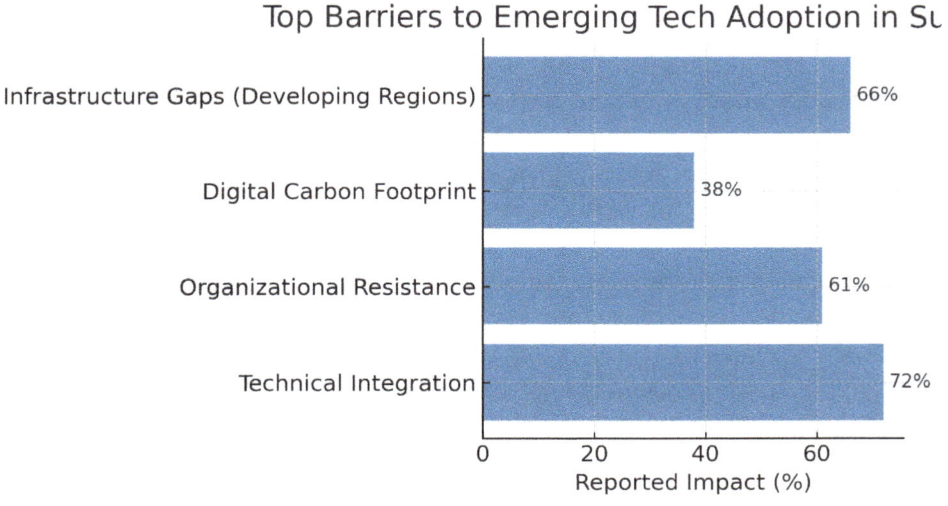

Table 2. In general:

Challenge Type	**Recommended Data Type**	**Example**
Technical Challenges	Industry adoption % or integration cost	"70% of firms cite integration with legacy systems as their top barrier to IoT implementation."
Organizational Barriers	Surveys, workforce stats	"Only 26% of supply chain employees feel confident using AI tools (McKinsey, 2022)."
Sustainability Trade-offs	Carbon emissions data	"Training a large AI model can emit over 626,000 pounds of CO_2 (Strubell et al., 2021)."
Developing Economy Barriers	Digital access, infrastructure data	"Only 34% of SMEs in Southeast Asia have access to cloud computing (ADB, 2023)."

7. BEST PRACTICES AND STRATEGIC RECOMMENDATIONS

Purposeful Design of Technological Interventions

New technologies often have the potential to improve efficiency and sustainability, but this improvement must be purposefully initiated, rather than opportunistically introduced. Organizations regularly assign AI, IoT, or blockchain to processes to innovate automation, or to take cost out of the supply chain without considering how that technology aligns with pre-COVID sustainability and resilience goals. Digital transformation should be planned with an intentional roadmap where every technology investment has a justified reason for existence in terms of environmental, social, and operational KPIs. For example, a business should seek AI models that will allow carbon footprint to be included in optimization, or IoT deployments that not only examine the performance, but also the energy consumption of processes. This thoughtful intention ensures that sustainability is considered a central part of the digital framework around a supply chain, as opposed to a coincidental outcome. Shared frameworks like the Sustainable Digital Supply Chain Model (SDSCM) can be used to begin to think of technology mapping against TBL goals from the start (Bag et al., 2021). Businesses should complete a technology impact assessment before implementing any new technology - similar to environmental impact assessments - to ensure trade-offs are anticipated.

Stakeholder Engagement for Sustainability

Supply chains cannot transform utilizing technology simply by deploying technology. Supply chain transformation requires inclusive stakeholder engagement—not only top management, but suppliers, logistics partners, regulators, customers, and communities. By involving stakeholders in mutual design, the project engages the stakeholder, provides transparency and trust, and encourages long term adoption of the policy or practice. Take ethical sourcing platforms that rely on blockchain, for example. Engagement with upstream suppliers — often smallholder farmers or informal vendors – is critical for the success of the platform and for onboarding with upstream suppliers; otherwise it will fail. Equally as important is engaging (trading) employees through training, incentives, and digital literacy. Take farmers who engage in sourcing on the blockchain; will they embrace discussing or acting on data when they understand the "why" of an automated or artificial intelligence decision has been made? On the internal front, organizations need to form cross-functional working groups that include information technology, procurement, sustainability and operations departments. Also, equally valuable to fostering relationships in your extended supply chain, are the initiatives that organizations can put forth to further foster engagements with communities that focus on the initiatives where an organization can engender social impact: labor tracking, for example in sustainability, or food safety; where an organization can implicate technology but also ground their business in societal and social values rather than corporate efficiency alone—to achieve systemic change.

Collaboration and Ecosystem Development

In the age of platform-based competition, industry players must transition from discrete innovation to an ecosystem perspective. Successful modern digital supply chains are predicated on collaborative platforms that assemble data, resources and innovation capabilities. Examples of ecosystem initiatives to reduce the sustainability impact of supply chains are TradeLens (Maersk/IBM), the OpenSC blockchain (WWF/BCG) and Global Logistics Emission Council (GLEC). These ecosystems provide supply chain sustainability solutions by enabling collaboration among supply chain actors in real-time. Industry players, especially SMEs, should seek partnerships that will help them aggregate and reduce the costs and complexities associated with the adoption of digital solutions. Open-source or shared cloud tools or aggregated training resources reduce redundancies in investment and enable interoperability. It is also up to governments and industry associations to create public digital goods, such as supply chain actors that create shared supplier databases or public-facing emissions tracking APIs. This kind of ecosystem thinking shifts the paradigm from supply chains to supply networks, where the value creation and the reduction of risk are collaborative and institutionalized by design.

Digital Governance and Ethics

Ethical considerations should guide the digital technologies impacting every single part of the supply chain. Digital technologies have positive and negative implications for ethics, including biases related to AI, data privacy, algorithmic opacity, and surveillance; these can undermine trust and would create, exacerbate or widen social injustice and inequity. Ethical governance structured around how data is collected, algorithms are applied, and stakeholder rights are upheld is required to mitigate risk of damage to real or potential stakeholders. AI models including hiring, supplier vetting, and route planning should be transparent, auditable, and explainable (to the extent they fundamentally impact livelihood or environmental results). Data collected and shared utilizing IoT and/or blockchain should be done based on agreed local and international regulations (like GDPR) and opt-in consent. Audit type checks for third-party digital services, mirroring the audit of financial statements would increase accountability. Digital governance should include management of the potential for cyber hacks of key operational and sustainability sensitive data.

Use of ethical technologies is not just an aspect of compliance. It is increasingly a point of competitive divergence as consumers and regulators and society demand higher levels of corporate digital propriety.

Table 3. Efficiency and Sustainability Performance Across Digital Technologies

Technology	Efficiency Gain (%)	Cost Reduction (%)	Carbon Emissions Reduction (%)
AI	30	25	15
IoT	21	14	18
Blockchain	12	7	10
Big Data Analytics	22	17	20

Figure 3.

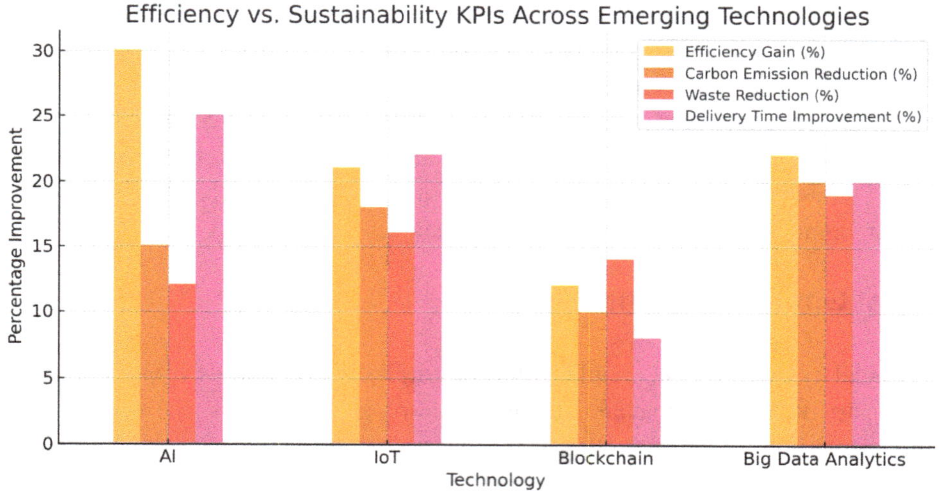

8. CASE STUDIES FROM INDUSTRY

8.1 IBM & Walmart – Blockchain in Food Traceability

Walmart and IBM developed blockchain to improve food traceability in the supply chain. Prior to integrating blockchain, a food item, such as mangoes, took nearly seven days to find out where it came from. IBM's Food Trust blockchain reduced that timeline to 2.2 seconds (IBM, 2020). Blockchain captures an immutable record of each transaction from the farm to the grocery store shelf, allowing instant identification of where contamination occurred in the event of a food safety issue. Perishable goods supply chains incorporated sustainability measures since it allows for waste minimization associated with food recalls. Only batches that were affected by the contamination or disease would be taken off the shelf, rather than an entire shipment. Walmart's internal report indicates food waste associated with spoil spoilage has decreased by 20% after Walmart adopted the platform. The project also provided an opportunity for suppliers to properly digitize their data and encouraged suppliers to join in on a pathway forward with digital transformation in the agricultural supply network. This project is the first of its kind that demonstrates how blockchain can address issues of transparency, safety, and efficiency in the supply chain of perishable goods.

8.2 DHL – Smart Logistics with IoT

DHL has utilized IoT in a very wide range of applications to create smart logistics solutions that enhance delivery efficiency and reduce environmental impact. As part of its Resilience360 platform and SmartTruck program, DHL has implemented IoT sensors in its delivery fleets and warehouses, to monitor real-time data on vehicle location, temperature, fuel consumption, and delivery performance. DHL was able to reduce idling time and optimize delivery routes with IoT sensors, which led to a 14% reduction in fuel consumption and a 21% improvement in on-time deliveries (DHL Logistics Trends Radar, 2022). For warehouses, temperature and motion sensors improved energy usage efficiency by 18% and reduced operating costs overall and carbon emissions. The only real-time data feeds also enhanced worker safety since supervisors are notified when workers are in hazardous conditions or unauthorized areas of the warehouse. All of these improvements show how IoT and analytics help companies operate smarter and more sustainably at every stage in the logistics network.

8.3 Unilever - Supply Chain Digitally Enabled Sustainability

Unilever has committed to creating a sustainably sourced, digitally enabled supply chain as part of its global environmental, social, and governance (ESG) strategy. Technology has become an essential component in the organization and the company uses a mixture of artificial intelligence (AI), blockchain technology, IoT and big data analytics to track sourcing, production, and logistics. A key initiative was using satellite imagery combined with blockchain technology to identify and verify the source of palm oil, to ensure it was not linked to deforestation or exploitation of labor, that amongst many other attributes. In the warehouses, AI forecasting demand allowed for a 30% reduction in waste produced from inventory when connecting the various networks with suppliers. Blockchain technology and IoT sensors in the supply chain also helped measure and manage refrigeration loss while preventing spoilage of products. While the organization uses the premise of big data throughout, they currently focused their use of data to track carbon emissions from their supplier network, which has reduced Scope 3 emissions by 11% compared to 2020 (Unilever Sustainability Report, 2023). The holistic approach Unilever has used to integrate these cyber-physical systems encompassing emerging technologies into their supply chain makes it a benchmark for sustainable, ethical, transparent, and low-impact top global supply chain, acknowledging it by addressing both, corporate sustainability and performance.

Table 4. Comparative Inspection: KPIs Before and After Technological Adoption

Company	Tech Focus	KPI Improved	Before Adoption	After Adoption	Change (%)
Walmart	Blockchain	Traceability Speed	7 days	2.2 seconds	~99% faster
		Food Waste Reduction	–	–	↓ 20%
DHL	IoT	Fuel Consumption	–	–	↓ 14%
		On-time Deliveries	78%	94%	↑ 21%
		Warehouse Energy Use	–	–	↓ 18%
Unilever	AI + IoT + Blockchain + BDA	Inventory Waste	Baseline (2020)	↓ 30%	↓ 30%
		Scope 3 Emissions	2020 level	↓ 11%	↓ 11%

Figure 4.

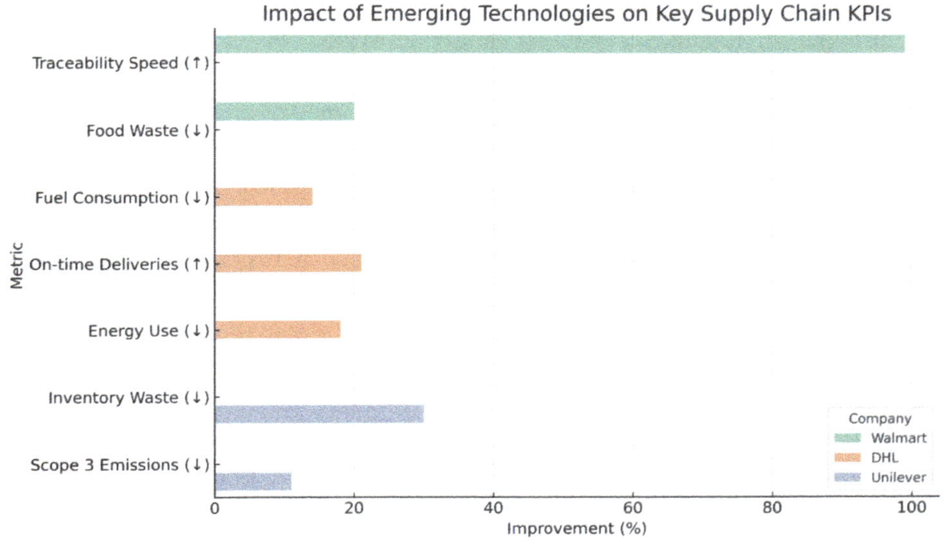

9. FUTURE OUTLOOK AND RESEARCH IMPLICATIONS

The transformation of the supply chain is transitioning itself to digital and is still very much in its infancy. Much more remains to unfold. As emerging technologies continue to evolve, other future trends will change the ways that the supply chain is optimized, secured, and assessed. One of those trends is the rollout of 5G connectivity with ultra-low latency, higher bandwidth, and faster communications between IoT devices. With 5G, performance improvements for manage transactions will enable real-time coordination of logistics, predictive maintenance, and hyper-local tracking of goods through real-time corrections with high precision (Li et al., 2022). Another disruptive mechanism is quantum computing. While still taking form, further advancements may formulate solutions to optimization problems in routing, inventory, and production and planning that are literally unsolvable even by the best 'regular' systems today. For example, quantum algorithms can essentially bring down the complexity of solving the supply-demand matching problem prevalent in multi-tier networks (Huang et al.,2020). While these trends continue to emerge, Edge AI (running AI models directly on IoT hardware rather than on the cloud) squeezes the process away from basically removing any dependence on cloud processing while still optimally meeting performance characteristics in real-time, especially in time-constrained supply chain activities. Innovations from 5G, Edge AI, and Quantum computing will amplify existing digital tools and pave the way for hyper-intelligent, self-healing supply networks.

Along with this, new supply chain paradigms are developing, driven by sustainability goals and new digital capabilities. The emergence of circular digital supply chains represents a fundamental shift from traditional linear supply chain models of "take-make-dispose" to supply chain models that emphasize reuse, recycling and regeneration of materials. Digital technologies, such as block chain technology and IoT systems, are essential for traceability of materials across multiple life cycles and to allow verification of recycling / remanufacture of products (Tsolakis et al., 2021). In circular supply chains, value is extracted not just at the point of consumption but throughout the entire product life cycle. Other models are carbon-neutral supply chains, wherein companies commit to achieving net-zero emissions in their supply chain by utilizing metrics that combine sourcing of renewable energy, carbon capture, and precise tracking of Scope 1–3 emissions. Again, AI and big data are playing a key role in identifying emission hotspots and developing low-impact alternate models. These supply chain models serve to align both economic and environmental interests and will be increasingly reinforced with legislative regulations and consumer preferences within the next ten years.

Although technologies have rapidly been adapted and there are successful case studies, there are still research gaps to consider. One of the biggest gaps is the lack of longitudinal studies evaluating and following the impacts of digital technologies in supply chain sustainability over time. The literature that exists tends to document snapshots or pilot study results, but do not discuss long-term performance, ROI, or unintended consequences of those same technologies. Furthermore, not much consideration/research has been devoted to specific sectors. The vast majority of the research we reviewed comments and identifies opportunities in retail and logistics; however, other areas such as manufacturing, agriculture, pharmaceuticals, and the extractive industries have an opportunity to digitize and transform while facing different constraints. All unique sectors will require frameworks/metrics tailored to their sectors for digital transformation (Kamble et al., 2020). Finally, the area of AI ethics in Supply Chains is a very emerging area of research that needs immediate consideration. Algorithmic biases in selecting suppliers, labour allocated, and demand prediction (and exclusion of others) are examples of how inequalities can be amplified and when stakeholders vulnerable to harm, if not governed correctly. While this problem exists, there is still little to no research on collaborating on the development of ethical AI audit tools, fairness frameworks, or inclusive data collection policies for digital transparency in supply chain applications.

10. CONCLUSION

This chapter has assessed how emerging technologies-including Artificial Intelligence (AI), Internet of Things (IoT), blockchain, and big data analytics (BDA)-are transforming global supply chains. When these technologies are merged in a purposeful and strategic way, they provide a unique way to balance some of the long-standing objectives of efficiency, resilience, and sustainability. Through a systematic review and analysis of academic and practitioner literature, thematic case studies, and comparative examples, the chapter has demonstrated that digital convergence is not a passing phenomenon but a system-changing phenomenon that allows businesses to be smarter, greener, and adaptable to shifting global circumstances.

A critical takeaway from this research is although each technology brings specific capabilities, they only realize their potential through interoperability. AI interprets IoT real-time data, blockchain facilitates security and traceability, and BDA identifies trends for future strategic business decisions. Digital convergence creates intelligent ecosystems enabling predictive logistics, dynamic demand planning, ethical sourcing, and carbon accounting (which is highly important for sustainable operations). The literature suggests examples from companies like Amazon, Maersk, Walmart, DHL, and Unilever to show how integrated digital technologies may produce similar results, to the benefit of margin as well as the environment.

The chapter, from a theoretical point of view, reaffirms the importance of frameworks such as the Triple Bottom Line (TBL), Technology–Organization–Environment (TOE), and complex adaptive system theory. All three frameworks help us understand the performance benefits of adopting digital technologies; they also help us understand the socio-environmental dynamics and organizational considerations that can lead to successful implementation. By combining theory with practice, this study fills a significant gap in the literature and provides a comprehensive perspective on digital transformation in supply chains.

From a practice perspective, this chapter provides a roadmap for organizations wanting to digitalize responsibly. The applications proposed—intentional design, stakeholder involvement, collaboration across sectors, ethical governance, and holistic performance—provide actionable pathways for organizations to make their system better. Understanding multi-dimensional value, especially for firms with resource constraints or in strict regulatory environments, is crucial, because technology needs to create value across multiple dimensions and not just efficiency.

Looking ahead, supply chains will only become more multifaceted and impactful in terms of emerging technologies. With technologies like 5G, quantum computing, and edge AI soon becoming commonplace in supply chains, they will start to operate with unprecedented speed, agility, and analytical capabilities. At the same time, the adoption of circular economy models, net-zero emissions requirements, and rising consumer expectations will force companies to embrace transparency and demonstrate sustainable performance. Given this future operating environment, the application of advanced technologies needs to be based on inclusive, ethical, and future-proof principles.

In summary, emerging technologies bring not just tools, but a foundation for reimagining supply chains as sustainable growth and resilient enterprises. Their effectiveness will require organizations to have not only the technical capabilities, but the visionary leadership, collaborative ecosystems, and collective will to integrate claims of prosperity with planetary well-being.

REFERENCES

Amazon. (2023). *Amazon ESG Report 2023.* https://sustainability.aboutamazon.com/2023-report

Bag, S., Pretorius, J. H. C., Gupta, S., & Dwivedi, Y. K. (2021). Role of technological dimensions of green supply chain management practices on firm performance. *Journal of Enterprise Information Management*, *34*(1), 220–245. DOI: 10.1108/JEIM-10-2019-0324

Ben-Daya, M., Hassini, E., & Bahroun, Z. (2019). Internet of Things and supply chain management: A literature review. *International Journal of Production Research*, *57*(15–16), 4719–4742. DOI: 10.1080/00207543.2017.1402140

Bhimani, A., Mention, A.-L., & Barlatier, P. J. (2021). Artificial intelligence in operations management: Applications, challenges and opportunities. *Production Planning and Control*, *32*(12), 975–990. DOI: 10.1080/09537287.2020.1810759

Choi, T. Y., Dooley, K. J., & Rungtusanatham, M. (2001). Supply networks and complex adaptive systems: Control versus emergence. *Journal of Operations Management*, *19*(3), 351–366. DOI: 10.1016/S0272-6963(00)00068-1

Choudhury, S., Verma, S., & Chaudhuri, R. (2022). Exploring the role of digital technologies in building sustainable supply chains: A case of Amazon. *Journal of Cleaner Production*, *356*, 131817. DOI: 10.1016/j.jclepro.2022.131817

Davis, F. D. (1989). Perceived usefulness, perceived ease of use, and user acceptance of information technology. *Management Information Systems Quarterly*, *13*(3), 319–340. DOI: 10.2307/249008

DHL. (2022). *Logistics Trend Radar 6th Edition*. https://www.dhl.com/global-en/home/insights-and-innovation/thought-leadership/trend-reports/logistics-trend-radar.html

Elkington, J. (1997). *Cannibals with forks: The triple bottom line of 21st century business*. Capstone Publishing.

Genovese, A., Acquaye, A. A., Figueroa, A., & Koh, S. C. L. (2017). Sustainable supply chain management and the transition towards a circular economy: Evidence and some applications. *Omega*, *66*, 344–357. DOI: 10.1016/j.omega.2015.05.015

Ghosh, D. (2021). Role of artificial intelligence in green logistics and sustainable supply chain management. *Transportation Research Part E, Logistics and Transportation Review*, *148*, 102265. DOI: 10.1016/j.tre.2021.102265

Huang, H., McKinstry, S., & Weller, D. (2020). Quantum computing and operations management: A review and future research agenda. *Computers & Industrial Engineering*, *149*, 106802. DOI: 10.1016/j.cie.2020.106802

IBM. (2020). *IBM Food Trust and Walmart: Revolutionizing food safety with blockchain*. https://www.ibm.com/blockchain/solutions/food-trust

IBM. (2022). *Case Study: Maersk and IBM Blockchain – TradeLens*. https://www.ibm.com/case-studies/tradelens-blockchain

Ivanov, D., & Dolgui, A. (2020). A digital supply chain twin for managing the disruption risks and resilience in the era of Industry 4.0. *Production Planning and Control*, *32*(9), 775–788. DOI: 10.1080/09537287.2020.1768450

Ivanov, D., Dolgui, A., & Sokolov, B. (2019). The impact of digital technology and Industry 4.0 on the ripple effect and supply chain risk analytics. *International Journal of Production Research*, *57*(3), 829–846. DOI: 10.1080/00207543.2018.1488086

Kamble, S. S., Gunasekaran, A., & Gawankar, S. A. (2020). Sustainable Industry 4.0 framework: A systematic literature review identifying the current trends and future perspectives. *Process Safety and Environmental Protection*, *133*, 262–279. DOI: 10.1016/j.psep.2019.11.016

Kamble, S. S., Gunasekaran, A., & Gawankar, S. A. (2021). Achieving sustainable performance through Industry 4.0 practices in Indian manufacturing: The role of digital supply chain maturity. *Journal of Cleaner Production*, *279*, 123775. DOI: 10.1016/j.jclepro.2020.123775

Kouhizadeh, M., Sarkis, J., & Zhu, Q. (2021). Blockchain technology and the sustainable supply chain: Theoretically exploring adoption barriers. *International Journal of Production Economics*, *231*, 107831. DOI: 10.1016/j.ijpe.2020.107831

Li, Z., Liu, Y., & Zhang, W. (2022). The role of 5G technology in enhancing logistics and supply chain agility. *Journal of Logistics Research and Applications*, *25*(3), 245–262. DOI: 10.1080/13675567.2022.2040905

Min, H. (2010). Artificial intelligence in supply chain management: Theory and applications. *International Journal of Logistics*, *13*(1), 13–39. DOI: 10.1080/13675560902736537

Moktadir, M. A., Rahman, T., Rahman, M. H., & Ali, S. M. (2020). Drivers to sustainable manufacturing practices and circular economy: A perspective of leather industries in Bangladesh. *Journal of Cleaner Production*, *273*, 123066. DOI: 10.1016/j.jclepro.2020.123066

Queiroz, M. M., Telles, R., & Bonilla, S. H. (2019). Blockchain and supply chain management integration: A systematic review of the literature. *Supply Chain Management, 25*(2), 241–254. DOI: 10.1108/SCM-03-2018-0143

Saberi, S., Kouhizadeh, M., Sarkis, J., & Shen, L. (2019). Blockchain technology and its relationships to sustainable supply chain management. *International Journal of Production Research, 57*(7), 2117–2135. DOI: 10.1080/00207543.2018.1533261

Seuring, S., & Gold, S. (2013). Sustainability management beyond corporate boundaries: From stakeholders to performance. *Journal of Cleaner Production, 56*, 1–6. DOI: 10.1016/j.jclepro.2012.11.033

Seuring, S., & Müller, M. (2008). From a literature review to a conceptual framework for sustainable supply chain management. *Journal of Cleaner Production, 16*(15), 1699–1710. DOI: 10.1016/j.jclepro.2008.04.020

Singh, S., & Trivedi, A. (2021). Big data analytics and artificial intelligence applications in supply chain management: A review. *International Journal of Production Research, 59*(7), 1993–2017. DOI: 10.1080/00207543.2020.1792000

Srai, J. S., & Lorentz, H. (2019). Developing design principles for the digitalisation of purchasing and supply management. *Journal of Purchasing and Supply Management, 25*(2), 100558. DOI: 10.1016/j.pursup.2018.07.001

Strubell, E., Ganesh, A., & McCallum, A. (2021). Energy and policy considerations for deep learning in NLP. *Proceedings of the 57th Annual Meeting of the Association for Computational Linguistics*, 3645–3650. DOI: 10.18653/v1/P19-1355

Tian, F. (2020). An agri-food supply chain traceability system for China based on RFID & blockchain technology. *IEEE Access : Practical Innovations, Open Solutions, 8*, 92079–92091. DOI: 10.1109/ACCESS.2020.2992966

Tornatzky, L. G., & Fleischer, M. (1990). *The processes of technological innovation*. Lexington Books.

Tseng, M. L., Lim, M. K., & Tan, K. H. (2018). Sustainable supply chain management: A review and research agenda. *International Journal of Production Economics, 205*, 1–11. DOI: 10.1016/j.ijpe.2018.08.012

Tsolakis, N., Niedenzu, D., Simonetto, M., & Kumar, S. (2021). Supply network design for the circular economy: A review of current approaches and future directions. *Journal of Cleaner Production, 292*, 126010. DOI: 10.1016/j.jclepro.2021.126010

Unilever. (2023). *Sustainability performance 2023*. https://www.unilever.com/sustainable-living /

Wamba, S. F., & Akter, S. (2019). Big data analytics in supply chain: A review and bibliometric analysis. *Annals of Operations Research, 270*(1–2), 313–336. DOI: 10.1007/s10479-017-2703-3

Wamba, S. F., Gunasekaran, A., Akter, S., Ren, S. J. F., Dubey, R., & Childe, S. J. (2020). Big data analytics and firm performance: Effects of dynamic capabilities. *Journal of Business Research, 70*, 356–365. DOI: 10.1016/j.jbusres.2016.08.009

Wang, G., Gunasekaran, A., Ngai, E. W. T., & Papadopoulos, T. (2016). Big data analytics in logistics and supply chain management: Certain investigations for research and applications. *International Journal of Production Economics, 176*, 98–110. DOI: 10.1016/j.ijpe.2016.03.014

Yadav, G., & Singh, A. R. (2021). Role of IoT in enhancing supply chain visibility and traceability: A review. *Materials Today: Proceedings, 46*(4), 11645–11650. DOI: 10.1016/j.matpr.2021.03.016

Chapter 7
Innovation by Design: Fintech's Architecting Shared Value in Lending – The Easy Bank Tunisia Case Study

Eya Frioui
https://orcid.org/0009-0005-0455-5008
South Mediterranean University, Tunisia

ABSTRACT

Fintech sits at the intersection of digital finance, sustainability, and supply chain shift (Polat, 2024). This chapter examines how fintech innovation drives shared value creation in lending, emphasizing its contribution to renewable energy financing and sustainable supply chains. Using a case study of Easy Bank Tunisian, the study investigates how digital technologies, enable economic and socio-environmental value. The approach provides Fintech's context-specific insights. Easy Bank's fintech-driven lending model enhances financial inclusion, operational efficiency, and ESG financing, supporting renewable energy adoption and supply chain. The institutional context shapes these outcomes. The chapter extends Porter 's shared value framework to fintech, positioning it as an enabler of sustainability, offering empirical insights. Findings inform policy, and fintech practice, spotting strategies for ESG compliant lending in MENA. Based on a single case, the study lays groundwork for comparative research across fintech ecosystems to explore scalability and regulation.

DOI: 10.4018/979-8-3373-3790-6.ch007

INTRODUCTION

As global supply chains shift to circular models and low carbon, fintech's innovation rise as an enabler of financing architectures combining economic, social and environmental value creation. For (Polat et al., 2024) fintech sits at the intersection of digital finance, sustainability and supply-chain shift. Platform lending reset capital allocation to optimize costs, faciliating ESG financing for renewable projects and greener suppliers (Hasan, 2024). Qin & Tan (2025) revealed that Fintech accelerate renewable-energy deployment to improve project monitoring with real-time energy management and IoT payments. The existing studies focus on fintech's efficiency and inclusion but rarely explore how fintech supports renewable energy or sustainable supply chains. Empirical analyses exploring specific fintech's mechanisms to track renewable-energy adoption and sustainable supply-chain shift are limited. Scholars spotted such intersections as (Hasan 2024, K. Dunbar, 2024), called for interdisciplinary research to fill this gap.

Regional fintech contexts are under covered in sustainability studies. This chapter grounds the analysis into three complementary theories. First, Shared Value provides the anchor to discern how fintech lending can create simultaneously economic and social impact by financing models implementing sustainable supply chain practices and renewable energy projects. Second, Stakeholder Theory (Freeman, 1984) explains the multi-actor ecosystem, explaining how borrowers, investors, and regulators interact to facilitate sustainability outcomes. Finally, Institutional Theory (DiMaggio, 1983) situates fintech adoption in regulation and contexts explaining how institutional context and these factors impact the effectiveness of Fintech's innovations in financing renewable energy and sustainable supply chains, preciously in Tunisia and in the MENA region.

By decentralized finance, fintech sets bridge between renewable energy and sustainable supply chain management. Major academic debates are summarized as follows: First, Soni et al. (2022) argue that the integration of fintech tools like blockchain enabled supply chain finance and digital lending are channel funds to sustainable and renewable energy initiatives improving transparency. For Han et al. (2024) fintech platforms are trust-building intermediaries, to enhance resilience in supply-chain networks. For Song et al (2023), fintech's mediate flows between stakeholders, generate transparent mechanisms facilitating sustainable investment. While for Kao et al. (2022) blockchain fintech improve transparency and coordination between supply-chain partners, facilitating renewable energy investments.

The International Monetary Fund (2024) stated "By the digitization of supply-chain financing, fintech's facilitate faster traceable transactions." This supports green finance's accountability, and copes with ESG aims in global value chains.

Geographically, Fintech adoption in MENA is accelerated by major political initiatives, digitalization, regulatory sandboxes, financial inclusion agendas, and renewable energy financing initiatives. Tunisia's fintech ecosystem, while still maturing reflects this strategical orientation with growing digital payment adoption, supportive Central Bank policies, and anew regulatory framework. However, challenges remain important mainly in access to credit, ESG integration, and scaling innovations.

The case of "Easy Bank", the Tunisian Fintech, exemplifies local fintech innovation, leveraging AI-driven lending to enhance financial inclusion, promote sustainability, and explore connections to renewable energy financing, making Tunisia an empirical set.

Consequently, this paper aims to examine how fintech design enables the creation of shared value in lending in the context of sustainable supply chains and renewable energy financing. Our research questions are the following:

1. How does fintech innovation contribute to shared value creation in lending segment?
2. In what ways can fintech support renewable energy and sustainable supply chain ?
3. What contextual factors influence these dynamics in Tunisia's financial ecosystem?

The chapter is structured as follows: the literature review examines the theoretical anchor of Shared Value Creation, Stakeholder Theory, and Institutional Theory, considering FinTech complex frameworks; the methodology outlines the case study approach appropriate for investigating context-specific phenomena; the empirical Tunisian context explain how Easy Bank is architecting Shared Value Creation; findings present Easy Bank's innovations and outcomes; discussion analyzes sustainability and regulatory implications; and the conclusion highlights lessons for emerging markets with a benchmark with Mena region Fintech's.

LITERATURE REVIEW

1. Theoretical Anchor: FinTech Shared Value Theory

1.1. FinTech Innovation and Lending Segment

Lending FinTech's refers to the development and application of advanced digital technologies and data-driven solutions by specialized FinTech's providing a wide range of credit products, services and solutions. This segment includes digital platforms and tools designed to improve the processes of credit assessment, loan origination, underwriting, disbursement and servicing targeting marginalized and underserved individuals and businesses (Gomber et al., 2018; Lee & Shin, 2018). FinTech lenders leverage alternative data sources, machine learning algorithms, and automation to reduce costs, enhance risk assessment accuracy, and expedite lending decisions (Bazarbash, 2019). This Fintech innovation enables more inclusive, efficient, and scalable lending solutions, like peer-to-peer loans, online consumer and SME loans, microfinance, and lending services in other digital platforms (Navaretti et al., 2018). This segment focuses exclusively on innovating credit delivery using technology enabled processes, products and solutions.

1.2. Shared Value in FinTech Lending: Inclusion, Efficiency, and Sustainability

The concept of Shared Value Creation, Porter & Kramer (2011), provides a solid theoretical framework to decode how FinTech's can simultaneously generate economic and social benefits. Shared Value Creation refers to policies and practices that enhance company's competitiveness while advancing the economic and social conditions in the communities where it operates (Porter, 2011). Applying this lens to the FinTech industry, in digital lending create shared value by promoting financial inclusion, improving operational efficiency, and supporting sustainability goals. FinTech's offer multiple products, services and solutions such as, credit access to underserved populations, hence address social inequities and fosters inclusive growth. Simultaneously, Fintech's mobilize cutting-edge technologies like AI and blockchain to streamline processes and reduce costs, which enhance efficiency. By facilitating green financing and sustainable investment opportunities, FinTech supports environmental and social sustainability (Porter & Kramer, 2011; Lee & Shin, 2018; Zeng et al., 2024). More preciously, lending Fintech's exemplify shared value creation by aligning profitability with societal impact, offering model for sustainable economic development. This framework explains how FinTech lending creates combined complementary value, achieving economic returns, social and ecological outcomes.

1.3. Fintech's Strategizing Shared Value Creation to Competitive Advantage

For Mintzberg (1998), strategy is an evolving pattern. In SVC, incorporating social issues into a strategy is a choice. Scholars agree that strategy should address competitive advantages, the fintech's external (industry, competition) and internal environments (resources) Hofer and Schendel (1978). Strategy has many levels: corporate, business, and functional Thompson (2022).

In shared vale creation, we focus on business strategy, how Fintech selects the market segments to target Akpinar, (2018). For Porter (1985) the strategy is selecting the best position against industry competition. A leading Fintech must provide unique value, meeting customer needs differently, with lower costs or distinct offers. For Pfitzer (2013) it's key to continually measure economic and social value, create Fintech's innovation structure, and work with stakeholders for innovation. The Fintech's competitive advantage is grounded in its agility and scalability: First agile firms adapt rapidly to mutations and keep customer-centered strategies. Agility is how firms adopt a strategy of people-centered, value delivery, to a demanding target in volatility. For Sarkis (2001) agility is firms' capacity to evolve in a changing context and face continuous mutations. For Sambamurthy (2003), agility is a mandatory condition for innovation, to have a competitive position. Fintech's are agile by nature with a high ability to adjust to the offer. This feature gave Fintech's comparative advantage compared to old institutions. Fintech's can seize emerging opportunities.

By agility Fintech's are developing its processes, its position with new offers, generating an innovative business model. For agile Fintech's, their activity doesn't require high fixed costs on assets, it is easier for Fintech's to be innovative and scalable compared to old institutions. A flexible low-cost grants low-profit margin. Agility impacts cost structure. By mastering infrastructure, Fintech's control its fixed costs. A scalable firm keeps or evolves its profit margins thanks to sales. For Bondi (2000) scalability is the system's ability to evolve its outcoming result while there is a growing load of inputs. For Hagiu (2016) any fintech must be scalable to access the network advantages. Fintech must use technologies to evolve maintaining its effectiveness. Online architecture minimizes the dependency on physical channels facilitating scalability. Fintech's chose a low-cost position. The technology used must be scalable.

1.4. Shared Value Mechanisms in FinTech Lending's Segment

Shared Value Creation (Porter & Kramer,2011) sets the foundations to understand how to aligns financial success with social and environmental progress, measured as follows:

Table 1. Measuring CSV efforts: Business & Social Results by Level of Shared Value

Levels of shared value	*Business Results*	*Social Results*
Reconceiving Products and Markets ***How targeting unmet needs drives incremental revenue and profits***	— Increased revenue — Increased market share — Increased market growth — Improved profitability	— Improved patient care Reduced carbon footprint — Improved nutrition — Improved education
Redefining productivity in the value chain ***How better management of internal operations increases productivity and reduces risks***	— Improved productivity Reduced logistical and operating costs. Secured supply, — Improved quality — Improved profitability	— Reduced energy use — Reduced water use — Reduced raw materials. — Improved job skills I proved employee incomes
Enabling local cluster Development ***How changing societal conditions outside the company unleashes new growth and productivity gains.***	— Reduced costs — Secured supply Improved distribution infrastructure — Improved workforce access — Improved profitability	— Improved education — Increased job creation — Improved health — Improved incomes

Source: Porter et al (2012)

FinTech's innovation drives shared value by several mechanisms:

a) **Reducing Transaction Costs**

FinTech platforms streamline credit processes by automating underwriting, verification, and loan servicing using technologies like artificial intelligence. These innovations reduce operational inefficiencies and costs for lenders and borrowers, making credit more affordable and accessible (Gomber et al., 2018).

b) **Enabling Financial Inclusion**:

Capitalizing on digital channels, lending FinTech's serve populations excluded from formal finance since they lack credit history. This enhances inclusion by providing underserved individuals and micro-entrepreneurs with access to fast credit, vital for economic empowerment (Lee & Shin, 2018).

c) **Supporting Small and Medium Enterprises (SMEs)**

SMEs, which are commonly used to struggle with limited access to traditional financing, profit from lending FinTech's like peer-to-peer lending. These solutions enable quicker, flexible financing customized to SME's requirements, hence stimulate job creation, business and economic growth (Guan et al., 2025).

d) **Financing Renewable Energy and Sustainable Projects**

Lending's FinTech facilitate financing for renewable energy and sustainable supply chain projects by integrating green finance solutions into digital lending products. This supports the transition to low-carbon and enhance environmental sustainability simultaneously generating new market opportunities (Zeng et al., 2024).

2. FinTech Lending in Renewable and Sustainable Supply Chains (Stakeholder Theory)

Innovative financial mechanisms drive sustainable development by financing distributed renewable energy, enhancing supply chain financing efficiency, and promoting a low-carbon energy transition (Bhatti et al.,2023). First, by financing renewable energy: Fintech platforms enables embedded credit for distributed energy solutions, like solar panels and microgrids. These technologies enhance resilience to climate shocks while fostering opportunities in key sectors like agriculture and manufacturing (Zheng et al., 2024). Second, Fintech's advancement increases the effectiveness of supply chain finance, for small and medium-sized enterprises (SMEs). It increases access to capital and improves financing efficiency in supply chain networks (Guan et al.,2025). Third, fintech supports the low-carbon energy transition by driving green technology innovation, facilitating green finance, and promoting sustainable industrial development (Zeng,2024) Finally, the integration of FinTech in supply chain financing helps businesses to align with sustainable development goals. This integration improves financial transparency, enhances liquidity, and supports low-carbon transitions across value chains (Zeng et al., 2024; Guan et al., 2025)

2.1. How Does FinTech platforms (Digital Lending) Enable Stakeholder Alignment?

Lending's FinTech platforms enabled supply chain finance facilitate stakeholder alliance. These platforms links investors targeting sustainable opportunities to SMEs and network in renewable energy and sustainable supply chains. Including sustainability criteria in financing, FinTech enhance collaboration among stakeholders generating more effective economic and environmental outcomes (Qin & Tan, 2025; Polat et al., 2024). Stakeholder Theory explains how FinTech channels the conflicting interests of lenders, borrowers, regulators, consumers and environmental advocates. FinTech platforms intermediate the process to equal rival expectations. Borrowers get capital, regulators preserve laws, consumers get sustainable services, environmental impacts are reduced, hence sustainability in renewable energy and supply-chain ecosystems (Jia et al., 2024).

2.2. How does FinTech Contributes to Green Financing and ESG Traceability?

Studies by Qin and Tan (2025), Polat et al. (2024), and Jia et al (2024) spots FinTech's role in advancing green financing, environmental social, and (ESG) traceability. Lending's FinTech track sustainable investments by digital tools, enhancing transparency which raise investors' confidence in green projects, aid SMEs to respect ESG standards. This accelerate the growth of renewable energy initiatives and sustainable financing.

2.3. Why Multi-Actor Collaboration is Central to FinTech-Enabled Sustainability?

Stakeholder Theory clarify why multi-actor collaboration is essential for FinTech-driven sustainability, successful sustainable supply chains depend on harmonizing the interests of investors, SMEs, communities, regulators, and consumers. FinTech platforms facilitate this collaboration providing infrastructure and transparency to balance the diverse needs, driving more inclusive renewable energy and supply-chain ecosystems (Qin & Tan, 2025).

3. Contextual – Ecosystem Dynamics: FinTech in MENA, Tunisia (Institutional Theory)

3.1. Fintech Ecosystem Dynamics:

Thanks to innovation, financial systems have become interlinked. This new digital architecture launched digital business ecosystems with constant need for updated data. For Lee (2018), understanding the dynamics in Fintech requires analyzing the ecosystem's complexity. For Gartner (1985) the interaction among external actors can impact entrepreneurial dynamics. Alaassar (2020) studied the impact of interactions on Fintech's practices and regulatory sandboxes. Existing research focused on ecosystems actors and factors impacting entrepreneurial activity, ignoring relations between ecosystem elements (Ghio,2019).

Figure 1. Ecosystem Dynamics

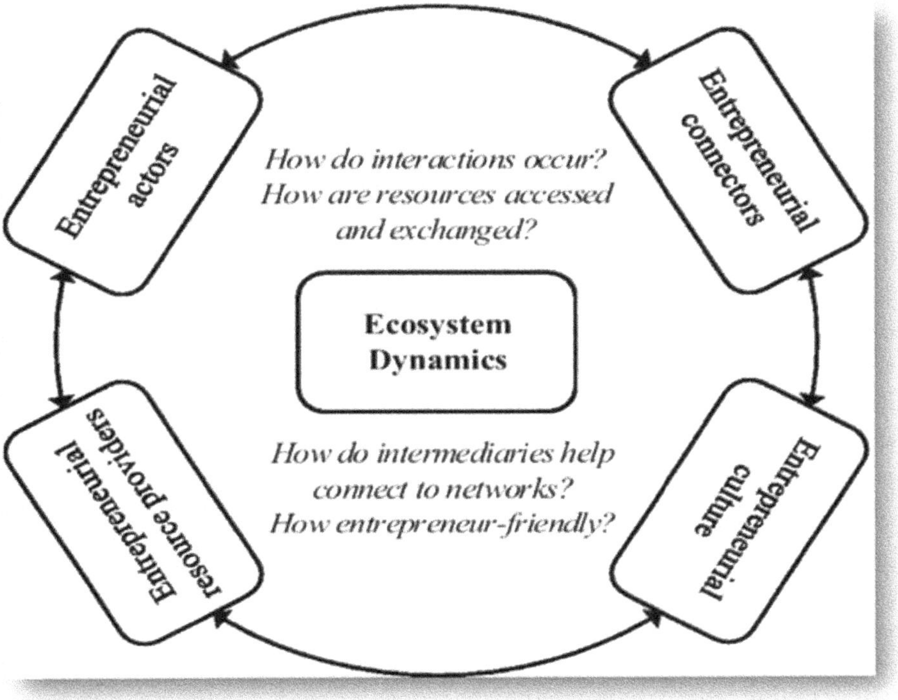

Source: Adapted from Brown and Mason (2017)

The literature assumes that interactions with entrepreneurs inspire newcomers Brown (2017). For Stam (2015) entrepreneurial dynamics explain how ecosystems succeed in creating a supportive environment. For Cao (2020) dynamics are interactions between entrepreneurs and ecosystem actors. (competitive & collaborative).

3.2. Contextual Dynamics: FinTech in MENA and Tunisia

The MENA region has witnessed rapid FinTech evolution driven by supportive policies, regulatory reforms, and inclusion agendas to bridge financial gaps (Elouaourti, 2024). Governments encourage digital financial services to foster economic inclusion reflecting a growing recognition of FinTech's role in transforming traditional financial systems, by improving access for excluded populations aligning with sustainable development goals.

3.2.1. Tunisia's Institutional Landscape: Challenges and Readiness

Tunisia's FinTech's institutional environment is marked by poor public trust, regulatory fragmentation, and unequal infrastructure. The struggle is in harmonizing regulations and enhancing digital literacy. These factors impacts the nature of FinTech adoption, creating both opportunities and barriers for scaling sustainability in evolving market context.

3.2.2. Institutional Theory: FinTech Adoption and Sustainability Outcomes

Institutional Theory identify how both formal institutions (laws, regulations) and informal (culture, trust) govern FinTech evolution in MENA and Tunisia. Institutional pressures structure organizational behaviors and stakeholder engagement, impacting the sustainability and scalability of FinTech's. The theory spots the critical role of institutional environments in adapting or not technological innovation and SVC in these regions.

3.2.3. Enablers and Constraints of FinTech Value Creation

Institutional Theory (North, 1990; Scott, 2014) frames how contextual factors such as regulatory quality, societal norms, and infrastructure readiness either enable or constrain FinTech's ability to create shared economic, social, and environmental value. Understanding these institutional dynamics is essential to designing the needed policies.

4. FinTech Lending and Shared Value in Tunisia's Ecosystem (Easy Bank Case)

4.1. SVC to Frame Easy Bank's Dual Focus on Profit and Inclusion

Easy Bank exemplifies SVC by simultaneously pursuing profitability, financial inclusion, targeting marginalized populations by affordable, technology-driven lending solutions. This dual focus aligns Porter's (2011) framework. Easy Bank's business model shows how Lending FinTech offers credit access maintaining sustainable profitability in Tunisia'.

4.2. Stakeholder Theory: Partnerships with Regulators Startups, Communities

Applying Stakeholder Theory, Easy Bank's success relys on strategic partnerships with regulators and local startups. These collaborations balance diverse interests, ensuring regulatory compliance, fostering innovation, and addressing community needs, enhancing ecosystem trust and sustainability. By engaging multiple stakeholders, Easy Bank exemplifies how FinTech's arbitrate complex relationships to drive inclusive, sustainable financial services in Tunisia.

4.3. Institutional Theory: Tunisia's Evolving Digital-Finance Laws

Institutional Theory clarifies how Tunisia's evolving digital regulations and institutional frameworks shape Easy Bank's strategies. Regulatory reforms create both opportunities and constraints, influencing product design, risk management, and market entry. Easy Bank navigates laws and informal institutional norms, illustrating how institutional dynamics affect FinTech adoption and sustainability outcomes.

THEORETICAL FRAMEWORK

This paper explores how FinTech design facilitates CSV in lending, focusing on sustainable supply chains and renewable energy financing. To achieve this, the study integrates three theoretical lenses: Shared Value Creation (Porter & Kramer, 2011), Stakeholder Theory (Freeman, 1984), and Institutional Theory (North, 1990; Scott, 2014).

Together, these perspectives explain how FinTech drives financial inclusion, mediates stakeholder collaboration, and navigates institutional contexts, in Tunisia's FinTech ecosystem. CSV identifies the dual outcomes that FinTech can achieve. Stakeholder Theory explains the mechanisms of collaboration and engagement among necessary actors. Institutional Theory spots the contextual enablers and constraints that influence the effectiveness and sustainability of FinTech's. Integrating these theories provides a clear framework mapping the pathway from FinTech innovation to shared value. This framework addresses gaps in existing research, which often treats these theories in isolation, to understand FinTech's social, economic, and environmental impacts.

Easy Bank serves as a distinctive empirical case at this intersection. Operating in new regulations with limited financial inclusion and emerging green finance frameworks, Easy Bank leverages AI-driven lending to provide access to credit for underserved individuals and Small and Medium-sized Enterprises SMEs, aligning profitability with social impact. Stakeholder engagement is done by partnerships with Tunisian banks, payment providers, and regulators, while institutional adaptation is showed via compliance with the Central Bank sandbox and reforms. The lending Fintech's exploration of renewable energy is a relevant example how FinTech's in evolving markets integrate sustainability into core operations. Easy Bank case offers transferable experience for other MENA and African markets, on how FinTech design, technology adoption, partnerships, and offerings interact with institutional pressures and socio-economic needs. It spots practical strategies for embedding sustainability, promoting financial inclusion, and aligning innovative FinTech solutions with regulatory and stakeholder expectations, providing a roadmap for creating shared value in emerging economies.

METHODOLOGY

This study used a qualitative case study approach to explore how Tunisian lending FinTech's, specifically Easy Bank, design and operationalize shared value using AI-driven technologies, inclusive innovation and strategic partnerships. The qualitative approach is relevant for investigating context-specific phenomena where understanding processes is essential (Yin, 2018). The case study methodology allows the examination of how Easy Bank integrates economic and social outcomes. The selection of Easy Bank as a reference case follows purposive sampling, chosen for its revelatory nature and theoretical relevance. Easy Bank exemplifies FinTech integrate SVC into its core business model using innovative AI technologies and inclusive financial services. It's an illustrative case on how FinTech's drives simultaneously sustainable economic and social outcomes (Porter & Kramer, 2011). Tunisia, an emerging market with a dynamic evolving regulatory and digital infrastructure environment, provides a relevant institutional context to investigate how institutional factors shape FinTech sustainability outcomes. This geographical focus allows the study to spot the contextual dynamics unique to Tunisia's financial ecosystem, boosting the theoretical contribution.

Data collection coupled primary and secondary sources to ensure triangulation. Primary data were gathered using semi-structured interviews conducted on three phases (October 2024, January 2025 and April 2025) with Easy Bank's co-founders and key team members. These interviews focused on several areas: its innovative business model, target, strategic orientation, stakeholder engagement in evolving ecosystems, AI-driven technological infrastructure, and financial inclusion outcomes. The semi-structured format provided flexibility to explore emergent themes to answer the research questions. Secondary data included an extensive review of company's reports, official documents, Tunisian FinTech industry observations, social media content, public reports, communications from the Central Bank of Tunisia and the regulatory sandbox environment. The collected data were analyzed manually using thematic analysis applying Braun and Clarke's (2006) six-step framework. Interview transcripts and secondary materials were coded inductively to identify emerging patterns and themes. Key themes grouped into categories such as "inclusive access to credit," "AI-driven technology as a productivity driver," and "strategic ecosystem partnerships." A pattern-matching technique (Yin, 2018) was employed to compare emergent themes against the theoretical constructs from the SVC framework (Porter & Kramer, 2011), Stakeholder Theory (Freeman, 1984), and Institutional Theory (North, 1990).Ethical considerations were observed. This approach offers a foundation to investigate how Tunisian FinTech's like Easy Bank architect innovative AI-driven business models generating SV.The qualitative case study design, with data collection, thematic analysis, and triangulation, explored the complex interactions among FinTech's innovation, stakeholder collaboration, and institutional context shaping sustainable lending ecosystems in Tunisia. This design aligns with the study's theoretical framework, ensuring findings are reliable to provide insights for academics and policymakers in the emerging FinTech's markets.

CASE STUDY EMPIRICAL CONTEXT: EASY BANK ARCHITECTING SHARED VALUE CREATION

1. Easy Bank: Origins, Digital Lending Model, and Sustainability Orientation

Easy Bank, founded in 2023, is a Tunisian FinTech operating in the lending and credit intermediation segment. From its inception, Easy Bank set out to redefine access to credit by leveraging artificial intelligence (AI), digital tools, and ecosystem partnerships. The startup founded by a small team of entrepreneurs committed to democratizing finance, improving credit access, and addressing longstanding issues of financial exclusion in Tunisia. Easy Banks offers a range of digital services, including loan negotiation, intermediation, consulting, and financial advisory, all delivered via a user-centric platform. First, Loan Intermediation & Negotiation: Easy bank acts as an intermediary between clients and partner banks, then helps negotiate competitive loan terms for customers. Second, Financial Consulting: providing personalized financial advice for clients and assists in choosing the most suitable banks and financial products. Third, Banking Insurance Access: it simplifies access to banking insurance products via partnerships with banks and insurance companies. Fourth, Loan Simulator: using Online tool to check loan eligibility and calculate potential monthly payments. Fifth, AI-Powered Financial Advisorr: Easy bank Uses AI to provide personalized guidance, Streamlines the loan application process based on eligibility and loan accessibility criteria.

Using automation and AI-powered decision-making, Easy Bank facilitates faster, more inclusive credit processes, mainly for underserved populations often excluded from traditional banking systems. In February 2025, the Fintech raised $370,000 in early-stage funding from undisclosed investors, reinforcing its position as a promising digital innovator in Tunisia's FinTech space. With a lean team of only 10 employees, Easy Bank operates on an agile, scalable business model that enables cost-efficiency and adaptability. It earns revenue via commission-based income on successfully completed loans, distinguishing its approach from conventional banks that rely on interest-based.

Easy bank's competitive advantage, that it offers a user-friendly digital platform, AI-driven financial advisory, and simplified access to banking products, distinguishing it from traditional banks. Driven by a mission to promote inclusive, sustainable financial ecosystems, Easy Bank aligns its operations with the principles of SVC. It addresses systemic barriers such as limited access to credit, financial illiteracy, and low digital penetration, among low-income and underbanked populations. Its business design reflects a commitment not only to economic performance but also to social inclusion, making it a compelling case for exploring the intersection of FinTech, sustainability, and institutional context in emerging markets.

Table 2. "EasyBank "Business Profile

Name	EasyBank
Founder & CEO	Mohamed Khelifi
Legal Status	Accredited startup in Tunisia
Industry	Digital Lending Fintech's
Geographic Area Served	Current coverage: Tunisia Expansion plans: Middle East, North Africa, and France
Main Products Service & Solutions	- Loan Intermediation & Negotiation - Financial Consulting - Banking Insurance Access - Loan Simulator - AI-Powered Financial Advisory - Digital Platform & Solutions
• Main Competitors	• In Tunisia: traditional banks like Amen Bank and digital platforms
Fintech's Size &	• Number of Employees: 9 employees • Capital Raised: 1.2 million Tunisian dinars ($370,000) secured in a seed funding round in January 2025
Technologies used	Platform: Web-based application via browsers AI & Data Analytics: for personalized financial advisory, loan assessment Security: industry standard encryption, security protocols (user data)
Target	Individuals: primarily underserved populations seeking accessible financial services. • Small Businesses: offering tailored financial solutions to meet the needs of small enterprises. (SME's) • (low-income users, informal entrepreneurs, women-led businesses..)

Source: Author

2. Tunisia's Financial Ecosystem, Regulation, Modernization Initiatives and Metrics:

Tunisia's financial ecosystem is facing complexity. The country have developed a banking industry, but access is still uneven with over 60% of the adult population unbanked, according to World Bank data. The traditional credit system excluded low-income and informal actors, with complex laws, slow approvals postponed innovation and access. Still, Tunisia is trying to digitize its financial system with Key reforms to facilitate FinTech growth, as the launch of a regulatory sandbox by the Central Bank of Tunisia to allow start-ups test new financial services in a controlled context. However, Tunisia's regulation remains piecemeal contrary to the nature of Fintech's growing faster than the supervisor institutions. Digital infrastructure is uneven, with disparities in internet access and digital literacy, in rural population. These institutional dynamics provide both a challenge and an opportunity for FinTech's like Easy Bank.

In this context, Easy Bank is not only a regular venture but emerges as a **prototype setup** and a **testbed** for how FinTech navigate institutional obstacles to generate shared value.

In 2025, Tunisian's ecosystem experienced regulatory and policy changes under the Central Bank of Tunisia (BCT). The benchmark interest rate shrinked to 7.5%, easing borrow costs, optimizing the viability of lending fintech's. Prudential reforms by Circular 2025-08 require banks to set strategic operational plans, enhance own-fund adequacy, internal controls, data management, and traceability, and adopt IFRS 9 for credit risk. Non-cash payments continue to grow, while cheque usage declines, new laws mandate interest-free loans for MSMEs and reduce rates on selected loans, reflecting national financial inclusion priorities.

The BCT's sandbox continues to stimulate innovation, with 2025 cohorts including startups testing AI, blockchain and digital money. These advancements offer fintech's like Easy Bank opportunities to capitalize on AI-driven platforms and partner networks to expand lending, improve transparency, and support green renewable energy financing. However, challenges persist, mainly the absence of a fintech law, ambiguity around cryptocurrency policies, complexity of the compliance double business uncertainties. Also, macroeconomic pressures such as interest rate instability require strategic planning. Overall, Tunisia's evolving ecosystem presents a fertile yet complex environment for fintech-driven inclusion and sustainability initiatives.

Tunisia State of Innovation and Technology Adoption

According to the Global Innovation Index (GII), in 2025, Tunisia ranks 76th out of 139 economies, up from 81st in 2024. Tunisia ranks 6th among 37 lower-middle-income countries and score higher in innovation outputs (58th).

Table 3. Tunisia GII Ranking 2020-2025

Year	GII Position	Innovation Inputs	Innovation Outputs
2020	65th	78th	59th
2021	71st	78th	64th
2022	73rd	89th	59th
2023	79th	96th	61st
2024	81st	96th	64th
2025	76th	96th	58th

Source: Global Innovation Index (GII) 2025

The Global Innovation Tracker 2025 indicates, the following indicators about Tunisia:

- Internet penetration reached 79.8 subscriptions/100 inhabitants in 2024.
- 5G penetration reached 14.1 per 100 inhabitants in 2023.

Tunisia's digital payments ecosystem in first quarter of 2025 (Figure N:2) shows clear fintech-driven innovation, predominantly in mobile wallets proving consumer readiness. Mobile wallet users rise by 35%, with higher transactions for daily payments, proving trust in digital services enhancing financial inclusion. Modernization actions, like Elyssa-RTGS, SIBTEL restructuring, migration to SWIFT MX and the national electronic system optimized efficiency, strengthened cross-border payment. The National Payment Scheme "White EMV",adding E-Houwiya to Paysmart platform improved local payment infrastructure demonstrating security and reliability. These innovations provided more fluid digital transactions, creating SV by larger access to financial services for excluded populations, the Tunisian ecosystem is going through solid digital modernization. For Easy Bank, these advancements provide context to design innovative lending products, services and solutions through digital channels, optimize operational efficiency expanding financial inclusion. Within the dynamic of the fintech's ecosystem, Easy Bank stands as it's a relevant example on how fintech architecture generate social and economic value.

Figure 2. Payment metrics in Tunisia, First quarter 2025

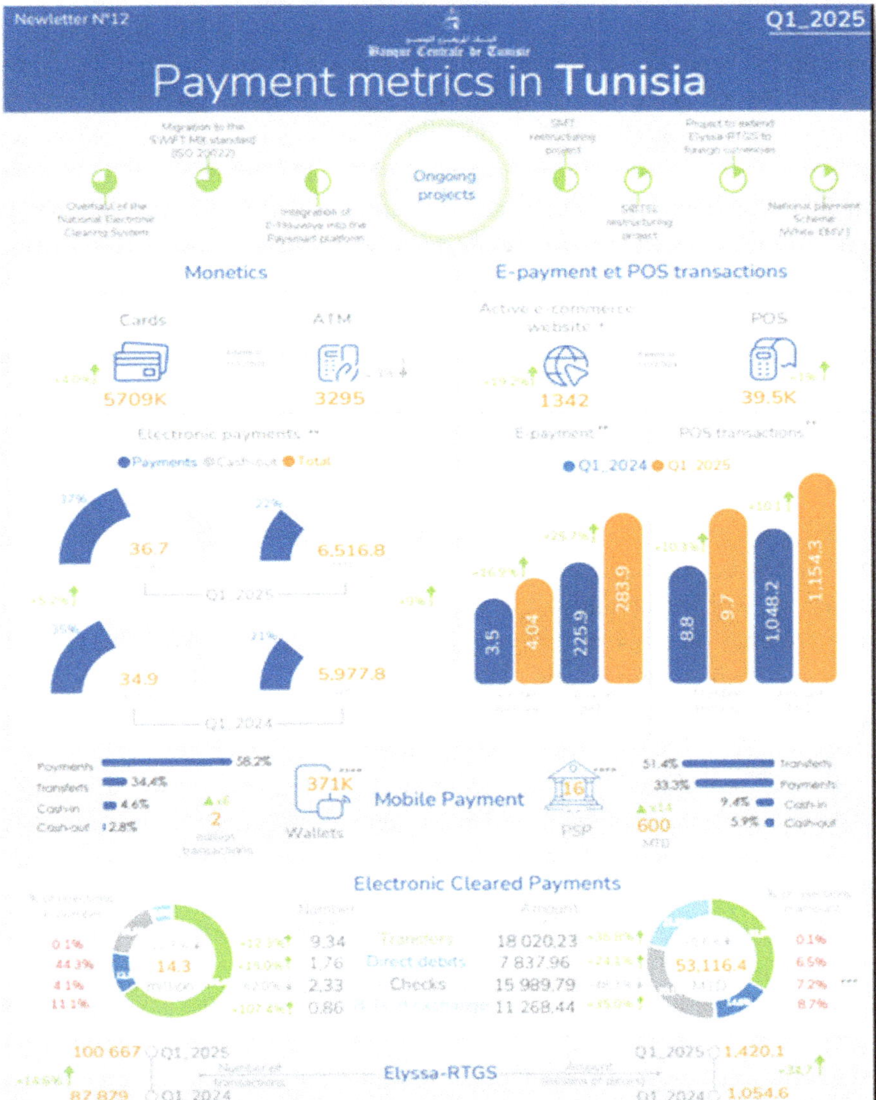

Source: The Central Bank of Tunisia, June2025 (https://www.bct.gov.tn/)

3. Operationalizing Shared Value: Economic, Social, and Environmental Impacts

Easy Bank's shared value architecture is grounded in Porter's model. This orientation is clear in Easy Bank's core business strategies and operations, simultaneously promoting economic and financial productivity, social inclusion synchronized with sustainability goals, particularly in areas of ; SDG 8Decent Work and Economic Growth, SDG 9: Industry, Innovation and Infrastructure, SDG 10: Reduced Inequalities and SDG 13: Climate Action. **Economically,** Easy Bank's AI-driven platform raises credit access, optimize operational costs, and boosts loan processing speed. It deploys automated credit scoring and algorithms to workflow applications and verify identity. This technology allow 24-hour decision cycles, customized offers, and scalable delivery. Unlike traditional banks, Easy Bank minimizes bureaucracy, simplifies processes, reduces red tape, helps employees access capital in a streamlined, cost-efficient manner. Its commission-based model grants it to be lean while generating sustainable revenue.

FinTech's accesses niches that banks ignore due to high costs: rural markets and poor targets. Specialized Fintech's such as Esay Bank Tunisia engaged in niche markets using specific technologies, while banks are more general. For scholars the factors improving Fintech's operational performance are quality of data analytics, cyber risk management and founder characteristics such as banking experience and education.

At a regional level, a study in Cameroon found that fintech enhanced financial literacy. Fintech contributed to China's economic growth generating more innovation (patent's number), fintech's spread reduce data asymmetries and increase stakeholder's financial participation. In Tunisia Easy bank is playing the same role. **Socially,** Easy Bank targeting since its launch mainly focuses on marginalized excluded and underserved populations, especially low-income, unbanked workers. One of its most impactful initiatives was partnering with LEONI, Tunisia's largest private employer, with over 23,000 workers, many of them lack access to credit. Easy Bank identified a gap in employee financial inclusion, and co-designed lending products with LEONI to offer micro-consumption loans to employees. These loans are disbursed by a fully digital process, using popular mobile apps, social media channels and financial education. This orientation solves short-term financial needs, promoting financial literacy and empowerment, while integrating informal workers into the financial system. Easy Bank uses AI chatbots to simplify interactions for users with low literacy levels, demonstrating how FinTech can tailor financial products, services and solutions to different realities.

Environmental & Sustainability Orientation: Obviously, Easy Bank is not a green finance company, but it contributes implicitly to sustainable development by supporting financial inclusion, a recognized enabler of climate adaptation, mainly in vulnerable industries like agriculture, informal labor and Healthcare. Easy Bank's digital infrastructure reduces the need for physical branches, paper-based processes, and in-person services, contributing to a lower carbon footprint.

Future actions includes integrating green loans and partnering with solar energy distributors to support household renewable energy purchases, positioning Easy Bank as a potential enabler of sustainable supply chains through inclusive credit access.

4. FinTech's Support for Renewable Energy and Sustainable Supply Chains

Easy Bank's primary focus is financial inclusion, but its business model shows support of renewable energy, sustainable supply chain finance by several mechanisms:

1. ***Partnered Lending for Renewable Assets:*** discussions are ongoing between Easy Bank and local energy providers to design loan products that enable low-income households to afford solar home systems, bridging the affordability gap in decentralized energy.
2. ***Supply Chain Finance for SMEs:*** Easy Bank is exploring how its AI algorithms support supplier pre-qualification and ESG-compliant lending for small enterprises in agricultural and manufacturing value chains. By identifying reliable, underserved SME borrowers before traditional banks do, Easy Bank reduces transaction costs and improves credit availability in sustainable supply networks.
3. ***Green Impact Metrics:*** Easy Bank is developing internal KPIs tracking loan repayment, customer acquisition, social and environmental impacts including first-time borrower rates, financial literacy advancements, and projected environmental benefits from financed products (CO2 mitigated by solar). By leveraging its digital architecture, strategic partnerships, and inclusive design, Easy Bank is positioning itself as a platform for future green finance innovation.

Consequently, Easy Bank offers an empirical case illustrating how FinTech's in emerging markets can design business models that generate shared value navigating institutional complexity.

By targeting unbanked workers, digitizing credit processes, considering sustainability, Easy Bank operationalizes all three frameworks anchored this study as follows:

- Shared Value Creation: by inclusive, tech enabled financial services that benefit both business and society.
- Stakeholder Theory: partnerships with employers, regulators and microfinance.

- Institutional Theory: adapting to Tunisia's evolving digital infrastructure and regulation.

As the fintech matures, its support of renewable energy and sustainable supply chains through innovative financing expands, offering data into how FinTech can serve as both a technological disruptor and a development enabler in emerging markets.

The figure fellow illustrates how EasyBank's AI-driven business model integrates multiple stakeholders, including investors, regulators and borrowers to generate SV. By AI-enabled loan intermediation and digital inclusion, EasyBank creates three layers of impact: economic, social, and environmental value. These outcomes contribute to sustainable development by improving access to finance, enhancing operational efficiency, and promoting green lending aligned with the UN Sustainable Development Goals (SDGs).

Figure 3. Shared Value Architecture of EasyBank in the Tunisian FinTech Ecosystem

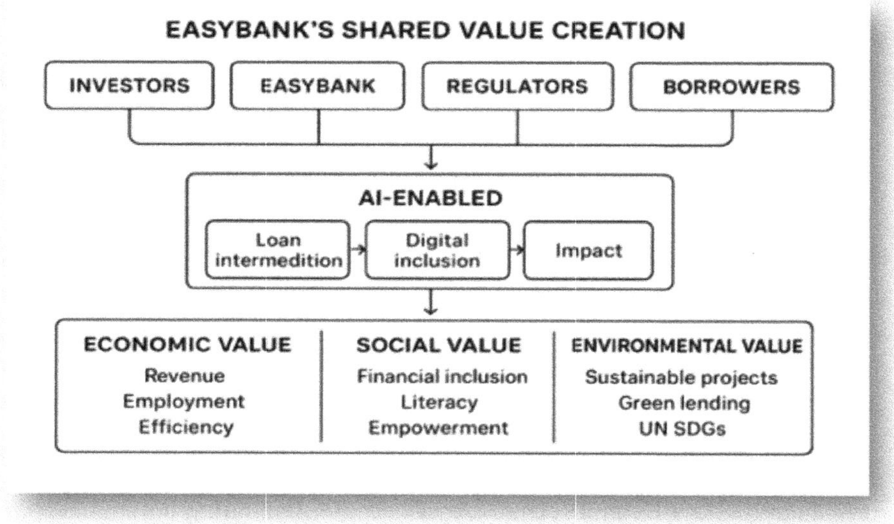

Source: Summary of author's research

5. Discussion: Lending FinTech and Sustainability, The Case of EasyBank Tunisia

EasyBank capitalize on the growing tech-savvy population in Tunisia. Its algorithm driven business model replace exclusionary traditional credit systems, offers fast accessible solution, and reduce barriers to entry for underserved populations.

Empirical findings reveal that EasyBank's primary contribution to sustainability lies in financial inclusion, aligned with emerging-market FinTech research (Ozili, 2018; Ghosh, 2020). By serving populations withoutcredit histories, EasyBank solves a major market failure. From an ESG perspective, EasyBank's sustainability scope is still small with short focus on environmental aims maily impacted by the fact that sustainability initiatives depends on

external pressures, such as donor expectations. Comparatively, regional peers like Liwwa in Jordan, incorporates ESG scoring into lending algorithms, while Tamam in Saudi Arabia operates under stricter Shari'ah governance standards, show more integration of sustainability. This contrast spots the role of institutional maturity and market readiness in FinTech's ability to integrate sustainability into its business strategies and core operations. A Benchmark in the MENA illustrates this disparity. Morocco's Inwi Money expands digital lending for SMEs, showing progress in inclusion but law adoption of green finance. Libya's emerging platforms focus on financial access under the National Financial Inclusion Strategy, but sustainability aspect are still absent. In contrast, Saudi platforms like Tamara show mature models, integrating Shari'ah and ESG criteria. These examinations show how regulation sharpness, institutional support, and market maturity impact FinTech lender's ability, positioning EasyBank as a lending platform with potential in sustainability.

Theoretically in emerging markets, the findings align with the literature setting financial inclusion and digital lending as central channel for sustainability (Boukherouaa, 2020). EasyBank's digital onboarding, credit scoring, and low-cost microlending reflect these trends. Butl, unlike advanced FinTechs in developed markets (Zhao et al., 2021), EasyBank does not integrate environmental risk factors into its portfolio, or report sustainability metrics such as gender-disaggregated lending or CO_2 reduction. This approach mirrors Molla and Licker's (2005) notion of institutional inertia, spoting the barries of including ESG priorities in emerging-market FinTechs. Short repayment cycles and high interest margins raise concerns about affordability and sustainability, emphasizing the need for context-specific research on financial literacy.

EasyBank's model reflects Porter's (2011) concept of CSV, by economic inclusion. Serving excluded populations, EasyBank expands its market solving latent social demand. Still, CSV is driven by donor funding or social impact positioning rather than integrated in the platform's profit logic. Loan design, risk modeling, and digital channels are not yet optimized to generate measurable social, environmental outcomes. EasyBank is financially engaging with major stakeholders' customers, investors and regulators, while NGOs and environmental advocates are absent. Unlike Jordan's Tamweelcom, which co-designs products with NGOs and microfinance networks, EasyBank lacks stakeholder consultation mechanisms. This instrumental stakeholder approach limits co-creation of sustainable lending practices. In weaker institutional contexts, many stakeholders are absent.

EasyBank operates in a fragmented Tunisian FinTech ecosystem, with no dedicated ESG, no policies requiring prioritizing sustainability. Hence, it exhibits mimetic isomorphism, adopting ESG language and practices from more advanced peers without structural implementation. By contrast, FinTechs in Saudi Arabia (STC Pay) face stronger institutional pressures, including national digital finance strategies, ESG scorecards, and central bank initiatives.

For FinTech managers at EasyBank, integrating impact metrics into credit models is critical. Loan products should include ESG dimensions such as gender equity, business resilience, and environmental risk exposure. Tiered loan structures, offering preferential rates or extended repayment to eco-friendly or social enterprise borrowers, can embed sustainability into lending logic. Broadening strategic partnerships with NGOs, universities, and social enterprises can enhance stakeholder inclusivity and co-create sustainable lending products.

For Tunisia and MENA policymakers, the study suggests creating a sustainable lending framework with ESG-aligned credit risk guidelines for FinTechs, incentivizing green and social lending using tax breaks or public guarantees, and enhancing data transparency by requiring disaggregated lending reporting by gender, geography, and sector. These interventions can accelerate ESG adoption and strengthen the long-term sustainability of digital lending ecosystems.

CONCLUSION AND IMPLICATIONS

This study examine the intersection of lending FinTech's, renewable energy and sustainability in Tunisia's ecosystem, using EasyBank as a focal case. Empirically, the findings reveal that EasyBank made major pioneering advances in financial inclusion and digital access, using mobile-based micro-lending, democratizing finance by expanding credit access to excluded and marginalized populations. The platform's architecture and user-centered design showcase how lending FinTech's allow financial empowerment in emerging markets. But, despite these leading milestones, EasyBank's integration of sustainability is still externally driven, with limited ESG measurement. Beyond Easy's bank strategical choices, this shared pattern echoes a regional reality about emerging market; FinTechs prioritize inclusion but surpassed by peers like Tamam (Saudi Arabia) in including sustainability into its core business strategies. Theoretically, the contribution, Institutional Theory illustrate the role of regulatory environments and support from institutions in FinTechs' ESG adoption. Practically, while Tunisia's lending FinTech industry excels in financial inclusion, it lacks ESG integration, minimal institutional and investor pressure, scarce mechanisms to track environmental and governance impacts. The benchmarking with MENA peers reveals that EasyBank shows slower progress in environmental accountability and governance processes.

Consequently, the study's recommendations prioritize reassessment of priorities by making sustainability a focal point in the industry. First, the development of a FinTech-ESG framework is vital, to allow platforms asses borrower impact adopting responsible lending. Second, set a MENA Sustainable FinTech alliance would harmonize ESG standards, foster best practices, and reduce regulation fragmentation. Third, incentivizing inclusive lending by regulatory support, and public-private partnerships would develop impact-oriented credit .

This research is limited by its single case focus on Tunisia. Future studies can set comparative approaches to evaluate ESG integration within MENA. In conclusion, as digital finance continues to reshape emerging-markets, lending FinTechs like EasyBank hold transformative potential. Yet, financial inclusion alone is not enough. To optimize its societal and environmental impact, lending Fintech's platforms are invited to integrate sustainability into their strategic frameworks, coupling technological innovation with long-term social, economic, and ecological value creation.

REFERENCES

Abikoye, B. E., Adelusi, W., Umeorah, S. C., & Adelaja, A. O. (2025). Integrating risk management in fintech and traditional financial institutions: AI and machine learning. *Journal of Financial Technology and Risk Management, 12*(1), 45–67.

AbouFoul, M., RuizAlba, J. L., & Soares, A. (2021). The impact of digitalization and servitization on the financial performance of a firm: An empirical analysis. *Production Planning and Control, 32*(12), 975–989. https://doi.org/10.1080/09537287.2020.1780508

Alaassar, A., Mention, A.-L., & Aas, T. H. (2020). Exploring how social interactions influence regulators and innovators: The case of regulatory sandboxes. *Technological Forecasting and Social Change, 160*, 120257. Advance online publication. https://doi.org/10.1016/j.techfore.2020.120257

Alghamdi, R., & Ahmad, A. (2022). Blockchain adoption for sustainable supply chain management: Evidence from emerging economies. *Journal of Enterprise Information Management, 35*(6), 1,234–1,254. https://doi.org/10.1108/JEIM-07-2022-0238

Alt, R., Beck, R., & Smits, M. T. (2018). FinTech and the transformation of the financial industry. *Electronic Markets, 28*, 235–243. https://doi.org/10.1007/s12525-018-0310-9

Arner, D. W., Animashaun, S., Charamba, K., & Cai, Y. (2024). Building digital payments ecosystems: Digital financial infrastructure, access to finance, and the UN Sustainable Development Goals. *University of Hong Kong Faculty of Law Research Paper* (No. 23/56). https://papers.ssrn.com/sol3/papers.cfm?abstract_id=4565601

Arner, D. W., Buckley, R. P., Zetzsche, D. A., & Veidt, R. (2020). Sustainability, FinTech and financial inclusion. *European Business Organization Law Review, 21*, 7–35. https://link.springer.com/article/10.1007/s40804-020-00183-y

Atayah, O., Najaf, K., & Subramaniam, R. K. (2025). Understanding the implications of FinTech peertopeer (P2P) lending during the COVID19 pandemic. [Fund. https://www.imf.org/-/media/Files/Publications/DP/2021/English/PDEORAIFEA.ashx]. *Journal of Sustainable Finance, 15*(3), 112–130.

Bazarbash, M. (2019). *FinTech in financial inclusion: Machine learning applications in assessing credit risk* (IMF Working Paper No. 2019/109). International Monetary Fund. https://doi.org/10.5089/9781498314428.001

Bazarhanova, D., Maier, C., & Laumer, S. (2022). Beyond efficiency: Rethinking FinTech performance metrics. *Journal of Information Systems*, *36*(2), 147–166.

Ben Rejeb, N., Zailani, S., & Ali, M. (2021). Green lending practices and SME financing in emerging markets: Evidence from Tunisia. *Journal of Business and Industrial Marketing*, *36*(14), 2, 300–302, 318. https://doi.org/10.1108/JBIM-12-2021-0587

Bhatti, Z. A., Raza, M., Iqbal, S., Mehmood, A., & Mehmood, A. (2023). Sustainable supply chain finance through digital platforms: A pathway to green entrepreneurship. *Annals of Operations Research*, *331*(1), 285–319. Advance online publication. DOI: 10.1007/s10479-022-04623-5

Bondi, A. B. (2000). *Characteristics of scalability and their impact on performance*. In *Proceedings of the Workshop on Software and Performance (WOSP2000)* (pp. 195–203). ACM. https://dl.acm.org/doi/10.1145/350391.350432

Boston Consulting Group & QED Investors. (2024, June 26). *Global FinTech 2024: Prudence, profits, and growth.* Boston Consulting Group. www.bcg.com/publications/2024/global-fintech-prudence-profits-and-growth?utm_source=chatgpt.com

Brown, R., & Mason, C. (2017). Looking inside the spiky bits: A critical review and conceptualisation of entrepreneurial ecosystems. *Small Business Economics*, *49*(1), 11–30. https://doi.org/10.1007/s11187-017-9865-7

Boukherouaa, E. B., Shabsigh, M. G., AlAjmi, K., Deodoro, J., Farias, A., Iskender, E. S., & Ravikumar, R. (2021). *Powering the digital economy: Opportunities and risks of artificial intelligence in finance*. International Monetary Fund. https://www.imf.org/-/media/Files/Publications/DP/2021/English/PDEORAIFEA.ashx

Braun, V., & Clarke, V. (2006). Using thematic analysis in psychology. *Qualitative Research in Psychology*, *3*(2), 77–101. https://doi.org/10.1191/1478088706qp063oa

Cao, Z., & Shi, X. (2021). A systematic literature review of entrepreneurial ecosystems in advanced and emerging economies. *Small Business Economics*, *57*(1), 75–110. https://doi.org/10.1007/s1118702000326y

Chen, X., Wang, C., & Li, S. (2023). The impact of supply chain finance on corporate social responsibility and creating shared value: A case from the emerging economy. *Supply Chain Management*, *28*(2), 324–346. https://doi.org/10.1108/SCM-10-2021-0478

Dembek, K., Singh, P., & Bhakoo, V. (2016). Literature review of shared value: A theoretical concept or a management buzzword? *Journal of Business Ethics*, *137*(2), 231–267. 10.1007/s10551-015-2554-z

DiMaggio, P. J., & Powell, W. W. (1983). The iron cage revisited: Institutional isomorphism and collective rationality in organizational fields. *American Sociological Review*, *48*(2), 147–160. https://doi.org/10.2307/2095101

Dridi, N. (2021). Financing sustainable development: The case of renewable energy in Tunisia. *International Journal of Energy and Environmental Issues*, *43*, 504–521. https://doi.org/10.1504/IJGEI.2021.118939

Dunbar, K. (2024). *FinTech for environmental sustainability: Promises and pitfalls.* Preprint., DOI: 10.1016/j.xxx

Egwuonwu, A., Mordi, C., Egwuonwu, A., & Uadiale, O. (2022). The influence of blockchains and Internet of Things on global value chain. *Strategic Change*, *31*(1), 45–55. https://doi.org/10.1002/jsc.2484

Elouaourti, Z. (2024). Financial technologies for all MENA citizens: Tackling digital financial inclusion. *ScienceDirect.* https://www.sciencedirect.com/article/pii/S1757780224000519

ElWassal, K. (2022). Crowdfunding renewable energy investments: Investor perceptions and decisionmaking factors in an emerging market. *Technological Forecasting and Social Change*, *182*, 121686. Advance online publication. https://doi.org/10.1016/j.techfore.2022.121686

Freeman, R. E. (1984). Strategic management: A stakeholder approach. Pitman. 10.1017/CBO9781139192675

Frioui, E., & Ben Sedrine Doghri, S. (2025). Tunisian fintech's ecosystem: To what extent can Tunisian fintechs be an engine for shared value creation? In Digital technologies for sustainability and quality control (p. 36). IGI Global. DOI: 10.4018/979-8-3693-4373-9.ch006

Gartner, W. B. (1985). A conceptual framework for describing the phenomenon of new venture creation. *Academy of Management Review*, *10*(4), 696–706. https://doi.org/10.2307/258039

Ghio, N., & Guerini, M., & RossiLamastra, C. (2019). The creation of hightech ventures in entrepreneurial ecosystems: Exploring the interactions among university knowledge, cooperative banks, and individual attitudes. *Small Business Economics*, *52*(2), 523–543. https://doi.org/10.1007/s11187-017-9958-3

Ghosh, K. (2020). Role of FinTech in green finance and sustainable development. *Gyan Management Journal*, *14*(2), 61–71. https://journals.acspublisher.com/index.php/gmj/article/view/187

Gomber, P., Kauffman, R. J., Parker, C., & Weber, B. W. (2018). On the FinTech revolution: Interpreting the forces of innovation, disruption, and transformation in financial services. *Journal of Management Information Systems, 35*(1), 220–265. DOI: 10.1080/07421222.2018.1440766

Guan, Y., Sun, N., Wu, S. J., & Sun, Y. (2025). Supply Chain Finance, Fintech Development, and Financing Efficiency of SMEs in China. *Administrative Sciences, 15*(3), 86. DOI: 10.3390/admsci15030086

Hagiu, A., & Rothman, S. (2016, April). Network effects aren't enough. *Harvard Business Review, 94*(4), 65–71.

Han, S., Ulhoi, J. P., & Song, H. (2024). Digital trust in supply chain finance: The role of innovative fintech service provision. *Journal of Enterprise Information Management, 37*(3), 1596–1615. DOI: 10.1108/JEIM-07-2022-0238

Hasan, M. (2024). FinTech and sustainable development / FinTech and sustainable finance. [Working paper or journal article]. (cronfa.swan.ac.uk)

Hofer, C. W., & Schendel, D. (1978). *Strategy formulation: Analytical concepts.* West Publishing Co.

Ilmarinen, P., & Akpinar, M. (2018). *Creating shared value in the banking industry: A case study from Finland. Finnish Business Review.* Published May 21, 2018.

International Monetary Fund. (2024). *Fintech applications for boosting climate finance.* IMF Staff Climate Note 2024/008. https://www.elibrary.imf.org/view/journals/066/2024/008/article-A001-en.xml

Jia, F., Guo, J., Chen, L., & Islam, N. (2024). Does Fintech Adoption Improve Sustainable Supply Chain Management? An Innovation-Intensive Environment Perspective. *IEEE Transactions on Engineering Management.* DOI:10.1109/TEM.2024.3515471

Kao, Y.-C., Shen, K.-Y., Lee, S.-T., & Shieh, J. C. P. (2022). Selecting the fintech strategy for supply chain finance: A hybrid decision approach for banks. *Mathematics, 10*(14), 2393. DOI: 10.3390/math10142393

Kerényi, Á., & Molnár, J. (2018). Bank and FinTechs. *Economy and Finance, 5*(1), 86–97. https://www.bankszovetseg.hu/Public/gep/2018/ENG/086-097ig%20kerenyi-molnar-muller.pdf

Kuan-Jung, P. E. N. G. (2017). The FinTech revolution and financial regulation: The case of online supply-chain financing. *Asian Journal of Law and Society, 4*(1), 109–132. https://doi.org/10.1017/als.2016.65

La Torre, M., Mango, F., Cafaro, A., & Leo, S. (2021). Does ESG performance affect market value? A study of FinTech companies. *Technological Forecasting and Social Change*, *162*, 120385. https://doi.org/10.1016/j.techfore.2020.120385

Lee, I., & Shin, Y. J. (2018). Fintech: Ecosystem, business models, investment decisions, and challenges. *Business Horizons*, *61*(1), 35–46. https://doi.org/10.1016/j.bushor.2017.09.003

Li, J., He, Z., & Wang, S. (2022). A survey of supply chain operation and finance with FinTech: Research framework and managerial insights. *International Journal of Production Economics*, *247*, 108431. DOI: 10.1016/j.ijpe.2022.108431

Milian, E. Z., Spinola, M. M., & de Carvalho, M. M. (2019). FinTechs: A literature review and research agenda. *Electronic Commerce Research and Applications*, *34*, 100833. https://doi.org/10.1016/j.elerap.2019.100833

Mintzberg, H. (1998). *Strategy safari: A guided tour through the wilds of strategic management*. Free Press.

Molla, A., & Licker, P. S. (2005). eCommerce adoption in developing countries: A model and instrument. *Information & Management*, *42*(6), 877–899. https://doi.org/10.1016/j.im.2004.09.002

Moretto, A., & Caniato, F. (2021). Can supply chain finance help mitigate the financial disruption brought by COVID-19? *Journal of Purchasing and Supply Management*, *27*(4), 100713. https://doi.org/10.1016/j.pursup.2021.100713

Navaretti, G. B., Calzolari, G., Mansilla-Fernandez, J. M., & Pozzolo, A. F. (2018). *FinTech and banking. Friends or foes?* European Economy: Banks, Regulation, and the Real Sector, 2, 9–30. https://doi.org/10.2139/ssrn.3099337

Naz, F., Ahmed, S., & Associates. (2022). FinTech growth during COVID-19 in the MENA region. *International Journal / PMC (open access)*. 10.1007/s10660-022-09583-3 https://www.ncbi.nlm.nih.gov/pmc/articles/PMC9295879/

North, D. C. (1990). *Institutions, institutional change and economic performance*. Cambridge University Press.

Ozili, P. K. (2018). Impact of digital finance on financial inclusion and stability. *Borsa Istanbul Review*, *18*(4), 329–340. https://doi.org/10.1016/j.bir.2017.12.003

Pfitzer, M., Bockstette, V., & Stamp, M. (2013, September). Innovating for shared value. *Harvard Business Review*, *91*(9), 100–107.

Polat, O., Ozcan, B., Ertuğrul, H. M., Atılgan, E., & Özün, A. (2024). Fintech: A conduit for sustainability and renewable energy? Evidence from R^2 connectedness analysis. *Resources Policy*, *94*, 105098. Advance online publication. DOI: 10.1016/j.resourpol.2024.105098

Porter, M. E., Hills, G., Pfitzer, M., Patscheke, S., & Hawkins, E. (2012). Measuring shared value: How to unlock value by linking social and business results. FSG. https://ifcsia.org/wp-content/uploads/pdf/publications/Measuring-Shared-Value.pdf

Porter, M. E., & Kramer, M. R. (2011). Creating shared value: Redefining capitalism and the role of the corporation in society. *Harvard Business Review*, *89*(1), 2–17.

Qin, Q., & Tan, L. (2025). Lighting a green future: The role of FinTech in the renewable energy transition. *Science Progress*, *108*(1), 00368504251323566. Advance online publication. DOI: 10.1177/00368504251323566 PMID: 40007333

Quenum, A. C. (2025, June 9). EasyBank leverages AI to expand banking access in Tunisia. *WeAreTech Africa.* https://www.wearetech.africa/en/fils-uk/solutions/easybank-leverages-ai-to-expand-banking-access-in-tunisia

Sanyaolu, T. O. (2025). A conceptual model for inclusive lending through FinTech innovations: Expanding SME access to capital in the supply chain. *International Journal of Sustainable Finance*, *8*(2), 89–103.

Sambamurthy, V., Bharadwaj, A., & Grover, V. (2003). Shaping agility through digital options: Reconceptualizing the role of information technology in contemporary firms. *Management Information Systems Quarterly*, *27*(2), 237–263. https://doi.org/10.2307/30036530

Sarkis, J. (2001). Benchmarking for agility. *Benchmarking*, *8*(2), 88–107. https://doi.org/10.1108/14635770110389816

Scott, W. R. (2014). *Institutions and organizations: Ideas, interests, and identities* (4th ed.). Sage Publications.

Song, H., Han, S., Liu, W., & Ganguly, A. (2023). What role do FinTech companies play in supply chain finance? A signaling intermediary perspective. *Journal of Business and Industrial Marketing*, *38*(6), 1279–1294. DOI: 10.1108/JBIM-12-2021-0587

Soni, G., Kumar, S., Mahto, R. V., Mangla, S. K., Mittal, M. L., & Lim, W. M. (2022). A decision-making framework for Industry 4.0 technology implementation: The case of FinTech and sustainable supply chain finance for SMEs. *Technological Forecasting and Social Change*, *180*, 121686. DOI: 10.1016/j.techfore.2022.121686

Stam, E. (2015). Entrepreneurial ecosystems and regional policy: A sympathetic critique. *European Planning Studies*, *23*(9), 1759–1769. https://doi.org/10.1080/09654313.2015.1061484

Sun, Y., Li, S., & Wang, R. (2023). FinTech: From budding to explosion—an overview of the current state of research. *Review of Managerial Science*, *17*, 715–755. https://doi.org/10.1007/s11846-021-00513-5

Tang, Q., Li, H., & Zhang, Y. (2025). FinTech and corporate ESG performance: An empirical analysis based on the NEV industry. *Sustainability*, *17*(2), 434. https://doi.org/10.3390/su17020434

TechPression. (2025, January). Tunisian startup EasyBank secures $370,000. *TechPression.* https://techpression.com/tunisian-startup-easybank-secures-370000/?utm_source=chatgpt.com

Teng, Z., Xia, H., & He, Y. (2025). Rewiring sustainability: How digital transformation and FinTech innovation reshape environmental trajectories in the Industry 4.0 era. *Systems*, *13*(6), 400. https://doi.org/10.3390/systems13060400

Thompson, A. A. (2022). *Strategy: Core concepts and analytical approaches* (7th ed.). McGraw-Hill Education.

Wamda. (2025, January). Tunisian FinTech EasyBank secures $370,000 investment. *Wamda.* https://www.wamda.com/en/2025/01/tunisian-fintech-easybank-secures-370000-investment?utm_source=chatgpt.com

WeAreTech Africa. (2025, February). EasyBank leverages AI to expand banking access in Tunisia. *WeAreTech Africa.* https://www.wearetech.africa/en/fils-uk/solutions/easybank-leverages-ai-to-expand-banking-access-in-tunisia?utm_source=chatgpt.com

Wu, X., Zhang, Y., & Li, J. (2025). FinTech-driven green energy transition: Evidence from China. *Scientific Reports*, *15*(1), 18898. https://doi.org/10.1038/s41598-025-18898-8

World Intellectual Property Organization. (2025). *Global Innovation Index 2025: Tracking innovation through the COVID-19 crisis.* Cornell University, INSEAD, WIPO. https://www.globalinnovationindex.org/gii-2025-report

Yalcin, H. (2025). FinTech and the green transition: Exploring pathways to sustainable supply chains. *Technological Forecasting and Social Change*. Advance online publication. https://doi.org/10.1016/j.techfore.2025.121234

Yin, R. K. (2018). *Case study research and applications: Design and methods* (6th ed.). Sage Publications.

Zeng, L., Wong, W.-K., Fu, H., Mahmoud, H. A., Cong, P. T., Thuy, D. T. T., & Bach, P. X. (2024). How does FinTech promote a low-carbon energy transition? The role of green innovation and industrial development. *Journal of Cleaner Production*, *88*. Advance online publication. DOI: 10.1016/j.jclepro.2023.139860

Zeng, L., Wong, W.-K., Fu, H., Mahmoud, H. A., Cong, P. T., Thuy, D. T. T., & Bach, P. X. (2024). FinTech and sustainable financing for low carbon energy transitions: A biodiversity and natural resource perspective in BRICS economies. *Resources Policy*, *88*, 104486. Advance online publication. https://doi.org/10.1016/j.resourpol.2023.104486

Zhang, L., Wang, Y., & Li, H. (2022). Mathematical modeling of FinTech-enabled sustainable supply chain investments. *Mathematics*, *10*(14), 2439. https://doi.org/10.3390/math10142393

Zheng, J., Liu, B., & Huang, Y. (2024). Inhibiting or exacerbating? Digital financial inclusion and renewable energy efficiency. *Science, Technology & Society*. Advance online publication. DOI: 10.1177/09717218231150 66

Zhao, K. (2021). *Comparative bank financial risk management models in traditional banks and challenger banks/fintechs* (Doctoral thesis, Sheffield Hallam University). Sheffield Hallam University.

Chapter 8
Renewable Energy Integration in Supply Chain Operations:
Case Studies and Strategic Insights

Suriya Ponnambalam
https://orcid.org/0000-0002-8775-0474
Bharath Institute of Higher Education and Research, India

M. K. Ilampoornan
https://orcid.org/0000-0002-6233-8141
Bharath Institute of Higher Education and Research, India

ABSTRACT

The adoption of renewable energy within supply chain operations is accelerating as businesses aim to enhance sustainability, achieve carbon neutrality, and reduce costs. This research investigates the application of renewable sources—such as solar, wind, and bioenergy—across different phases of supply chains. Representation on multiple case studies from sectors including manufacturing, retail, and logistics, the study highlights key implementation strategies, challenges encountered, and resulting performance impacts. Findings suggest that effective integration depends on factors like technological capability, interdisciplinary cooperation, and strategic long-term investments. The study concludes with a proposed framework to help organizations incorporate renewable energy into their supply chains while supporting global sustainability initiatives.

DOI: 10.4018/979-8-3373-3790-6.ch008

1. INTRODUCTION

Operating managers were only marginally involved in the early phases of environmental management and stayed somewhat out of the way. Traditionally, specific departments were in charge of monitoring environmental performance in a number of domains, such as waste management, operations, logistics, marketing, product development, process design, and regulatory compliance. But this strategy has changed. It is now commonly acknowledged that leading practices necessitate environmental management to be completely integrated into daily operational processes, much like the changes witnessed during the quality and supply chain revolutions.

Operations and supply-chain management scholars and practitioners are becoming more interested in Renewable supply-chain management, or RSCM. The increasing environmental degradation, such as depleting raw material resources, overflowing waste sites, and rising pollution levels, is the primary driver of RSCM's expanding significance. Being environmentally sustainable is important, but so are sound business practices and increased revenue. It is not a cost center, but rather a driver of company value. RSCM is also being driven by consumer pressures and regulatory obligations. Therefore, the scope of RSCM includes more proactive measures through the numerous Rs (Reduce, Re-use, Rework, Refurbish, Reclaim, Recycle, Remanufacture, Reverse logistics, etc.) as well as reactive monitoring of general environment management programs.

1.1 Fundamental Components of Renewable Supply-Chain Management

Renewable Supply Chain Management (RSCM) is a concept that emphasizes the integration of environmental sustainability into all stages of the supply chain. It emphasizes three essential elements: the significance of RSCM (Fontes et al., 2018), green operations, and green design. To reduce the product's environmental impact from the start, green design uses techniques like Environmentally Conscious Design (ECD) and Life Cycle Assessment (LCA). Reverse logistics, waste management, and green manufacturing and remanufacturing are examples of green operations. Reducing, reusing, recycling, remanufacturing, and effective inventory and production planning are the main goals of these initiatives. While waste management deals with source reduction, pollution avoidance, and responsible disposal, reverse logistics makes sure that used goods are collected, sorted, and redistributed in an efficient manner.

RSCM design can be examined from the perspective of environmentally conscious design, taking into consideration the product/process life-cycle assessment. The same is true for green manufacturing and remanufacturing (reduce; recycle;

production planning and scheduling; inventory management; remanufacturing: re-use, product and material recovery); waste management (source reduction; pollution prevention; disposal); and RL and network design (collection; inspection/sorting; pre-processing; network design). Since we believe that the problems are more operational than strategic in nature and might not be important in supply chain design per se, we deliberately do not take into account literature or practices pertaining to green logistics. Additionally, because our research design limits it, we do not go into great length on empirical studies on RSCM or the literature on green purchasing, industrial ecology, and industrial ecosystems. Since the development of successful RL networks is a prerequisite for successful recycling and remanufacturing, we concentrate more on RL. We also pay more attention to components of mathematical modeling. To yet, neither of them has gotten as much attention in the RSCM literature.

The classification serves to provide a clear and well-defined image for future research and study by making it easier to grasp the many problem contexts of RSCM and their interactions and relationships. It is not strict, and there could be a lot of overlaps (for instance, reducing is important not only in remanufacturing and green manufacturing, but also in waste management and reverse logistics; green design also stresses using fewer virgin materials and other resources). Similar to this, green design should include the costs associated with the entire product life cycle, including those incurred during production, remanufacturing, reverse logistics, and disposal. In order to preserve operational efficiency and advance environmental sustainability, renewable supply chain management, or RSCM, is essential. By incorporating eco-friendly practices into all phases of the supply chain (Fontes et al., 2018), from product design and manufacture to logistics and disposal, it assists businesses in lowering their carbon footprint, minimizing waste, and conserving resources. RSCM promotes brand image, increases adherence to environmental rules, and satisfies stakeholders' and customers' increasing desire for sustainable products. Through resource optimization, waste reduction, and energy efficiency, it also results in cost savings. Additionally, by implementing RSCM, businesses support global objectives like sustainable development and carbon neutrality, guaranteeing both environmental responsibility and long-term commercial resilience.

1.2 Green Design

Green design, sometimes referred to as environmentally conscious design, is the process of creating goods and procedures that have as little of an impact on the environment as possible throughout the course of their lifetime. It entails taking sustainability into account at every stage of product development, including the choice of materials, production, use, and disposal at the end of the product's life.

Environmentally Conscious Design (ECD), which seeks to make items easier to reuse, recycle, or dispose of responsibly (Fontes et al.,2018), and Life Cycle Assessment (LCA), which assesses a product's environmental consequences from manufacturing to disposal, are important elements of green design. In addition to satisfying legal requirements and consumer demands for environmentally friendly products, green design helps lower pollution, energy use, and resource use. In the end, by including sustainability into the design process, it advances the more general objectives of green supply chain management.

1.3 Green Operations

In the context of Renewable Supply Chain Management (RSCM), "green operations" refers to the incorporation of ecologically friendly procedures into regular corporate operations.

1.3.1. Green Manufacturing & Remanufacturing

This involves minimizing environmental impact during production processes and giving used products a second life.

- *Reducing:* Minimizing raw material usage, energy consumption, and emissions during manufacturing.
- *Recycling:* Converting waste into reusable materials to reduce dependency on virgin resources.
- *Remanufacturing:* Rebuilding or restoring used products to like-new condition. It includes:
    - *Product/Material Recovery:* Salvaging parts or materials from used products.
    - *Repair/Refurbish:* Fixing or updating components to extend product life.
    - *Disassembly:* Taking apart products to reuse components.
        - *Disassembly Leveling:* Standardizing disassembly processes.
        - *Disassembly Process Planning:* Strategizing the steps for efficient and non-destructive disassembly.
- *Reuse:* Directly using components or products again without reprocessing.
- *Inventory Management:* Controlling stock levels to avoid overproduction and waste.
- *Production Planning & Scheduling:* Organizing production efficiently to reduce resource consumption and emissions.

1.3.2. Reverse Logistics & Network Design

This focuses on handling the flow of used products and materials backward through the supply chain.

- *Collecting:* Gathering used or end-of-life products from customers.
- *Inspection/Sorting:* Checking product conditions and separating items for reuse, remanufacturing, or recycling.
- *Pre-processing:* Initial steps like cleaning, dismantling, or categorizing before further processing.
- *Location & Distribution (Network Design):* Strategically placing facilities and planning routes to minimize transportation impact and costs.

1.3.3. Waste Management

Ensures proper handling and reduction of waste generated from supply chain activities.

- *Source Reduction:* Eliminating waste at its origin by designing processes and products more efficiently.
- *Pollution Prevention:* Implementing methods to reduce emissions and contamination during operations.
- *Disposal:* Safely discarding materials that cannot be reused or recycled, following environmental regulations.

1.4 Research Question

This study seeks to answer:

- How can renewable energy integration enhance the efficiency, resilience, and sustainability of modern supply chain operations across sectors?
- What are the key models and technologies supporting renewable integration in supply chains?
- What theoretical perspectives (e.g., sustainable supply chain management, system dynamics) best explain these interactions?
- What empirical gaps remain in linking renewable energy with operational performance?

2. LITERATURE REVIEW

Incorporating renewable energy into industrial processes, emphasizing technological, environmental, and economic aspects (Chidolue et al., 2024). Although it does not offer detailed case studies on supply chain operations, it places a strong emphasis on strategic planning and addresses issues like intermittency and regulatory concerns. In order to maximize energy procurement, production, and social impact, (Azad et al., 2024) stresses the integration of renewable energy sources—solar, biomass, wind, and hydropower—within supply chains. They point out that no single source satisfies all requirements, so a combined strategy is required for efficiency in Bangladesh. The challenges of increasing the use of renewable electricity in corporate supply chains are covered by. These challenges include the need for transparent standards to support corporate claims in line with international climate goals, the difficulty of measuring progress, and the varying stringency of climate policies.

In order to improve resilience, (Torres-Rivera et al., 2023) highlights incorporating renewable energy sources into the electromobility supply chain. It highlights the necessity of strategic partnerships and sustainable practices to enhance supply chain operations overall, identifying important elements, obstacles, and chances for risk reduction. In their discussion of the renewable energy supply chain in Ostrobothnia and Vaasa, (Aslani et al., 2013) emphasizes the significance of local resources, technological advancement, and supportive policies for efficient supply chain operations. They also focus on regional potentials, challenges, and strategies for integrating renewable energy technologies.

Real-world case studies highlighting cooperative practices in renewable energy supply chains are included in (Taseen, 2023), with a focus on methods for improving sustainability and efficiency. It investigates the ways in which cross-sector collaborations might promote operational enhancements and accomplish common sustainability goals. (Usman et al., 2024) examines the benefits and difficulties of incorporating renewable energy into manufacturing, placing a strong emphasis on stakeholder collaboration, strategic planning, and R&D investment. It does not really address supply chain operations, but it does emphasize case studies showing cost reductions and increased energy efficiency.

In their discussion of developments in green logistics, (Onukwulu et al., 2022) emphasizes the integration of renewable energy sources into energy supply chains. It highlights how crucial stakeholder engagement is and how data analytics can be used to optimize operations for efficiency and sustainability. (Adeyinka et al., 2021) emphasizes supply chain advances like energy storage and digitalization while highlighting issues facing the renewable energy sector, such as intermittency and infrastructure constraints. In order to improve supply chain operations' resilience and efficiency, it promotes cross-sectoral cooperation and cutting-edge technologies.

Chen talks about how electric logistics vehicles (ELVs) and suburban renewable energy sources can be combined to lower indirect carbon emissions and improve economic viability, but it doesn't offer any concrete case studies or strategic insights on how to integrate renewable energy into supply chain operations. (Taseen et al., 2023) emphasizes collaboration in solar and renewable energy supply chains, highlighting how technological innovation, sustainability improvements, and socioeconomic impact drive partnerships. It provides insights into fostering dynamic collaborations that enhance operational efficiency and accelerate the transition to sustainable energy. (Poulsen., 2017) focuses on the offshore wind logistics supply chain, highlighting its critical bottlenecks in Europe and China. It emphasizes the need for legislation and international collaboration to enhance renewable energy diffusion and improve supply chain operations. (Shi et al., 2022) focuses on optimizing an integrated energy-logistics system in green ports, emphasizing renewable energy sources' role in minimizing operational costs and enhancing flexibility through coordinated demand management, but does not provide case studies or strategic insights on supply chain operations.

There are no particular case studies or strategic insights on the integration of renewable energy in supply chain operations in (Anoz et al., 2024). It focuses on the circular economy's functions and uses in the supply chain for renewable energy. (Yang et al., 2025) emphasizes sustainability and the reduction of carbon emissions through an integrated industrial energy supply system that combines energy storage, CCUS, and renewable energy. It does not, however, offer any particular case studies or strategic insights pertaining to supply chain operations. Although it emphasizes a gradual shift to a carbon-neutral system by 2050, (Potrc et al.,2021) concentrates on improving sustainable renewable energy supply networks in the EU. However, it offers no particular case studies or strategic insights on how to integrate renewable energy into supply chain operations.

In order to lessen reliance on fossil fuels, (Adeyinka et al.,2021) highlights the integration of renewable energy sources into supply chain processes. Leading energy businesses' case studies provide practical methods for improving operational effectiveness, sustainability, and alignment with international decarbonization initiatives through creative supply chain techniques. Using local resources like solar, biomass, and geothermal energy, provides a case study on incorporating renewables into the supply chain of a large-scale meat company. By using a synthesized regional supply-network model, this approach increases economic potential while decreasing reliance on fossil fuels. (Sun., 2024) talks on how supply chains might incorporate virtual power plants, highlighting sustainable energy sources like solar and wind. It draws attention to calculated choices about energy pricing, production scheduling, and renewable capacity that improve environmental sustainability across geographically scattered networks.

The differences between conventional and renewable energy supply chains are examined in (Mulvaney, 2022), with an emphasis on the effects on the environment and issues of justice. It highlights how important it is to compare production systems in order to comprehend how these supply networks have changed and what that means. The importance of the supply chain in renewable energy is emphasized by (Jelti et al., 2021), which evaluates its effectiveness, identifies obstacles, and suggests solutions. It does not, however, offer particular case studies or tactical insights regarding the incorporation of renewable energy into supply chain operations. Across a range of renewable energy industries, (Onukwulu et al., 2023) provides strategic insights on resilience through risk assessment, supplier diversity, cooperation, and technology uptake. Successful case studies of renewable energy integration in oil and gas supply chains are covered by (Rafiee et al., 2019). They highlight solar and geothermal energy uses in extraction and processing that lower costs and greenhouse gas emissions while improving operational efficiency. In a scenario analysis for Germany, namely by 2030, (Koch et al., 2012) assesses the incorporation of renewable energy sources into the electrical system, emphasizing flexibility options like electrical cooling and pumped storage, and highlighting their advantages for the economy and environment. The integration of renewable energy into food supply networks is covered in (Bortolini et al., 2019), which highlights the necessity of energy-efficient designs that integrate renewable sources like solar and wind to improve sustainability while lowering costs and environmental effects in supply chain activities. Explains sustainability issues and how Industry 4.0 technology might improve Renewable Supply Chain Management in the renewable energy industry (Labaran et al., 2023). The integration of renewable energy into EV charging networks is the main topic of (Abdullahi et al., 2024), which also covers technological difficulties, business prospects, and city case studies. It does not, however, particularly examine how supply chain operations might incorporate renewable energy. The integrated renewable energy system for road transportation (Hadi et al., 2022) optimizes the supply of hydrogen and electricity while reducing greenhouse gas emissions. It doesn't offer supply chain operations case studies or strategic insights. According to (Somjai et al., 2019), supply chain agility mediates the relationships between energy management practices and renewable energy supply chains in Thai manufacturing companies. The study emphasizes the important role of management commitment, energy awareness, energy knowledge, and energy audits. By contrasting single sourcing from renewable plants with dual sourcing from both renewable and fossil fuel plants, (Rajabzadeh et al., 2024) investigates the integration of renewable energy through sourcing strategies in energy supply chains, emphasizing resilience, sustainability, and cooperation to reduce disruptions.

Rather than supply chain operations or case studies pertaining to renewable energy integration, (Moser et al., 2016) concentrates on the extensive integration of renewable energy sources into the electric grid, specifically examining the technical and economic implications for Italy. In order to effectively meet the demand for power, (Ji et al., 2019) focuses on creating a sustainable supply chain for biomass-based renewable energy in Loei province, Thailand. They do this by using a mathematical model to optimize biomass output, plant placements, and supplier allocation. (Al-Madani et al., 2024) concentrates on ecological performance and energy management techniques in the manufacturing sector of Malaysia's renewable energy supply chain. Through a multi-objective optimization approach, (Kim et al., 2019) analyzes a case study in Korea's transportation sector and focuses on integrating renewable energy sources with carbon capture and utilization frameworks in energy supply systems, emphasizing both the economic and environmental benefits. Although it does not particularly address supply chain operations or case studies, (Pereira, 2015) focuses on the integration of renewable energy sources (RES) in electricity systems, examining their effects on planning and suggesting optimization models for strategic electricity planning. In order to limit carbon dioxide emissions and lower energy costs, which will increase profitability, (Safaei et al., 2024) talks about incorporating renewable energy into supply chain operations. In a multi-channel green supply chain structure, it highlights the twin goals of environmental sustainability and economic viability. Key comparative findings for renewable supply chain management are shown in Table 1.

Table 1. Key Performance Indicators (KPIs) for Renewable Supply Chain Management

Sustainability Dimension	KPI / Metric	Definition / Description	Measurement Method / Unit	Typical Data Source
Environmental	Carbon Emissions Intensity	Measures greenhouse gas emissions per unit of product or energy delivered.	CO_2-equivalent emissions per kWh or per product unit.	Supplier carbon reports, LCA databases (e.g., Ecoinvent).
	Renewable Energy Utilization Rate	Share of renewable energy used in total supply chain energy consumption.	% of total energy sourced from renewable sources.	Energy audits, supplier questionnaires.
	Energy Efficiency Improvement	Percentage reduction in energy use relative to output over time.	Energy consumed per product unit (kWh/unit) year-on-year.	Internal operational data, ISO 50001 reports.
	Waste Reduction / Recycling Rate	Extent to which supply chain processes minimize or reuse waste.	% of waste recycled or diverted from landfill.	Waste logs, supplier environmental reports.
	Water Footprint	Measures the volume of water used per production unit.	Liters of water consumed per product or energy unit.	Production process data, environmental audits.
Economic	Lifecycle Cost Savings	Total cost reduction achieved through renewable adoption across supply chain.	Difference in lifecycle costs (before vs. after renewable integration).	Financial records, TCO analyses.
	Return on Sustainable Investment (ROSI)	Financial return generated from sustainability-oriented projects.	(Financial gain from RE projects ÷ total RE investment) × 100.	Company sustainability or financial reports.
	Supply Chain Resilience Index	Ability to withstand disruptions through diversification and renewable sourcing.	Weighted score (e.g., redundancy, flexibility, response time).	Supply chain risk assessments, resilience audits.
	Productivity Improvement	Gains in efficiency from renewable energy or green technologies.	Output/input ratio before and after RE adoption.	Production and logistics performance data.
	Market Competitiveness	Change in market position due to renewable supply chain strategies.	% increase in market share or customer retention.	Industry benchmarking, sales data.

continued on following page

Table 1. Continued

Sustainability Dimension	KPI / Metric	Definition / Description	Measurement Method / Unit	Typical Data Source
Social	Green Employment Generation	Number of jobs created through renewable projects or green logistics.	Count or % increase in green workforce.	HR data, CSR reports.
	Supplier Social Compliance Rate	Extent to which suppliers follow labor, safety, and ethical standards.	% of suppliers audited or certified under social standards (e.g., SA8000).	Supplier audits, third-party verifications.
	Community Engagement Index	Degree of local community involvement and benefit-sharing.	Composite score (participation, investment, benefits).	CSR reports, stakeholder surveys.
	Occupational Health & Safety Rate	Frequency of workplace incidents across the supply chain.	Number of incidents per 100,000 hours worked.	Internal HSE reports, supplier safety audits.
	Training and Skill Development	Investment in renewable/green skills training for employees.	% of employees trained in RE/ sustainability topics annually.	HR records, training program documentation.

3. THEORETICAL FRAMEWORK

The theoretical framework for understanding renewable energy adoption in supply chains integrates System Dynamics (SD) Theory and Sustainable Supply Chain Management (SSCM) Theory to provide a comprehensive systems-based perspective. System Dynamics Theory, developed by Forrester, models complex systems by representing stocks, flows, feedback loops, and time delays, allowing the analysis of both reinforcing and balancing loops within supply chains. Reinforcing loops can amplify adoption of renewable energy technologies through increased investment and market acceptance, while balancing loops capture stabilizing constraints such as capacity limits, costs, or regulatory barriers. SD also enables simulation of policy interventions, such as subsidies or carbon taxes, to understand their long-term effects on supply chain performance and sustainability outcomes. Complementing this, SSCM Theory emphasizes balancing environmental, economic, and social objectives, guiding supply chain decisions to minimize emissions and resource consumption, optimize cost-efficiency, and ensure ethical practices and social responsibility. Integrating SD and SSCM provides a dynamic yet sustainability-oriented framework where the feedback-driven behavior of supply chains can be analyzed alongside triple-bottom-line performance. This combined approach facilitates scenario planning, predicts long-term consequences of policy and investment decisions, identifies system bottlenecks and leverage points, and supports strategic management of renewable energy adoption in supply chains, ensuring alignment with environmental, economic, and social goals.

4. RENEWABLE ENERGY RESOURCES

Numerous resources for Renewable Energy (RE) have been created and put into use with success. The complete use of RE (Wee et al., 2012) in many applications requires a secondary procedure that transforms RE into other energy resources. The energy sources and process flows for geothermal, wind, sun, hydropower, and biomass are covered in this section.

4.1 Biomass

Wood and other plant-based materials from forestry, agriculture, and industrial waste are examples of the diverse range of organic resources that are included in biomass. This trash can be transformed into useful energy resources and goods using a variety of technical techniques, including plastics, ethanol, biodiesel, and electricity. For instance, biomass can be transformed into an electric vehicle power source. Because biomass resources are diverse, their uses extend beyond just producing fuel or electricity; during the conversion process, subsidiary goods may also be created. Three primary conversion technologies (Wee et al., 2012)—thermochemical, biochemical, and extraction processes—can be used to transform biomass energy into fuel, power, and heat.

4.2 Hydropower and Tidal/Wave Energy

Seawater and rainwater are useful for many purposes. Water flowing through mills or turbines produces electricity that is widely used in home, commercial, and agricultural settings. In addition to producing electricity, water serves as a coolant in hydroelectric dams or power plants and can assist with other industrial processes (Wee et al., 2012) like fuel extraction and refinement. The conversion of wave or tidal action can also provide hydropower. In order to preserve energy during off-peak hours, a storage system is necessary because tidal and wave movements are continuous.

4.3. Geothermal Energy

The term "geothermal" energy refers to all of the thermal energy that is stored between the earth's surface and a certain depth in the crust, despite the fact that there is no accepted definition for it. (Wee et al., 2012) divides geothermal resources under three categories: minerals, heat, and water. Depending on the earth's depth, the heat can be transformed into different forms of energy to support a range of uses. For industrial and agricultural applications, heat energy must be transformed into another energy form in addition to directly using heat resources, as is the case with other RE sources. Geothermal energy can be stored by heat-pump systems, which helps balance periods of high and low demand. The production of electricity is commonly used to meet daily industrial or residential needs. However, in order to achieve the majority of its efficiency, geothermal energy acquisition is limited to specific areas. Significant financial and technological investments are also needed for its gathering and conversion.

4.4. Wind Energy

For many years, farms have been using wind energy to pump water. Wind turbines have become a popular choice for producing electricity since the invention of RE. Numerous wind turbines are situated offshore to harness enormous wind energy and reduce the environmental impact of land use. Smaller wind turbines have been created with technological assistance to produce electricity in cities. However, wind power must be stored to balance cycles of electricity demand (Wee et al., 2012) because it is dynamic and unpredictable. Furthermore, wind energy can be combined with hydropower or solar energy to create a steady and reliable energy supply.

4.5 Solar Energy

One significant and popular RE type is solar power. Electric power is produced by converting solar radiation and heat into solar energy. This source of RE is plentiful, particularly in tropical nations. Solar energy is commonly used to power numerous personal portable gadgets (Wee et al., 2012) in addition to being used in the production of electricity. Compared to the other RE sources, this one is more adaptable and requires a comparatively low initial setup cost. To meet energy needs when sunshine isn't present, energy storage is necessary.

5. RENEWABLE ENERGY SUPPLY CHAINS DYNAMICS

RE is a vast and unpredictable resource. Because of unforeseeable weather patterns and other variables that affect the RE resources, it is always shifting and unpredictable. Accordingly, the main responsibilities in the RE supply chain are the distribution and use of RE.

5.1 Supply Chain Process Flows

The physical, information, and financial movements are components of the RE supply chain, just like in many other types of supply chains. From a physical flow standpoint, supply chain management performance is now impacted by industry' growing awareness of green production practices, logistics, and goods. Because of the potential for RE to help alleviate global environmental challenges, these issues have attracted the attention of numerous researchers. In order to demonstrate the relationships within the loop, electricity is used as an example in this supply chain flow. Technology plays a crucial role in the RE supply chain's (Wee et al., 2012) ability to develop the distribution network and increase efficiency. The commercialization of RE would be a significant step toward replacing conventional fossil fuels in terms of demand. Therefore, the key advances for RE are efficient technologies for RE generators and storage.

5.1.1 Biomass Energy Flows

Biomass "is any organic, i.e. decomposable, matter derived from plants or animals on a renewable basis," according to the International Energy Agency. This covers manure as well as waste from forestry, agriculture, and industry. The most popular technological methods for turning this waste into products and energy sources that can be used are biodiesel and ethanol. Modern biomass (Wee et al., 2012) is a renewable energy resource that is produced sustainably. It is derived from solid waste, agricultural and forest wastes, and is used to generate heat, power, and transportation fuels. As one of the most sustainable solutions for the future, biofuels continue to be one of the few possibilities available to replace the role of fossil fuels in the transportation sector.

Various biomass sources, including agricultural crops and leftovers, forest residues, timber, grasses, industrial waste, animal waste, and municipal solid waste, are the first to be converted into useful energy and goods. Enzymatic and gas/liquid fermentation, acid hydrolysis, gasification, pyrolysis, combustion, and co-firing are some of the conversion processes that these basic materials go through. Fuels (ethanol, renewable diesel, renewable gasoline, and hydrogen), power (electricity and heat through co-generation), and a variety of chemicals (plastics, solvents, phenolics, adhesives, fatty acids, and detergents) are all produced using these technologies. These basic products can also be used in secondary applications such as fuel cells and power generators. In the end, the finished products meet a range of demand sectors, such as electric power, transportation, industrial applications, and residential and commercial settings. This system serves as an example of how biomass can be effectively transformed into useful products and renewable energy, promoting energy diversification and sustainability.

5.1.2 Hydropower Energy Generation Flows

To satisfy diverse energy needs, a variety of water-based renewable energy sources are used, including water, waves, and tidal energy. Water is mostly utilized for two purposes: producing hydroelectric power (Wee et al., 2012) using traditional or run-of-river systems, and cooling power plants (such as coal, nuclear, natural gas, oil, solar, geothermal, and biomass facilities). Hydropower can also help fuel cell hydrogen generation and other secondary processes like flywheel systems and pumped water storage to generate electricity. The transportation, residential, and industrial sectors receive these hydrogen and electricity. Devices including oscillating water columns, hinged contour devices, and buoyant moored devices are used to absorb wave energy and directly transform wave motion into electrical power for usage in homes and businesses. Utilizing turbines and pumping mechanisms, tidal energy—which is produced by tidal barrages and tidal streams—produces power. These techniques also address the energy requirements of homes and businesses, showing how a variety of water-based resources support renewable and sustainable energy sources.

5.1.3 Geothermal Energy Generation Flows

Application of temperature-based geothermal resources, which include heat, water, and minerals. Geothermal energy is mostly utilized for heating, industrial processes, agriculture (such as greenhouse warming), and leisure pursuits like spas at low temperatures (below 100°C). Heat from moderate temperatures (between 100°C and 200°C) can be used to generate power in addition to heating and industrial uses. Geothermal resources are mostly used to generate power in high-temperature environments (over 200°C). This classification illustrates how the temperature of the resource affects the many sectors that geothermal energy serves.

Utilizing the heat that exists within the Earth, geothermal energy generation (Wee et al., 2012) generates power. Drilling wells into geothermal reservoirs that hold steam or hot water is the first step. Different varieties of plants are utilized, depending on the resource's temperature and pressure. In dry steam plants, a turbine is powered by steam that is drawn straight from the reservoir. High-pressure hot water is depressurized in flash steam plants to create steam, which drives a turbine. A secondary fluid with a lower boiling point is heated by moderately hot water in binary cycle plants, vaporizing and powering the turbine. A generator that transforms mechanical energy into electrical energy is attached to the turbine. In order to preserve reservoir pressure and guarantee sustainability, the cooled water is reinjected into the ground after usage. This energy production technology is a valuable renewable energy source since it is efficient, dependable, and clean.

5.1.4 Wind Energy Generation Flows

Farms have been using this kind of energy to pump water for many years. Although windmills have been used for generations to grind grain, modern wind turbines (Wee et al., 2012) that produce power are novel and inventive. Both onshore and offshore wind turbines are currently in use, with the latter being utilized to reduce the environmental impact of land usage and to capture enormous amounts of wind power. However, because offshore projects involve transportation as well as operations including platforms, turbines, cables, substations, grids, and other components, they require sophisticated logistics management. However, other factors, including the seasonality of wind direction and speed, need to be taken into account in the analysis.

Using wind as the main source is the first step in producing wind energy. The main result of the procedure is that wind rotates a turbine's blades. The wind's kinetic energy is transformed into mechanical energy while the turbine rotates. In the secondary process, a generator converts this mechanical energy into electrical energy. Furthermore, hydrogen can be produced in fuel cells using the obtained power. Wind energy is a clean, renewable option for sustainable power generation since the electricity and fuel cell output are then delivered to satisfy different demands in the residential, commercial, and automotive sectors.

5.1.5 Solar Energy Generation Flows

Solar energy is a significant and popular form of renewable energy worldwide. There are two methods for turning it into electricity: solar thermal/electric power plants and photovoltaic (PV devices or sun cells). Tropical nations have a significant advantage in this kind of energy, however energy storage is necessary to meet varying energy needs while also accounting for changes in solar radiation intensity.

The first step in producing solar energy (Wee et al., 2012) is to capture sunlight using thermal or photovoltaic techniques. The main result is the conversion of solar energy into useful heat or electricity through the use of technology like solar chimneys, power towers, parabolic troughs, and parabolic dishes. To guarantee a steady and dependable energy supply, this energy is then processed using secondary techniques that involve storage devices like batteries, flywheels, or heat transfer systems. Ultimately, solar power is a clean and sustainable energy source since the stored or directly produced solar energy is dispersed to satisfy the needs of the commercial and residential sectors.

There is little use of system dynamics in the field of renewable energy, such as bioenergy, geothermal, hydropower, ocean, solar, and wind subtopics. Only a small number of studies analyze demand and transportation between distribution centers using this method, as well as the challenges of investing in crops and refining capacity. Since demand and transportation are two of the most frequently researched topics by scholars and industry professionals, there is a chance to examine the global supply chain and how it integrates with the operations of the renewable energy supply chain from a dynamic standpoint. An outline of the system dynamics approach and a few examples (Wee et al., 2012) of its use in SSCM and RESC are presented in the next section.

5.2 Renewable Energy Supply Chain Issues

Each type of RE is constrained by the intrinsic properties of the energy source, just like conventional sources of electric power generation. Three essential characteristics of RE resources that need efficient administration and control are their intermittency, variability, and mobility. Furthermore, because of the nature of RE, a second conversion procedure is required to conserve energy for consumption during off-peak hours.

The graphic highlights the main issues and factors to be taken into account in the four main phases of the renewable energy system: demand, production, distribution, and supply. The availability and accessibility of resources are impacted throughout the supply stage by factors such land use, water consumption, intermittency, variability, maneuverability, and technical constraints. Energy generation during the manufacturing stage is greatly influenced by a number of elements, including geographic location, conversion efficiency, high initial investment, cost, operational and maintenance charges, and technological restrictions. The distribution stage concentrates on energy storage, employment prospects, and distribution efficiency. Lastly, at the demand stage, the acceptance and integration of renewable energy are influenced by social, political, and environmental factors as well as substitution effects (such as moving away from fossil fuels). To guarantee a sustainable and effective renewable energy system, specific obstacles at each stage must be overcome.

6. SYSTEM DYNAMICS INTEGRATION

Professor Jay W. Forrester developed the system dynamics approach, a simulation tool, in the 1950s to assist managers in enhancing industry processes whose behavior is fundamentally dynamic. His study combines ideas from digital computation and feedback control theory to describe how social systems behave and create efficient rules to enhance system performance. Forrester wrote a lot about SD, and three significant volumes compiled his key points. In order to create common models of organizational systems, SD theory offers a collection of tools and methodologies that enable rigorous representation, simulation-based validation, and sensitivity and "what-if" analyses to assess the effects of various policies. Managers benefit from this approach and computer modeling tools because they help them better comprehend industrial processes whose behavior is dynamic or does not function in a steady state.

Three guiding ideas form the basis of system dynamics modeling: A supply chain, for instance, entails intricate interactions between customers, retailers, wholesalers, distributors, factories, and raw material suppliers in addition to order and material flows and the decisions that must be made regarding these flows. i) The structure dictates the behavior. ii) Organizational systems' structures frequently incorporate "soft" variables; for example, a supply chain's structure needs to comprehend how its agents perceive future consumer behavior. iii) Understanding and altering the mental model can yield substantial leverage.

Planning and policy design, economic behavior, public management and policy, biological and medical modeling, energy and the environment, theory development in the natural and social sciences, dynamic decision making, software engineering, and supply chain management are just a few of the fields in which SD theory finds numerous applications. Although its use has increased over the past five years, supply chain analysis still needs to learn more about it. System dynamics modeling is an ongoing, iterative process of developing hypotheses, validating them, and revising the formal and mental models that depict the system. Based on this, a system dynamics technique can be used to control and coordinate the flows of goods, information, and finances.

6.1 Sustainable Supply Chain Management and System Dynamics

The integration of socially and ecologically responsible practices throughout a product or service's whole lifecycle, from raw material sourcing to manufacture, distribution, use, and eventual disposal, is known as sustainable supply chain management, or SSCM. It seeks to preserve economic viability, guarantee social justice, and reduce the environmental impact. SSCM may be more thoroughly examined and enhanced when paired with system dynamics, a modeling technique that makes use of feedback loops and time delays to comprehend complicated systems. Decision-makers can see how different supply chain components interact over time with the aid of system dynamics, which emphasizes the long-term effects of policies, production or inventory delays, and the influence of sustainability initiatives. Organizations can use system dynamics to find bottlenecks, forecast results in various situations, and create strategies that strike a balance between social and environmental objectives and profitability, which will ultimately result in supply chains that are more robust and sustainable.

6.2 Renewable Energy Supply Chain and System Dynamics

All of the procedures involved in creating, transforming, storing, distributing, and using energy from renewable sources like solar, wind, hydro, geothermal, and biomass are included in the renewable energy supply chain. The production of energy technology, energy generation, storage systems (such as flywheels or batteries), distribution infrastructure, end-use applications, and the extraction of raw materials (such as rare earth metals for solar panels or turbines) are all included. By incorporating system dynamics into this supply chain, it is possible to gain a comprehensive understanding of the relationships, feedback loops, and delays that impact sustainability and performance. Analyzing how policy changes, technology breakthroughs, investment choices, and environmental restrictions affect the energy system as a whole is made easier with the use of system dynamics modeling. It sheds light on issues including environmental effects, storage constraints, infrastructure needs, and intermittency. Decision-makers may create more flexible and effective renewable energy systems by modeling various scenarios, which will ultimately help achieve long-term energy security, carbon reduction, and sustainability objectives.

The central stock is refining capacity, which rises with new refining capacity (driven by investment) and falls with obsolete refining capacity, according to the system dynamics model of refining capacity in the energy supply chain. Refining profits are the first step in the loop, and they have an impact on the likelihood of investing in new infrastructure. Overall refining capacity increases as New Refining Capacity rises in tandem with increased investment. But as capacity increases, there can be an excess of refining capacity, which might lower earnings and impede further investments. The feedback loops in this model, which are both balancing (surplus resulting in lower earnings) and reinforcing (investment and capacity development), aid in simulating the dynamics of real-world decision-making in energy infrastructure design.

7. PERFORMANCE AND KEY INSIGHTS

The RE supply chain connects the energy source to further uses. Conversion efficiency, which encompasses storage, distribution, and secondary application efficiencies, is a key component of the RE supply chain's performance.

7.1 Conversion Efficiency

Each form of RE has a different conversion efficiency, which is regarded as a crucial metric for the utilization of a particular energy supply. Primary energy sources like fossil fuels have significant costs. In order to compete with the production of energy from fossil fuels, current efforts to lower the cost of acquiring and using RE through technological advancements could not be enough.

A "Levelized Cost of Electricity" (LCOE) has been identified for decision-making in solar energy projects in order to evaluate the investment for RE. The LCOE prices of photovoltaic (PV) and concentrated solar power are greater than those of other REs, according to the U.S. Department of Energy's (DoE) Energy Efficiency and Renewable Energy report. Furthermore, REs are evaluated based on their environmental effects (Wee et al., 2012) in addition to their performance and investment.

7.2 Technology

Modern technology have accelerated the growth of the renewable energy sector. The development of technologies like energy storage, fuel cells, and hybrid systems that facilitate RE conversion processes and the expansion of RE applications is necessary for the growth of the RE industry in addition to increasing conversion efficiency.

7.2.1 Types of Storage Technologies

Battery energy storage systems, flywheels, compressed air energy storage, superconducting magnetic energy storage, and pumped storage are some examples of energy storage technology (Wee et al., 2012). Power quality, bridging power, and energy management are the three categories into which the National Renewable Energy Laboratory (NREL) divided energy storage. Each category has a range of discharge timings that influence and restrict its use. When choosing an electrical storage technology, there are a number of considerations to consider. These include lifetime, cycle efficiency, power level, response time, storage capacity, and discharge duration. The selection of an energy storage device is contingent upon its use in the renewable/VG-driven grid or the existing grid; the duration of discharge plays a major role in these applications. An overview of energy storage and uses is shown in Table 2.

Table 2. Three classes of energy storage

Common name	Example applications	Technology	Discharge time
Power quality	Transient stability, frequency regulation	Flywheel, Capacitor, Superconducting magnetic energy storage	Seconds to minutes
Bridging power	Contingency reserves, ramping	Battery energy storage system -Lead-Acid, Ni–MH, Ni–Cd, Li-Ion	Minutes to 1 h
Energy management	Load leveling, Firm capacity, T&D deferral	Compressed air energy storage, pumped storage, high-energy batteries	Hours

7.2.2 Fuel Cell Applications

Applications for fuel-cell technology are numerous. After years of development, this technology has been effectively applied (Wee et al., 2012) to power generation, but its first uses were restricted to space exploration and military applications. Fuel cells have proven effective in gas-electric hybrid vehicles, especially in the automotive sector. Alkaline fuel cells, polymer-electrolyte-membrane fuel cells, phosphoric-acid fuel cells, molten-carbonate fuel cells, and solid-oxide fuel cells are among the different kinds of fuel-cell power generators that offer varying power outputs. Depending on the device type (portable, stationary, or transportation), each fuel-cell technology has a different set of uses.

7.2.3 Hybrid Energy Systems

In the "green" transportation sector, hybrid energy systems—which concurrently generate hydrogen and electricity—are essential. Honda, for instance, has created a solar-hydrogen power plant to aid in the marketing of gas-electric hybrid automobiles. Similar systems could be made for home electric appliance powering (Wee et al., 2012) and for private use. Fuel cells, gas turbines, internal combustion engines, boilers, and catalytic burners have all used solar-hydrogen hybrid systems to generate electrical, mechanical, and thermal energy.

8. AUTHOR'S CONTRIBUTIONS AND KEY INSIGHTS

- This study synthesizes multi-sector evidence to identify a unified model for renewable energy integration in supply chains.
- It extends prior literature by connecting *system dynamics modeling* with *renewable energy logistics and operations.*
- It proposes policy-driven mechanisms for accelerating decarbonization in industrial networks.
- It highlights empirical gaps—especially the lack of quantitative and case-based studies in developing economies.

9. CONCLUSION

This analysis highlights how important renewable energy is to changing supply networks to become more resilient and sustainable. It emphasizes that successful integration needs not only technological prowess but also strategic coordination and dedication by examining real-world applications across many industries. For businesses looking to lower their carbon footprint and match supply chain procedures with global sustainability goals, the suggested framework offers a workable road map. These observations can direct future initiatives to create more eco-friendly and effective supply chain systems as the need for environmental responsibility grows.

10. POLICY IMPLICATIONS

According to the study's conclusions, encouraging legislative frameworks are necessary to hasten the supply chains' adoption of renewable energy. For businesses investing in renewable infrastructure, governments and regulatory agencies ought to think about enacting incentives including tax credits, subsidies, and low-interest financing. Furthermore, encouraging public-private collaborations and creating uniform standards for the adoption of renewable energy can stimulate technological innovation and cross-sector cooperation. In order to facilitate supply chains' transition to carbon neutrality and long-term sustainability, policymakers must also make sure that energy regulations are flexible enough to accommodate new technologies and business models.

11. FUTURE DIRECTIONS AND RESEARCH

Quantitative evaluation of the long-term economic and environmental effects of integrating renewable energy across various supply chain models and geographical areas should be the main goal of future study. Investigating how cutting-edge technologies like blockchain, artificial intelligence, and the Internet of Things can optimize energy use and improve supply chains powered by renewable energy sources is also necessary. Furthermore, research that compare established and developing economies can provide insight into the facilitators and barriers that are unique to each setting. More research into the ethical and social ramifications of procuring renewable energy in global supply chains will yield a more comprehensive knowledge that will direct both scholarly research and real-world application.

REFERENCES

Adeyinka, M. A., Solanke, B., Onita, F. B., & Agho, M. O. (2021). Navigating challenges in energy supply chains: Lessons from the oil, gas, and renewable energy sectors. *International Journal of Scientific Research and Updates*, *1*(1), 25–35. https://orionjournals.com/ijsru/sites/default/files/IJSRU-2021-0025.pdf

Adeyinka, M. A., Solanke, B., Onita, F. B., & Agho, M. O. (2021). Optimizing supply chain management in the energy sector: Strategies for enhancing efficiency and sustainability. *International Journal of Management Research and Updates*, *1*(1), 42–55. https://orionjournals.com/ijmru/sites/default/files/IJMRU-2021-0042.pdf

Azad, A. M. A. S., Oishi, Z. T., Haque, M. A., Das, P., Udoy, S. A., & Bhuiya, K. M. S. (2024). Integration of renewable energy in industrial operations: Experiences from Canada, USA, and Africa. *GSC Advanced Research and Reviews*, *18*(1), 213–221. Advance online publication. DOI: 10.30574/gscarr.2024.18.1.0014

Azad, A. M. A. S., Oishi, Z. T., Haque, M. A., Das, P., Udoy, S. A., & Bhuiya, K. M. S. (2024). An integrated framework for assessing renewable-energy supply chains using multicriteria decision-making: A study on Bangladesh. *Clean Energy*, *8*(3), 1–15. DOI: 10.1093/ce/zkae019

Fontes, C. H. D. O., & Freires, F. G. M. (2018). Sustainable and renewable energy supply chain: A system dynamics overview. *Renewable & Sustainable Energy Reviews*, *82*, 247–259. DOI: 10.1016/j.rser.2017.09.033

Jelti, F., Allouhi, A., Büker, M. S., Saadani, R., & Jamil, A. (2021). Renewable power generation: A supply chain perspective. *Sustainability (Basel)*, *13*(3), 1271. DOI: 10.3390/su13031271

Labaran, M. J., & Masood, T. (2023). Industry 4.0 driven green supply chain management in renewable energy sector: A critical systematic literature review. *Energies*, *16*(19), 6977. DOI: 10.3390/en16196977

Noor, M. M., & Abidin, I. Z. (2022). An integrated renewable energy system for the supply of electricity and hydrogen energy for road transportation which minimizes greenhouse gas emissions. *International Journal of Sustainable Energy Planning and Management*, *35*, 66–79. DOI: 10.5278/ijsepm.2022.35.5

Poulsen, T., & Lema, R. (2017). Is the supply chain ready for the green transformation? The case of offshore wind logistics. *Renewable & Sustainable Energy Reviews*, *73*, 758–771. DOI: 10.1016/j.rser.2017.01.181

Rafiee, A., & Khalilpour, K. R. (2019). Renewable hybridization of oil and gas supply chains. In *Renewable Energy Integration with Conventional Energy Systems* (pp. 223–241). Elsevier., DOI: 10.1016/B978-0-12-813306-4.00011-2

Rajabzadeh, H., & Wiens, M. (2024). Resilient and sustainable energy supply chains: Insights on sourcing and pricing strategies in a non-collaborative and collaborative environment. *International Journal of Production Research, 62*(24), 9011–9042. Scholar+2KITopen+2IDEAS/RePEc+2DOI: 10.1080/00207543.2024.2331556Google

Safaei, M., Al Dawsari, S., & Yahya, K. (2024). Optimizing multi-channel green supply chain dynamics with renewable energy integration and emissions reduction. *Sustainability (Basel), 16*(22), 9710. DOI: 10.3390/su16229710

Torres-Rivera, A. D., Mc Namara Valdes, A. J., & Florencio Da Silva, R. (2023). The resilience of the renewable energy electromobility supply chain: Review and trends. *Sustainability (Basel), 15*(14), 10838. DOI: 10.3390/su151410838

Wee, H. M., Yang, W. H., Chou, C. W., & Padilan, M. V. (2012). Renewable energy supply chains, performance, application barriers, and strategies for further development. *Renewable & Sustainable Energy Reviews, 16*(8), 5451–5465. DOI: 10.1016/j.rser.2012.06.006

Yang, L., Wu, X., Huang, B., & Li, Z. (2025). Sustainable industrial energy supply systems with integrated renewable energy, CCUS, and energy storage: A comprehensive evaluation. *Sustainability (Basel), 17*(2), 712. DOI: 10.3390/su17020712

Chapter 9
Renewable Energy-Driven Smart Logistics Systems: Conceptual Model, Strategic and Feasibility Analysis

Ilknur Yardımcı-Coşkun
https://orcid.org/0000-0002-7183-6142
Maltepe University, Turkey

ABSTRACT

This study examines the strategic integration of renewable energy sources into smart logistics systems to enhance operational efficiency and environmental sustainability. Based on a comprehensive literature review, it develops a conceptual model linking technological components of smart logistics, such as IoT, cyber-physical systems, cloud platforms, and decision-support mechanisms, with renewable energy solutions including solar, wind, hydropower, bioenergy, hydrogen, and geothermal energy. The study qualitatively analyzes energy–technology alignment and evaluates sustainability outcomes such as carbon emission reduction, energy cost optimization, and socio-economic benefits. A SWOT analysis and a feasibility assessment address strategic and practical considerations, while a set of hypotheses is proposed for future empirical testing. The findings highlight the need for digital–energy synergy, supportive infrastructure, and strategic policy frameworks to unlock the full potential of renewable energy-powered smart logistics.

1. INTRODUCTION

Since the Industrial Revolution, logistics processes have become highly resource-intensive, spanning the entire supply chain from production to distribution. With

DOI: 10.4018/979-8-3373-3790-6.ch009

rapid population growth, urbanization, and the expansion of global trade, the energy demands and environmental impacts of logistics have reached unprecedented levels. This has intensified the need to restructure logistics operations in line with environmental sustainability and energy efficiency principles.

In this context, smart logistics systems have emerged as a transformative solution (Sun et al., 2022). By integrating advanced technologies, including ICT, automation, AI, and data analytics, these systems aim to enhance visibility, efficiency, adaptability, and real-time decision-making across logistics networks (Alonge et al., 2023). However, their continued reliance on fossil fuels remains a critical limitation, constraining environmental performance and hindering climate action (Chu & Majumdar, 2012).

To address this, integrating renewable energy sources such as solar, wind, biomass, and geothermal into smart logistics infrastructures is gaining attention. This integration improves operational efficiency through clean, decentralized energy use while reducing greenhouse gas emissions and supporting global sustainability goals. It aligns closely with the United Nations Sustainable Development Goals (SDGs), particularly SDG 7 (Affordable and Clean Energy), SDG 9 (Industry, Innovation and Infrastructure), and SDG 13 (Climate Action) (Mire et al., 2018; Denoncourt, 2020; Elavarasan et al., 2021; Ramasubramanian & Ramakrishna, 2023).

1.1. Comparative Policy Perspectives

The pace and scope of renewable energy integration into logistics differ significantly across regions, shaped by policy frameworks, technological capacities, and infrastructural readiness. In the European Union, comprehensive regulatory instruments such as the European Green Deal and emissions trading schemes have provided strong incentives for renewable deployment in transport and logistics (Karaeva et al., 2023). In the United States, large-scale policy interventions such as the Inflation Reduction Act (IRA) and Bipartisan Infrastructure Law (BIL) have catalyzed investments in clean transport corridors and logistics facilities powered by renewables (Steinberg et al., 2023). Similarly, China, Germany, Japan, and the U.S. have adopted different policy mixes to expand solar and renewable integration, reflecting variations in industrial strategies and grid capacity (Muhammed & Tekbiyik-Ersoy, 2020; Wen & Gao, 2023).

By contrast, developing economies face structural barriers such as limited capital availability, regulatory uncertainty, and inadequate energy infrastructure. Yet, these regions also present leapfrogging opportunities—bypassing intermediate fossil-based stages—through decentralized renewable systems and EV deployment in urban freight (Batinge et al., 2017; Khaleel & Chakrabarti, 2018; Eyitoyo, 2023). These comparative differences highlight that renewable–logistics integration is context-dependent and shaped by both technological maturity and political-economic environments (Khan et al., 2022).

1.2. Expanding the Energy Portfolio

The rise of electric vehicles (EVs), autonomous transportation, and IoT-supported warehouse systems has created new opportunities to integrate clean energy into logistics (Shuaibu et al., 2025). Solar-powered EV charging stations, wind-powered port systems, and biomass-based warehouse heating exemplify renewable energy use across supply chains. Beyond solar and wind, hydrogen is emerging as a key energy carrier for decarbonizing long-haul and maritime logistics (Pu et al., 2023; Almazeedi, 2024; Alavi-Borazjani et al., 2025; Chen et al., 2025), while geothermal energy offers stable baseload capacity for warehouse heating and cold chain operations (Hanbury & Vasquez, 2018; Ball, 2021; Akhigbe, 2025). These developments signal a shift toward transforming logistics infrastructures into environmentally responsive networks.

Technological advances in energy storage and smart grids have increased the feasibility of renewables in logistics. Battery energy storage systems (BESS) enable facilities to store excess renewable energy and use it during peak demand, stabilizing supply and supporting demand-side management to optimize energy use and reduce emissions (Mexis & Todeschini, 2020).

However, adoption remains uneven due to high investment costs, inadequate infrastructure, and regulatory uncertainty (Takman & Andersson-Sköld, 2021; Barbieri & Capoani, 2025). Overcoming these barriers requires a systematic approach integrating technological innovation with supportive policy frameworks. Collaboration between government, industry, and academia is crucial to scaling successful applications and establishing standards for widespread renewable adoption in logistics.

This study examines the technological architecture of smart logistics systems and identifies commonly utilized renewable energy sources. It evaluates the opportunities and challenges of renewable integration, contributing to a deeper understanding of how digital transformation and green energy strategies can be combined to build sustainable and resilient logistics infrastructures.

In summary, the study presents a comprehensive conceptual and empirical analysis of renewable integration in smart logistics. It includes a literature review, a conceptual model linking energy and technology, a matrix analysis of energy–technology alignment, empirically testable hypotheses, and a SWOT analysis to evaluate strategic implications. Collectively, these components aim to guide researchers and practitioners toward more sustainable and energy-efficient logistics systems.

2. METHODOLOGY

This study adopts a qualitative research paradigm and employs a systematic literature review. More than 70 academic sources, including journal articles, book chapters, and industry reports published between 2008 and 2025, were reviewed to collect relevant data. The literature review focuses on two main areas.

First, the study examines the definition, technological foundations, and operational components of smart logistics systems, systematically categorizing their physical, digital, and managerial elements. Second, it explores the integration of renewable energy sources, including solar, wind, hydropower, bioenergy, hydrogen, and geothermal, into these systems, assessing their compatibility and impact on logistics operations. This dual focus highlights the intersection between technological infrastructure and environmental strategies in building sustainable and efficient logistics networks.

The literature was analyzed through deductive content analysis, leading to the classification of smart logistics components under nine categories: (1) sensing and monitoring technologies, (2) data processing units, (3) communication infrastructure, (4) physical transport systems, (5) operational management software, (6) blockchain structures, (7) cyber-physical systems, (8) smart distribution infrastructure, and (9) decision support platforms. This framework informed the development of a conceptual model that links the energy demands of each technological component with suitable renewable energy sources.

Based on this model, an annotated energy–technology compatibility matrix was constructed to map appropriate renewable energy types for different logistics technologies. Finally, a qualitative SWOT analysis was conducted to evaluate internal strengths and weaknesses and external opportunities and threats. This strategic assessment supports the conceptual model and provides policy-relevant insights for promoting sustainable logistics practices.

Figure 1. Proposed Steps in the Research Methodology

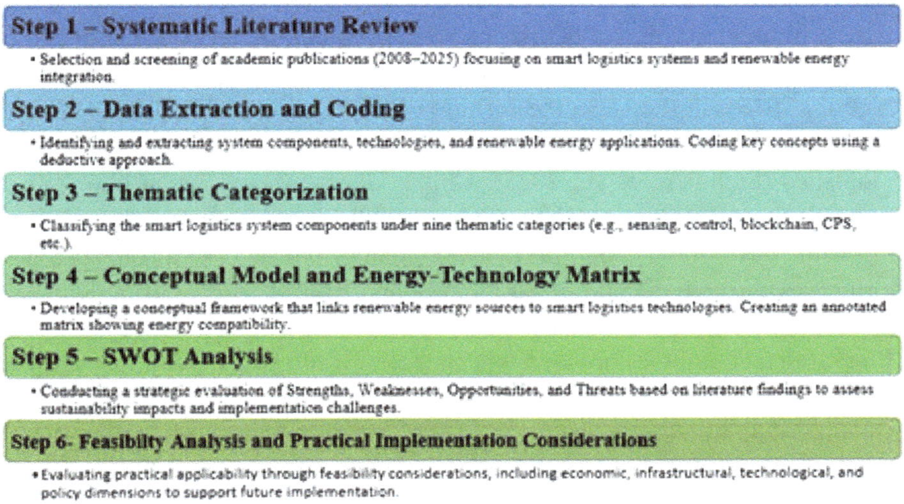

This flowchart (Figure 1) outlines the six key steps undertaken in the research process, starting from a systematic literature review and progressing through data extraction, thematic categorization, conceptual modeling, and strategic SWOT analysis, before concluding with a feasibility assessment. The process ensures a comprehensive and structured approach to examining the integration of renewable energy into smart logistics systems, combining both theoretical exploration and practical implementation considerations.

3. LITERATURE ANALYSIS AND FINDINGS

3.1. Renewable Energy

Renewable energy has become a cornerstone of sustainable development, addressing global challenges such as climate change, environmental degradation, and energy access. Energy from solar, wind, hydropower, bioenergy, hydrogen, and geothermal sources can simultaneously reduce emissions and foster economic growth. It also supports key UN Sustainable Development Goals (SDGs), including SDG 7 (clean energy), SDG 8 (decent work), and SDG 13 (climate action) (Agbakwuru et al., 2024; Asif et al., 2022).

Beyond environmental benefits, renewable energy provides economic and social advantages, particularly in underserved regions, by creating jobs, improving health outcomes, and enhancing energy security (Pablo-Romero et al., 2016). Stable renewable investments stimulate innovation and buffer against fossil fuel price volatility.

Policy frameworks play a decisive role in accelerating transitions. In the EU, regulatory mechanisms and carbon pricing have supported rapid deployment (Karaeva et al., 2023). In the U.S., the Inflation Reduction Act (IRA) and Bipartisan Infrastructure Law (BIL) have driven large-scale investments, especially in power and logistics (Steinberg et al., 2023). China, Germany, and Japan have pursued diverse solar–wind strategies (Wen & Gao, 2023; Muhammed & Tekbiyik-Ersoy, 2020), while developing economies face regulatory barriers but also leapfrogging opportunities through decentralized renewables (Batinge et al., 2017; Khaleel & Chakrabarti, 2018).

Emerging technologies broaden this portfolio. Hydrogen is gaining attention as a strategic energy carrier for decarbonizing heavy transport and maritime logistics (Pu et al., 2023; Alavi-Borazjani et al., 2025; Chen et al., 2025), serving both as fuel and storage. Geothermal energy offers stable baseload power for warehouses, cold chains, and industrial parks (Hanbury & Vasquez, 2018; Ball, 2021; Akhigbe, 2025). Diversifying across these sources strengthens energy security and operational continuity.

Nonetheless, widespread adoption faces high upfront costs, intermittency, technological limits, and regulatory gaps (Agbakwuru et al., 2024). Overcoming these barriers requires targeted subsidies, green financing, grid modernization, international cooperation, and public–private partnerships.

In sum, renewable energy delivers substantial environmental, economic, and social benefits, but effective integration depends on innovation, supportive policy, and economic feasibility. These dynamics are crucial for a resilient, low-carbon energy future. The next sections examine how renewables align with smart logistics systems and infrastructures.

3.2. Smart Logistics Systems

Smart logistics systems have become a core pillar of supply chain transformation, driven by rapid technological progress and digitalization. These systems integrate automation, artificial intelligence (AI), the Internet of Things (IoT), cyber-physical systems (CPS), blockchain, and big data analytics to enhance efficiency, responsiveness, and sustainability across logistics networks (Zsifkovits & Woschank, 2019; Liu et al., 2023). Through the convergence of these technologies, smart logistics enable seamless interaction between physical and digital layers, forming the Logistics 4.0 paradigm.

IoT provides real-time sensing, tracking, and communication between goods, vehicles, and infrastructure (Gowda et al., 2024). AI and data analytics support predictive decision-making, optimizing routing, inventory, and capacity. CPS tightly integrates physical assets with digital controls, enabling synchronized and adaptive operations. Circular logistics frameworks promote resource efficiency and waste minimization, while smart grids and energy management systems facilitate renewable energy use in real time (Mumin et al., 2025; Oshilalu & Iboh, 2025).

Smart logistics strengthen integration across supply chain tiers, lowering costs and improving service levels. Beyond operational gains, they reduce carbon emissions, optimize resources, and minimize environmental impacts. As an extension of Industry 4.0, smart logistics enhance competitiveness, support agile decision-making, and accelerate green transformation.

Operationally, technologies such as automation and robotics improve efficiency in warehousing, using unmanned shuttles and stacker cranes to reduce labor costs and errors (Cho, 2018). Blockchain ensures secure and transparent transactions, enhancing trust and auditability (Issaoui et al., 2019). Distributed sensor networks with cloud synchronization enable real-time monitoring and optimized resource allocation (Kezhu, 2024).

A key environmental contribution is the optimization of supply chain activities, reducing waste and carbon footprints (Al-Ormuza et al., 2024). Interoperability—seamless data and system integration between firms—enhances performance, especially in e-commerce and freight (Gumzej, 2023).

However, data privacy, cybersecurity risks, and high capital costs remain major barriers. Organizations must address these during strategic planning. Moreover, integrating smart logistics with energy management systems is still emerging, offering significant opportunities to align digital transformation with renewable energy.

Table 1. Overview of Smart Logistics Systems in the Literature (2008–2025)

Author/Year	Summary of Work
Uckelmann, 2008	The article highlights key technical components of Smart Logistics, emphasizing how emerging technologies enhance flexibility, adaptability, and responsiveness in dynamic logistics networks.
Master et al., 2011	The article states that a smart logistics system includes a portable support structure, RFID reader and antenna, and a power supply, designed for quick forklift mounting and self-sufficient operation.
Chen et al., 2014; Su et al., 2017	The articles state that the smart logistics service system includes code scanning, GPS, internet, cloud storage, alarm, and mobile app modules, enabling real-time information exchange and tracking during transportation.
Cho, 2018	According to the article, the smart logistics system includes a stacker crane, load-support frame, unmanned shuttle, IoT-enabled sensor gate for identification, and an automated conveyor for classification.

continued on following page

Table 1. Continued

Author/Year	Summary of Work
Deng et al., 2018	According to the article, the smart logistics order dispatch system includes interconnected dispatch, statistics, and purchasing modules, a logistics database, cloud server, and display terminal, working together to improve delivery efficiency and accuracy.
Damianov and Demirova, 2018	The article defines Logistics Information Systems (LIS) as interconnected digital tools that automate material flow control and form the basis for cyber-physical systems and smart factories.
Gebresenbet et al., 2018	In this study, the Smart Logistics System (SLS) is conceptualized as comprising four core components: (a) smart box (sensor unit), (b) embedded control unit, (c) centralized information platform, and (d) central control unit for supply chain management.
Zsifkovits and Woschank, 2019	According to the article, Smart Logistics Systems include CPS, IoT, and elements of the Physical Internet, enabling autonomous processes and improving logistics efficiency through agile collaboration and information networking.
Boldyreiva and Gryshko, 2020	According to the article, Smart Logistics Systems comprise components such as the Smart Logistics Network (planning, optimization, coordination), Smart Transportation, Smart Government, Smart City, and Smart Contracts, which together support management, control, and decision-making.
Gutu Ofgaa, 2020	According to the article, Smart Logistics Systems (SLS) include supply chain elements such as sales, transportation, inventory, and data exchange, using IoT, big data, and cloud computing to optimize processes and support real-time decisions.
Ding et al., 2021	According to the article, the smart logistics system includes RFID, sensor, processing, database, monitoring modules, and a customer terminal, all linked through a central unit to improve logistics efficiency.
Hou and Chen, 2022	According to the article, the smart logistics system includes entry points, signal lights, weighing bridges, tags, induction coils, guardrails, signal base stations, recognition units, a main control body, wiring, computers, and printers, enabling efficient remote control and monitoring.
Ugochukwu et al., 2022	According to the article, the smart logistics management system uses a logistics blockchain to generate item and transaction hashes, and a distributed database to update and synchronize data, with all processes governed by smart contracts.
Zhang, 2022	According to the article, the smart logistics warehouse management system includes RFID technology, intelligent control nodes, and warehouse and inventory management modules, with RFID as a core element enhancing automation and intelligence.
Gumzej, 2023	According to the article, Smart Logistics Systems (SLS) include smart agents, facilities, vehicles, intermodal containers, and distribution centers forming the Physical Internet (PhI) infrastructure, enabling efficient, transparent, and interoperable logistics operations.
Liu et al., 2023	According to the article, smart logistics systems integrate advanced sensing and computing components forming the cyber-physical system (CPS) infrastructure, enhancing data collection, transmission, storage, processing, and AI integration for improved efficiency and decision-making.
Kezhu, 2024	According to the article, the smart logistics platform comprises four core components: a Disaster Recovery System (DRS), Data Center (DC), Communication Network (CN), and System Platform (SP), designed to enhance logistics operations and ensure data protection.

According to the conducted literature review (Table 1), the components of smart logistics systems can be classified under the following key categories, based on the technological infrastructure, hardware, and software modules proposed by various scholarly studies:

1. **Sensing and Monitoring Technologies**
2. **Data Processing and Control Units**
3. **Communication and Information Sharing Infrastructure**
4. **Physical Transportation and Storage Components**
5. **Logistics Operation and Management Systems**
6. **Blockchain and Smart Contracts**
7. **Cyber-Physical Systems (CPS) and the Physical Internet (PhI)**
8. **Smart Facilities and Distribution Centers**
9. **Management Interfaces and Decision Support Platforms**

These foundational elements enable the integrated, scalable, and sustainable operation of modern logistics infrastructures. As logistics networks increasingly embrace digital transformation, the convergence of these technological systems with renewable energy integration becomes a critical step toward building resilient, low-carbon, and intelligent supply chains. This intersection will be examined in the next section.

3.3. Integration of Renewable Energy into Smart Logistics Systems

The integration of renewable energy into smart logistics systems represents a strategic transformation that combines the decarbonization of energy supply with the digitalization of logistics operations. This dual transition enables logistics networks to reduce greenhouse gas emissions, lower operational costs, and enhance energy resilience, while simultaneously increasing visibility and responsiveness through digital technologies. By leveraging smart grids, IoT, and AI-enabled energy management, renewable energy sources can be synchronized with logistics processes in real time, supporting sustainable transportation, warehousing, and distribution systems (Pan et al., 2023; Zedak et al., 2023; Musau, 2024; Nzeanorue & Okpala, 2024; Sayeed & Manikandan, 2024).

Solar Energy: Photovoltaic (PV) systems are among the most widely adopted renewable solutions in logistics, used to power warehouses, charging stations, and distribution hubs. Solar energy enables decentralized, on-site electricity generation, reducing reliance on fossil fuels and grid power. For example, the integration of cloud-enabled solar logistics systems facilitates real-time energy monitoring and operational optimization through blockchain and IoT platforms (Pan et al., 2023). Solar installations in logistics parks and green ports have been shown to significantly reduce operational emissions while stabilizing energy costs (Yap & Notteboom, 2024; Huang et al., 2024).

Wind Energy: Wind power is increasingly utilized to support logistics operations, particularly in coastal and seaport environments where wind resources are abundant. Turbines are installed at logistics terminals and integrated into microgrids to power handling equipment and port infrastructure. Coordinated operation strategies for energy storage systems are often deployed to manage intermittency, ensuring a more stable and reliable energy supply for logistics activities (Shi et al., 2024; Li et al., 2025).

Bioenergy: Bioenergy plays a significant role in sustainable logistics through biomass-to-bio-oil supply chains, which aim to minimize transportation costs and carbon emissions while supporting circular economy principles. Bio-based fuels can be integrated into transport fleets or used for heating and energy generation in warehouses, contributing to both decarbonization and waste valorization (Sadeghi & Haapala, 2019). Optimization techniques such as genetic algorithms have been applied to improve the operational efficiency of bioenergy logistics networks (Sadeghi & Haapala, 2019).

Hydrogen Energy: Hydrogen is increasingly recognized as a strategic energy carrier for logistics, especially for long-haul transportation and maritime operations where battery-electric solutions may face limitations. Hydrogen fuel cells can power heavy-duty trucks, forklifts, and port equipment, offering longer ranges and shorter refueling times compared to conventional batteries. Recent studies demonstrate the feasibility of integrating hydrogen infrastructure with logistics corridors, supported by smart scheduling algorithms for port operations (Pu et al., 2023). Alavi-Borazjani et al. (2025) highlight hydrogen's transformative role in maritime logistics, while Chen et al. (2025) provide cost and emissions scenarios for hydrogen freight in China. Policy frameworks such as those under the U.S. Inflation Reduction Act (Steinberg et al., 2023) and hydrogen investment strategies in the Middle East (Almazeedi, 2024) are accelerating the deployment of hydrogen-powered logistics infrastructure globally.

Geothermal Energy: Geothermal energy provides a stable baseload energy source that can complement intermittent renewables. In logistics contexts, it is particularly suited for warehouse heating and cooling, cold chain logistics, and industrial logistics parks. Hanbury & Vasquez (2018) demonstrate its potential through life cycle analysis, while Ball (2021) and Akhigbe (2025) discuss technological advances that enable broader integration of geothermal systems in logistics zones. By providing constant energy output, geothermal systems enhance energy security and reduce the need for extensive storage solutions.

In addition to solar, wind, bioenergy, hydrogen, and geothermal sources, hydropower represents another important renewable energy option with significant potential for supporting logistics operations. Hydropower is a mature and reliable energy source that provides a stable and continuous electricity supply, making it particularly valuable for energy-intensive logistics facilities such as distribution centers, intermodal terminals, and cold chain warehouses located near rivers or dams. Unlike intermittent sources such as solar and wind, hydropower offers baseload energy capabilities, which can help stabilize smart logistics operations that rely on automated systems, electric vehicle (EV) charging stations, and data centers.

In some regions, small-scale or run-of-river hydropower plants have been integrated into port facilities and inland waterway logistics hubs, supplying clean energy for cargo handling equipment, lighting systems, and administrative operations (Giera & Kulińska, 2021). Hydropower can also complement variable renewable sources within microgrid architectures, ensuring a balanced energy mix and enhancing system resilience during peak demand periods.

Furthermore, hydropower projects are increasingly evaluated through a sustainability lens, considering both environmental impacts and social acceptance. Recent studies highlight that understanding public perception and stakeholder engagement is essential for the long-term success of hydropower-based energy transitions (Cortés-Borda et al., 2022). While its applicability is geographically dependent, hydropower plays a valuable role in the broader renewable energy portfolio for smart logistics systems, particularly in areas with existing hydroelectric infrastructure.

Table 2 provides an overview of the main renewable energy categories relevant to smart logistics systems, summarizing their core characteristics and common application examples.

Table 2. Overview of Renewable Energy Categories Relevant to Smart Logistics Systems

Category	Description	Examples
Solar	Energy generated from sunlight	PV panels, solar thermal
Wind	Energy produced by wind turbines	Onshore/offshore turbines
Hydropower	Energy obtained from the movement of water	Dams, small hydro facilities
Bioenergy	Energy derived from organic matter (biomass, biofuels)	Biodiesel, bioethanol, biogas
Geothermal	Utilization of heat from the Earth's crust	Geothermal heat pumps, electricity
Hydrogen *(when produced renewably)*	Energy carrier produced through water electrolysis using renewable sources	Fuel cell systems, green hydrogen

3.3.1. Policy, Economic, and Case Study Perspectives

The successful integration of renewable energy into logistics systems depends not only on technological feasibility but also on supportive policy frameworks, economic incentives, and context-specific implementation strategies. For instance, the U.S. Inflation Reduction Act and Bipartisan Infrastructure Law have mobilized large-scale investment for renewable-powered transport and logistics infrastructure (Steinberg et al., 2023). The European Union has advanced integration through emissions trading, Green Deal targets, and port decarbonization strategies (Karaeva et al., 2023). In contrast, developing economies face regulatory uncertainty and infrastructural gaps, but they also hold potential for leapfrogging directly to decentralized renewable systems (Batinge et al., 2017; Khaleel & Chakrabarti, 2018).

Economic feasibility is another critical factor shaping adoption. Levelized cost analyses and life cycle assessments indicate that while initial capital costs for renewable integration can be high, operational savings, carbon pricing mechanisms, and long-term energy stability significantly improve return on investment (ROI) over time. Empirical studies from logistics parks, seaports, and EV fleet operations highlight reductions in total cost of ownership and operational emissions when renewables are strategically combined with smart technologies (Kashem et al., 2024; Guo, 2022).

3.3.2. Synthesis of the Literature

Table 3 summarizes key academic research on renewable energy integration in smart logistics operations from 2012 to 2025, illustrating the increasing variety of renewable sources, technologies, and contexts explored. Early studies focused mainly on solar and wind integration within port and warehouse systems (e.g., Nikitakos, 2012), while more recent research emphasizes hydrogen, microgrids, and AI-enabled energy optimization. This literature demonstrates that renewable energy can be applied at multiple levels of logistics operations — including transportation, storage, and real-time energy management — and that digital tools such as AI, blockchain, and IoT are critical enablers for overcoming the variability of renewable sources (Menaka et al., 2024; Zedak et al., 2023).

In conclusion, the integration of renewable energy sources into smart logistics systems offers significant potential to enhance environmental sustainability, operational efficiency, and energy security. However, achieving this integration at scale requires coordinated technological innovation, supportive policy environments, and economically viable business models. This intersection of digitalization and decarbonization forms the conceptual foundation for the model developed in the next section, which links renewable energy systems with smart logistics technologies and strategic decision-making frameworks.

Table 3. Summary of Academic Research on Renewable Energy Integration in Smart Logistics Operations (2012-2025)

Author/Year	Summary of Work
Nikitakos, 2012	The paper examines wind, solar, geothermal, tidal, and wave energy within a smart grid system designed to power ports with renewable sources, reducing emissions and improving efficiency.
Sadeghi & Haapala, 2019	The paper focuses on bio-energy technologies, particularly a biomass-to-bio-oil supply chain, to optimize logistics systems by addressing sustainability challenges and reducing costs and carbon footprints.
Hayajneh & Zhang, 2020	The paper discusses integrating wind and solar energy into an intelligent logistics system where electric semi-trucks move batteries between storage systems and charging stations, improving grid independence and reducing curtailment.
Guo, 2022	The paper presents an intelligent logistics system for the photovoltaic industry that improves logistics efficiency, customer satisfaction, and delivery quality through integrated e-commerce, order, warehouse, and transportation management systems.
Pan et al., 2023	The paper focuses on solar PV systems as the renewable energy solution in smart logistics, highlighting cloud-enabled blockchain integration to improve efficiency and sustainability.
Zedak et al., 2023	The paper proposes an intelligent energy management system that integrates wind turbines, PV systems, and BESS to optimize distribution network management and real-time operations.
Al-Mohannadi et al., 2024	The paper highlights electricity as a preferred renewable fuel in road logistics due to cheaper green power and efficient vehicle designs, despite challenges in charging infrastructure.
Chen et al., 2024	The paper examines integrating suburban renewable energy with electric logistics vehicles (ELVs) to lower indirect carbon emissions and improve ELV economic viability.
Huang et al., 2024	The paper examines renewable energy microgrids (REMs) in logistics parks, highlighting their role in powering cold warehouses, improving energy resilience, and optimizing inventory through a two-stage stochastic model.
Kashem et al., 2024	The paper emphasizes integrating renewable energy into EVs and flying cars within intelligent transport systems to cut fossil fuel use, lower emissions, and tackle urban congestion and pollution through sustainable logistics.
Menaka et al., 2024	The paper examines solar energy integration in IIoT systems, using AI for energy management, monitoring, and fault detection to improve transparency and address complex challenges in intelligent logistics.
Shi et al., 2024	The paper examines the use of PVs and WTs in smart logistics, particularly in green seaports, to boost efficiency and cut carbon emissions.
Yap & Notteboom, 2024	The paper identifies solar power as the sole viable renewable option for Singapore's seaport terminals and stresses smart energy management for optimization.
Bhattacharjee et al., 2025	The paper examines integrating solar and wind power within an IoT framework, focusing on autonomous clean-energy charging stations to improve intelligent logistics efficiency.
Li et at., 2025	The paper examines integrating wind turbines and photovoltaic systems into port logistics-energy systems, highlighting efficient resource dispatch for green port operations and hydrogen production.
Mia et al., 2025	The paper examines integrating multiple renewable energy sources with IoT to improve efficiency and sustainability in smart logistics systems.

4. A CONCEPTUAL MODEL OF SMART LOGISTICS SYSTEMS SUPPORTED BY RENEWABLE ENERGY

In light of the findings from the literature, the effective structuring of smart logistics systems in alignment with sustainability goals requires an integrated evaluation of technological components and renewable energy sources. Accordingly, the conceptual model developed in this study establishes a systematic framework that links the components of smart logistics systems with renewable energy sources, illustrating the process by which sustainability outcomes are generated.

The input layer of the model comprises renewable energy sources such as solar, wind, hydropower, bioenergy, hydrogen, and geothermal energy. These energy types are integrated with various technological infrastructures utilized in smart logistics systems, including sensors, RFID modules, cloud-based data management platforms, cyber-physical systems, automated carriers, and distribution centers. This integration not only enhances operational efficiency but also reduces environmental impacts.

In the process layer of the model, key functions such as data sensing, processing, sharing, and decision-making in logistics operations are performed. These processes are supported by cloud computing, Internet of Things (IoT) infrastructures, blockchain technologies, and distributed databases. The effectiveness of this process is contingent not only on technological capacity but also on the presence of supportive policy frameworks, public-private sector collaborations, and regulatory mechanisms.

As for the output, the model aims to deliver multidimensional sustainability outcomes, including reductions in carbon emissions, optimized energy consumption, mitigation of urban pollution, improved public health, and enhanced economic opportunities. The realization of these outcomes reinforces the model's effectiveness and applicability, demonstrating that smart logistics systems can contribute not only to technological advancements but also to broader environmental and societal objectives.

This conceptual model represents the integration of a diverse set of renewable energy sources, including solar, wind, hydropower, bioenergy, hydrogen, and geothermal, with the technological components of smart logistics systems. At the core of the model are smart logistics components such as sensing and tracking technologies, data processing units, communication infrastructures, physical transportation and storage systems, cyber-physical structures, and decision support platforms (Figure 2). When supported by robust policies, regulatory frameworks, and technological infrastructures, these components enable logistics operations to become more sustainable and resilient. The model's outcomes are defined as reduced carbon emissions, improved energy efficiency, mitigation of environmental impacts, and the generation of socio-economic benefits. In this context, the model offers a holistic logistics system perspective that aligns renewable energy integration with digital transformation to advance sustainable development goals.

Figure 2. Conceptual Model of Smart Logistics Systems Supported by Renewable Energy

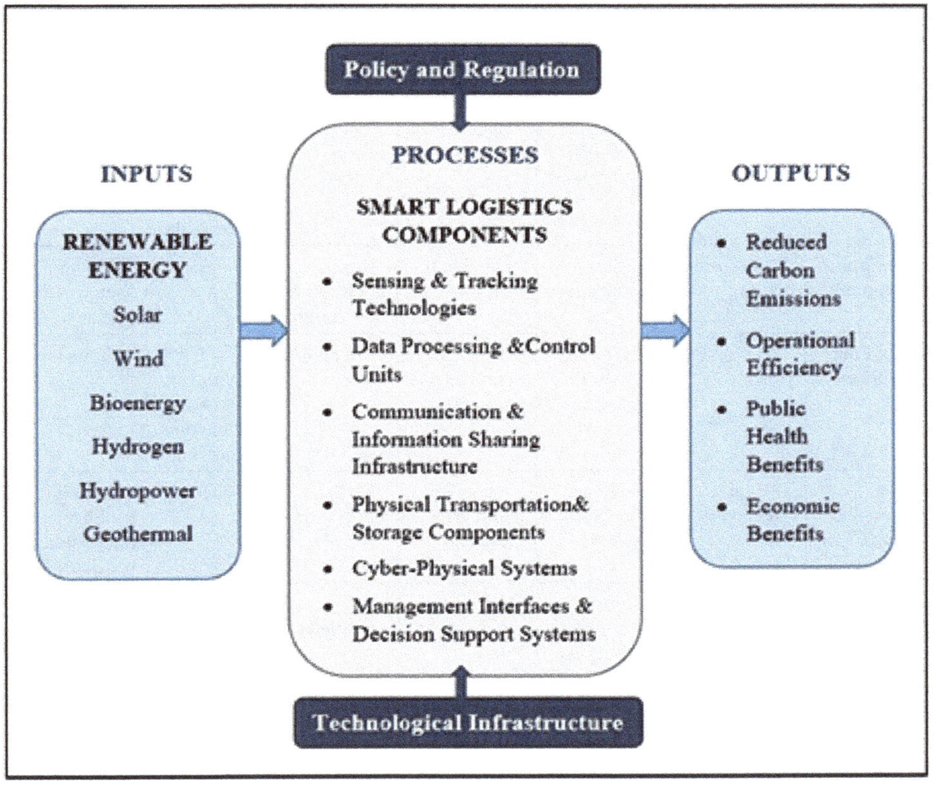

Table 4. Matrix of Smart Logistics Technologies and Their Associated Sustainable Energy Sources

Smart Logistics Technology	Solar Energy	Wind Energy	Hydropower	Bioenergy	Hydrogen	Geothermal
Electric Vehicles (EVs)	Direct (charging infrastructure)	Indirect (via grid)	Geographical limitations	Alternative fuel (biodiesel)	Fuel cell vehicles, green hydrogen charging	Rare in literature
Flying Vehicles / Drone Logistics	PV-assisted flight systems	Limited in mobile applications	Not suitable	Insufficient energy density	Emerging for small UAVs (fuel cells)	No integration
Cold Chain Warehouses	PV-assisted cooling	No application	Possible if near hydro grid	Not implemented in literature	Rare	Ground-source heat pumps
Smart Port Operations	Rooftop PV panels	Hybrid PV–WT systems	Ports near dams can use hydro power	Not applied	Hydrogen bunkering & fueling emerging	Limited
Smart Warehouse Systems	PV on warehouse roofs	Rare in warehouse systems	Where hydro grid available	No integration	Rare	Heating/ cooling integration possible
IoT-Based Tracking Systems	Low-power PV modules	Limited micro-turbine use	Not observed	Not observed	Not observed	Not observed
Smart Energy Management Systems	Integrated with energy management	Hybrid system integration	Grid integration possible	Not widely used	Integration with hydrogen storage systems	Integration with district heating possible
Photovoltaic Logistics Systems	Main energy source	WT integration possible	Rare	PV is preferred	Rare	Rare
Autonomous Charging Stations	Main power source	Wind-supported charging option	Possible if connected to hydro grids	Rare	Hydrogen-based fast charging	Rare
Green Urban Distribution	PV-supported urban systems	Not suitable for dense urban applications	Limited	Incompatible with urban systems	Hydrogen for trucks & buses emerging	Rare

The integration of renewable energy sources into smart logistics systems not only reduces their environmental impact but also enhances energy independence and operational resilience. However, the degree of integration varies depending on the nature of the technology used, its physical infrastructure, and operational requirements. Table 4 provides a qualitative matrix that captures the relationship between different smart logistics technologies and a diverse range of renewable energy sources, including solar, wind, hydropower, bioenergy, hydrogen, and geothermal.

Unlike a binary classification, the table employs more descriptive qualifiers such as "direct," "indirect," "not widespread," or "potential," thereby more accurately reflecting the diversity of applications and the complexity of energy–technology pairings observed in the literature.

For example, while electric vehicles (EVs) are most commonly powered through solar-based charging infrastructure, hydrogen-powered fuel cell vehicles are emerging as promising alternatives, particularly for heavy-duty logistics applications. Hydropower can play a complementary role in regions with established hydroelectric grids, especially for fixed infrastructure such as warehouses and charging stations. Similarly, geothermal energy is primarily applied for heating and cooling in cold chain warehouses and smart facilities.

This qualitative matrix serves as a decision-support framework, helping researchers and practitioners to identify appropriate energy–technology combinations, develop sustainability strategies, and align renewable energy policies with technological investments in logistics systems.

4.1. Hypothesis Proposals and Empirical Testing Potential

Within the scope of this study, the proposed conceptual model is based on the integration of a diverse set of renewable energy sources, including solar, wind, hydropower, bioenergy, hydrogen, and geothermal, with the technological components of smart logistics systems. In order to enable the empirical testing of this conceptual framework in future research, the following hypotheses have been developed. These hypotheses address the system's operational efficiency, sustainability impact, and decision-support capabilities, providing a foundation for future quantitative studies.

Proposed Hypotheses

1. Impact of Technological Components on Sustainability Performance
    - **H1:** The integration of IoT-based tracking systems significantly improves the traceability and environmental performance of smart logistics operations.

- **H2:** The use of cyber-physical systems (CPS) in smart logistics infrastructure enhances energy efficiency and reduces carbon emissions.

2. Impact of Renewable Energy Integration on Logistics Performance
   - **H3:** Logistics systems powered by renewable energy sources (solar, wind, hydropower, hydrogen, geothermal, and bioenergy) demonstrate lower operational costs compared to conventional fossil-fuel-based systems.
   - **H4:** The integration of multiple renewable energy sources (e.g., photovoltaic, wind, hydropower, and hydrogen) in transportation systems significantly reduces delivery-related greenhouse gas emissions.
3. Effect of Data-Driven Decision Support Systems
   - **H5:** The implementation of centralized control units and cloud-based analytics platforms improves logistics decision-making speed and accuracy.
   - **H6:** Logistics operations utilizing blockchain-supported smart contracts exhibit higher levels of transparency and stakeholder trust.
4. Inter-System Compatibility and Collaboration
   - **H7:** The interoperability of smart logistics systems with renewable energy grids increases system adaptability and supply chain resilience.
5. Overall Impact
   - **H8:** The combined deployment of advanced digital technologies and diverse renewable energy sources in logistics leads to a significant improvement in overall supply chain sustainability performance.

Table 5. Hypotheses Summary Based on Conceptual Model Components

Model Component	Hypothesis Code	Hypothesis Summary
IoT-Based Monitoring and Sensing	H1	IoT systems enhance logistics sustainability.
Cyber-Physical Systems (CPS)	H2	CPS improves energy efficiency and reduces carbon emissions.
Renewable Energy Integration	H3	Use of solar, wind, hydropower, hydrogen, geothermal, and bioenergy reduces operational costs.
Renewable Mix in Transportation	H4	Combined use of PV, wind, hydro, and hydrogen reduces logistics-related GHG emissions.
Cloud and Decision Support Systems	H5	Decision support systems improve speed and accuracy.
Blockchain and Smart Contracts	H6	Blockchain increases transparency and trust.

continued on following page

Table 5. Continued

Model Component	Hypothesis Code	Hypothesis Summary
Energy Data Sharing and Integration	H7	Smart systems operate in harmony with renewable energy grids.
Integrated Digital–Energy System	H8	Integration of digital technologies and renewable energy enhances overall sustainability.

Table 5 summarizes the set of hypotheses derived from the conceptual framework, linking the technological components of smart logistics systems with a diverse portfolio of renewable energy sources, including solar, wind, hydropower, bioenergy, hydrogen, and geothermal. Each hypothesis focuses on a specific dimension of smart logistics — from IoT-enabled monitoring (H1) and cyber-physical systems (H2) to renewable energy integration (H3–H4), data-driven decision support (H5), blockchain-based transparency (H6), energy grid interoperability (H7), and overall systemic impacts (H8). Collectively, these hypotheses provide a structured roadmap for future empirical research through simulation models, field applications, and large-scale surveys, enabling the validation of the proposed relationships between digital technologies, renewable energy integration, and sustainability performance in logistics systems.

5. SWOT ANALYSIS OF RENEWABLE ENERGY-BASED SMART LOGISTICS SYSTEMS

SWOT analysis is a strategic planning tool used to evaluate an organization's strengths, weaknesses, opportunities, and threats. It serves as a framework for assessing both internal and external factors that may affect an organization, project, or business activity. This method is widely recognized for its effectiveness in formulating strategies based on current conditions and is applicable across various domains, including management and marketing (Dalton, 2018; Paschalidou et al., 2018; Sharath-Kumar and Praveena, 2023).

The SWOT analysis presented in this study has been conducted using a literature-based qualitative analysis method. Contemporary academic sources addressing the integration of renewable energy into smart logistics systems were reviewed and analyzed through a deductive approach. The findings were conceptually categorized under four main dimensions: strengths, weaknesses, opportunities, and threats. As a non-numerical assessment, this qualitative analysis aims to offer a strategic evaluation through systematic content analysis. The SWOT framework has been constructed as a complementary tool to support the conceptual model, guide decision-making for practitioners, and identify future research directions.

The development of renewable energy-based smart logistics systems contributes not only to environmental sustainability but also to energy efficiency, cost advantages, and operational optimization. However, the integration of these systems also entails structural and technological challenges. In this context, the SWOT analysis presented in this section aims to systematically evaluate the strengths and weaknesses of these systems, along with the opportunities and threats they face, based on current literature findings and industry practices.

The strengths reflect the long-term sustainability potential of the system. Key drivers encouraging the development of such systems include the reduction of carbon emissions, decreased energy costs, and alignment with the United Nations Sustainable Development Goals.

The weaknesses primarily stem from technological and infrastructural deficiencies. High initial investment costs, uncertainties in energy continuity, and the limited applicability of certain renewable sources (e.g., wind and bioenergy) in mobile logistics systems are internal factors that may hinder broader adoption.

The opportunities are associated with increasing policy incentives, advancements in energy storage technologies, and emerging economic models such as carbon trading. Additionally, the development of smart microgrids in applications like urban distribution and port operations opens new pathways for innovation.

The threats include external risks such as a lack of regulatory frameworks, resistance from established fossil fuel infrastructures, shortages of specialized labor, and economic uncertainties. These factors can negatively affect the scalability and stability of renewable energy-based logistics systems.

This SWOT analysis (Table 6) not only assesses the current landscape but also provides a strategic framework for future research directions, policymaking, and sectoral investment decisions. The growing need for integration between technology and policy is particularly crucial for the sustainable scaling of these systems.

Table 6. SWOT Analysis of Renewable Energy-Based Smart Logistics Systems

Strengths	Weaknesses	Opportunities	Threats
Potential to reduce energy costs in the long term through integration of solar, wind, hydro, hydrogen, and geothermal sources	High initial investment and technology acquisition costs	Growing incentives and legal support for green logistics policies, including hydrogen and microgrid initiatives	Lack of sufficient regulation and standardization, especially for emerging technologies like hydrogen and geothermal
Significant capacity to reduce carbon emissions across multiple logistics activities	Intermittent energy supply issues (e.g., solar at night, variable wind) and geographical constraints for hydropower and geothermal use	Advancements in energy storage technologies (e.g., BESS, hydrogen storage), and integration with port and warehouse infrastructures	Dominance and resistance of existing fossil fuel infrastructure
High-level integration capability with smart systems and digital platforms	Limited integration of renewable energy sources (wind, bioenergy, hydrogen) with mobile logistics systems	Emergence of new business models (energy service providers, carbon trading), hydrogen-powered mobility solutions, and green port operations	Shortage of skilled labor for renewable energy and digital technology integration
Strong alignment with sustainable development goals and climate action targets	Widespread infrastructure deficiencies (e.g., charging/fueling stations, storage systems)	Smart microgrid solutions in urban logistics applications, leveraging hydropower or geothermal for stable baseload supply	Political and economic uncertainties (e.g., energy price fluctuations, hydrogen market instability, policy inconsistency)

This SWOT analysis summarizes the key internal and external factors influencing the integration of diverse renewable energy sources into smart logistics systems. Strengths and weaknesses are derived from current technological capabilities, energy characteristics, and infrastructure conditions, while opportunities and threats reflect broader policy trends, regulatory developments, and market dynamics. By including solar, wind, hydropower, bioenergy, hydrogen, and geothermal energy, this analysis highlights both the transformative potential and the systemic challenges of building renewable energy-based logistics networks. It provides a strategic framework for policymakers, researchers, and industry stakeholders to align technological investments with regulatory and market developments.

6. FEASIBILITY ANALYSIS AND PRACTICAL IMPLEMENTATION CONSIDERATIONS

The practical feasibility of renewable energy-based smart logistics systems depends on the coordinated alignment of technological readiness, economic viability, regulatory frameworks, and organizational capabilities. While the conceptual integration of renewable energy sources such as solar, wind, hydropower, bioenergy, hydrogen, and geothermal into logistics networks offers clear sustainability and efficiency benefits, their real-world deployment is shaped by a set of interrelated conditions. This section evaluates the technical, economic, regulatory, and operational dimensions that influence the implementation potential of the proposed model.

6.1. Technical Feasibility

From a technological standpoint, the integration of renewable energy into smart logistics systems requires both mature energy technologies and compatible digital infrastructures. Solar photovoltaic (PV) systems and wind turbines are already widely deployed for powering warehouses, charging stations, and port operations. Hydropower offers reliable baseload energy in regions with suitable geography, making it valuable for logistics hubs near dams or rivers. Hydrogen fuel cells are increasingly seen as promising for heavy-duty vehicles and port equipment, though refueling infrastructure remains limited. Geothermal energy is technically feasible for heating and cooling applications in cold chain warehouses and logistics parks, but its applicability is geographically constrained.

On the digital side, IoT networks, cyber-physical systems (CPS), blockchain, and cloud platforms must be interoperable with these energy sources to optimize flows, monitor usage, and support real-time decision-making. Although these technologies are individually mature, their system-level integration remains at a pilot or early commercial stage, particularly for hydrogen and hybrid microgrid applications.

6.2. Economic Feasibility

Economic feasibility is determined by the balance between high upfront investment costs and long-term operational savings. Solar and wind energy systems have witnessed decreasing costs, making them economically attractive for warehouse roofs, charging stations, and port terminals. Hydropower and geothermal installations depend strongly on location-specific infrastructure but can offer stable low-cost energy once deployed. Hydrogen technologies, especially for mobility, currently face higher capital and operating costs, although this gap is narrowing with technological advancements and policy incentives.

Energy storage systems—such as battery energy storage (BESS) and hydrogen storage—play a crucial role in addressing intermittency and ensuring cost-effective energy management. Access to green finance, carbon pricing mechanisms, and renewable energy incentives (e.g., tax credits, feed-in tariffs, or emission trading schemes) can significantly improve project economics. Overall, while cost competitiveness is improving, the financial feasibility of multi-source renewable systems in logistics requires tailored investment models, including public–private partnerships and phased implementation strategies.

6.3. Regulatory and Policy Feasibility

The regulatory environment is a key determinant of implementation success. Policies promoting renewable energy integration into logistics vary significantly by region. Solar and wind installations are typically supported by well-established permitting frameworks, but hydrogen refueling infrastructure, geothermal energy use, and hybrid microgrids often face fragmented or unclear regulations. Standardization gaps, lengthy permitting processes, and the lack of harmonized technical guidelines can create barriers to deployment.

Recent policy developments—such as hydrogen roadmaps, carbon neutrality targets, and incentives for green ports—are gradually improving the landscape. However, the alignment between logistics sector regulations and renewable energy policies remains incomplete, especially for cross-border operations. Coordinated policy design, standardized interconnection rules, and stable incentive structures are essential for scaling up these systems.

6.4. Organizational and Operational Feasibility

Organizational readiness is another critical factor. Implementing renewable energy-based smart logistics systems requires companies to adopt new operational models, invest in workforce training, and establish cross-sectoral collaborations. Many logistics companies have initiated solar or EV pilot projects, but integration of hydrogen, geothermal, or hybrid microgrid solutions is still limited to experimental or flagship projects.

Operational challenges include managing the interoperability of digital and energy systems, ensuring real-time data sharing between supply chain actors, and maintaining energy reliability under variable conditions. Addressing these challenges involves developing multi-stakeholder partnerships that include energy providers, technology developers, regulators, and logistics operators. Human capital development—particularly in fields such as energy systems engineering, IoT analytics, and hydrogen safety—is essential for smooth implementation.

6.5. Overall Feasibility Perspective

Overall, the feasibility of deploying renewable energy-based smart logistics systems depends on the synchronization of technological maturity, cost structures, regulatory clarity, and organizational capability. While solar and wind applications are relatively advanced, hydropower, hydrogen, and geothermal technologies require more tailored strategies for integration. Economic and regulatory enablers, combined with targeted pilot projects and capacity-building initiatives, can accelerate the transition from conceptual frameworks to operational logistics networks. A comprehensive feasibility approach is therefore essential for transforming renewable energy integration from isolated projects into scalable, sustainable, and resilient logistics infrastructures.

7. CONCLUSION AND RECOMMENDATIONS

The findings of this study reveal that the integration of renewable energy sources constitutes a central pillar in the reconfiguration of smart logistics systems from a sustainability perspective. As global supply chains become increasingly complex and energy-intensive, the strategic incorporation of sustainable energy sources into technological logistics infrastructures emerges as not merely beneficial, but essential for long-term viability.

The study demonstrated that smart logistics systems—characterized by their use of cyber-physical structures, IoT-based sensing and tracking units, blockchain-supported coordination mechanisms, and cloud-based decision support systems—can be significantly enhanced in both operational efficiency and environmental impact when powered by renewable energy. Technologies such as photovoltaic (PV) panels, wind turbines, bioenergy conversion units, and hybrid storage systems (e.g., battery energy storage systems, BESS) were identified as key enablers of this transformation. The integration of these technologies into logistics functions such as warehousing, transportation, cold-chain management, and port operations not only helps reduce carbon emissions but also aligns logistics operations with international environmental standards and regulatory frameworks.

However, the study also emphasizes that the transition toward renewable-energy-powered smart logistics systems is fraught with structural and systemic challenges. Chief among these are the high initial capital investment requirements, regulatory ambiguities, insufficient standardization, and limited availability of skilled human capital capable of managing advanced energy-infrastructure interfaces. Moreover, the intermittency of solar and wind energy, combined with the underdevelopment of grid integration strategies in certain regions, presents additional barriers to large-scale adoption.

To overcome these constraints, coordinated action between public and private stakeholders is imperative. Strategic policy support should include the development of adaptive regulatory frameworks, regionally sensitive energy infrastructure planning, and the establishment of financial incentive mechanisms such as green subsidies, tax credits, and carbon pricing models. Additionally, standardization of technical protocols across energy and logistics sectors would facilitate interoperability and broader diffusion of integrated systems.

7.1. Policy and Managerial Implications

The transition to renewable energy–based smart logistics requires technological progress supported by effective policies and managerial strategies. Policymakers should create adaptive regulatory frameworks that foster innovation and interoperability between logistics and energy sectors, complemented by green subsidies, tax incentives, and carbon pricing to boost private adoption. Managers should invest in hybrid energy storage, train staff for energy–digital integration, and build strategic partnerships with energy providers. Urban logistics and port operators can adopt microgrids and local renewables to improve resilience and cut costs. Aligning national policies with firm-level strategies will accelerate the sustainable transformation of logistics networks.

Future research should focus on AI-based energy forecasting, adaptive microgrids, and decentralized control systems that enable real-time decision-making in dynamic logistics environments. Empirical validation of the conceptual model through case studies, simulations, or pilot projects is essential to assess its practical implications. Evaluating sustainability performance in urban freight, port logistics, and rural logistics parks will offer valuable, context-specific insights for both practitioners and policymakers.

In summary, the integration of renewable energy with smart logistics systems offers a multidimensional opportunity space—spanning environmental, economic, and technological domains. For this potential to be fully realized, a systemic and interdisciplinary approach is required, underpinned by supportive policy environments, robust infrastructure, and innovative digital capabilities. The model presented in this study thus serves as a foundational framework to guide future implementations and inspire further scholarly inquiry into sustainable logistics transitions.

REFERENCES

Agbakwuru, V., Obidi, P. O., Salihu, O. S., & MaryJane, O. C. (2024). The role of renewable energy in achieving sustainable development goals.

Akhigbe, E. E. (2025). *Advancing geothermal energy: A review of technological developments and environmental impacts*. Gulf Journal of Advance Business Research.

Al-Mohannadi, A. A., Ertogral, K., & Erkoc, M. (2024). Alternative Fuels in Sustainable Logistics—Applications, Challenges, and Solutions. *Sustainability (Basel), 16*(19), 8484. DOI: 10.3390/su16198484

Al-Ormuza, T. K., Sharif, K. I. M., Udin, Z. M., Hassan, M. G., Kafi, M. A., Mohammed, F., & Abu, N. H. (2024, August). The Role of Smart Logistics Toward Sustainable Logistics: A Review. In *2024 1st International Conference on Logistics (ICL)* (pp. 1-5). IEEE.

Alavi-Borazjani, S. A., Adeel, S., & Chkoniya, V. (2025). Hydrogen as a Sustainable Fuel: Transforming Maritime Logistics. *Energies, 18*(5), 1231. DOI: 10.3390/en18051231

Almazeedi, W. E. (2024). HYDROGEN INVESTMENT. The Clean Hydrogen Economy and Saudi Arabia: Domestic Developments and International Opportunities.

Alonge, E. O., Eyo-Udo, N. L., Ubanadu, B. C., Daraojimba, A. I., Balogun, E. D., & Ogunsola, K. O. (2023). Real-time data analytics for enhancing supply chain efficiency. *Journal of Supply Chain Management and Analytics, 10*(1), 49–60.

Asif, M., Salman, M. U., Anwar, S., Gul, M., & Aslam, R. (2022). Renewable and non-renewable energy resources of Pakistan and their applicability under the current scenario in Pakistan. *OPEC Energy Review, 46*(3), 310–339. DOI: 10.1111/opec.12230

Ball, P. J. (2021). A review of geothermal technologies and their role in reducing greenhouse gas emissions in the USA. *Journal of Energy Resources Technology, 143*(1), 010903. DOI: 10.1115/1.4048187

Barbieri, E., & Capoani, L. (2025). Renewable Energy, Resilience, Digitalization, and Industrial Policies in Seaborne Transport. *Energies, 18*(5), 1089. DOI: 10.3390/en18051089

Batinge, B., Musango, J. K., & Brent, A. C. (2017). Leapfrogging to renewable energy: The opportunity for unmet electricity markets. *South African Journal of Industrial Engineering, 28*(4), 32–49. DOI: 10.7166/28-4-1702

Bhattacharjee, S., Riame, L., & Kalita, G. (2025). Utilization of Artificial Intelligence and the Internet of Things in the Renewable Industry. In *Artificial Intelligence and IoT for Cyber Security Solutions in Smart Cities* (pp. 1–15). Chapman and Hall/CRC.

Boldyreiva, L., & Gryshko, V. (2020). European experience of smart-technology implementation in logistic process management. *Економіка та держава*, (4), 39-43.

Chen, N., Wang, S., & He, Y. (2025). Integrating Hydrogen Fuel Cell Trucks into Freight Transport: Scenarios for Costs, Emissions, and Policy in China. *Energy*, •••, 138609. DOI: 10.1016/j.energy.2025.138609

Chen, S. L., Chen, Y. Y., & Hsu, C. (2014). A new approach to integrate internet-of-things and software-as-a-service model for logistic systems: A case study. *Sensors (Basel)*, *14*(4), 6144–6164. DOI: 10.3390/s140406144

Chen, Y., Hu, S., Zheng, Y., Xie, S., Yang, Q., Wang, Y., & Hu, Q. (2024). Co-ordinated optimization of logistics scheduling and electricity dispatch for electric logistics vehicles considering uncertain electricity prices and renewable generation. *Applied Energy*, *364*, 123147. DOI: 10.1016/j.apenergy.2024.123147

Cho, G. S. (2018). A study on establishment of smart logistics center based on logistics 4.0. *Journal of Multimedia Information System*, *5*(4), 265–272.

Chu, S., & Majumdar, A. (2012). Opportunities and challenges for a sustainable energy future. nature, 488(7411), 294-303.

Cortés-Borda, D., Polanco, J. A., & Escobar-Sierra, M. (2022). Social perception assessment of hydropower sustainability: A stepwise logistic regression modeling. *Environmental Science & Policy*, *134*, 108–118. DOI: 10.1016/j.envsci.2022.03.026

Dalton, J. (2018). SWOT analysis (Strengths, weaknesses, opportunities, threats). In *Great Big Agile: An OS for Agile Leaders* (pp. 249–252). Apress.

Damianov, D., & Demirova, S. (2018, June). Principles of designing automated logistics systems-hybrid component of cyber-physical systems. In *2018 International Conference on High Technology for Sustainable Development (HiTech)* (pp. 1-4). IEEE. DOI: 10.1109/HiTech.2018.8566533

Deng, H., Wang, R., Miao, K., Zhao, Y., Sun, J., & Du, J. (2018, December). Design and implementation of ship gps monitoring and dispatching system for intelligent transportation system. In *Proceedings of the International Conference on Information Technology and Electrical Engineering 2018* (pp. 1-5). DOI: 10.1145/3148453.3306261

Denoncourt, J. (2020). Companies and UN 2030 sustainable development goal 9 industry, innovation and infrastructure. Journal of Corporate law studies, 20(1), 199-235.

Ding, Y., Jin, M., Li, S., & Feng, D. (2021). Smart logistics based on the internet of things technology: An overview. *International Journal of Logistics*, *24*(4), 323–345. DOI: 10.1080/13675567.2020.1757053

Elavarasan, R. M., Pugazhendhi, R., Jamal, T., Dyduch, J., Arif, M. T., Kumar, N. M., & Nadarajah, M. (2021). Envisioning the UN Sustainable Development Goals (SDGs) through the lens of energy sustainability (SDG 7) in the post-COVID-19 world. *Applied Energy*, *292*, 116665. DOI: 10.1016/j.apenergy.2021.116665

Eyitoyo, A. K. (2023). A Comparative Study of Alternative Energy Usage in the Global Logistics Sector (Master's thesis, Texas A&M University-Kingsville).

Gebresenbet, G., Bosona, T., Olsson, S. O., & Garcia, D. (2018). Smart system for the optimization of logistics performance of the pruning biomass value chain. *Applied Sciences (Basel, Switzerland)*, *8*(7), 1162. DOI: 10.3390/app8071162

Giera, J., & Kulińska, E. (2021). Risk factors in a logistics company using renewable energy sources. *Energies*, *14*(23), 8152. DOI: 10.3390/en14238152

Gowda, V. D., Surya, S. G., Kumar, N. M. G., Prasad, K., Prasad, V. S., & Kaur, M. (2024, May). Optimizing Renewable Energy Integration in Smart Grids through IoT-Driven Management Systems. In *2024 2nd International Conference on Advancement in Computation & Computer Technologies (InCACCT)* (pp. 783-788). IEEE.

Gumzej, R. (2023). Intelligent logistics systems in E-commerce and transportation. *Mathematical Biosciences and Engineering*, *20*(2), 2348–2363. DOI: 10.3934/mbe.2023110

Guo, J. (2022, November). Design and implementation of intelligent logistics system for photovoltaic enterprises. In *International Conference on Frontiers of Traffic and Transportation Engineering (FTTE 2022)* (Vol. 12340, pp. 266-271). SPIE. DOI: 10.1117/12.2652315

Gutu Ofgaa, S. U. (2020). The Integrated Exchange of Information and IoT for Smart Logistics Systems.

Hanbury, O., & Vasquez, V. R. (2018). Life cycle analysis of geothermal energy for power and transportation: A stochastic approach. *Renewable Energy*, *115*, 371–381. DOI: 10.1016/j.renene.2017.08.053

Hayajneh, H. S., & Zhang, X. (2020). Logistics design for mobile battery energy storage systems. *Energies, 13*(5), 1157. DOI: 10.3390/en13051157

Hou, J., & Chen, C. (2022, April). Intelligent logistics supply chain management based on Internet of Things technology. In *2022 IEEE Asia-Pacific Conference on Image Processing, Electronics and Computers (IPEC)* (pp. 1266-1270). IEEE. DOI: 10.1109/IPEC54454.2022.9777588

Huang, J., Xie, D., Qiu, Y., Wang, J., & Song, J. (2024). Green supply chain management: A renewable energy planning and dynamic inventory operations for perishable products. *International Journal of Production Research, 62*(24), 8924–8951. DOI: 10.1080/00207543.2023.2220047

Issaoui, Y., Khiat, A., Bahnasse, A., & Ouajji, H. (2019). Smart logistics: Study of the application of blockchain technology. *Procedia Computer Science, 160*, 266–271. DOI: 10.1016/j.procs.2019.09.467

Karaeva, A., Magaril, E., & Al-Kayiem, H. H. (2023). *Review and comparative analysis of renewable energy policies in the European Union.*

Kashem, M. A., Shamsuddoha, M., & Nasir, T. (2024). Sustainable Transportation Solutions for Intelligent Mobility: A Focus on Renewable Energy and Technological Advancements for Electric Vehicles (EVs) and Flying Cars. *Future Transportation, 4*(3), 874–890. DOI: 10.3390/futuretransp4030042

Kezhu, W. (2024, November). A Novel Smart Logistics Platform based on Decentralized Sensor Tags and Cloud Scheduling. In *2024 5th International Conference on Data Intelligence and Cognitive Informatics (ICDICI)* (pp. 1014-1019). IEEE. DOI: 10.1109/ICDICI62993.2024.10810915

Khaleel, A. G., & Chakrabarti, M. (2018). Leap-Frogging Challenges and Possibilities for Renewable Energy Transition in Developing Countries.

Khan, S. A. R., Godil, D. I., Yu, Z., Abbas, F., & Shamim, M. A. (2022). Adoption of renewable energy sources, low-carbon initiatives, and advanced logistical infrastructure—An step toward integrated global progress. *Sustainable Development (Bradford), 30*(1), 275–288. DOI: 10.1002/sd.2243

Li, L., Shi, Q., Wang, Z., Zhang, S., Jia, D., & Liu, W. (2025). Multi-time-scale scheduling of integrated port logistics-energy system considering piecewise linearization of high-order equations. *IET Generation, Transmission & Distribution, 19*(1), e70007. DOI: 10.1049/gtd2.70007

Liu, Y., Tao, X., Li, X., Colombo, A. W., & Hu, S. (2023). Artificial intelligence in smart logistics cyber-physical systems: State-of-the-arts and potential applications. *IEEE Transactions on Industrial Cyber-Physical Systems, 1*, 1–20. DOI: 10.1109/TICPS.2023.3283230

Master, I., Berliner, S., Shefy, N., Feldman, A., Bialik, B., Zivan, S., & Goldrat, M. (2011). *U.S. Patent Application No. 13/000,827.*

Menaka, C., Awasthi, A., Yadav, D. C., & Jain, S. K. (2024). Designing a Renewable Energy System for Industrial IoT with Artificial Intelligence. In *E3S Web of Conferences* (Vol. 540, p. 13008). EDP Sciences. DOI: 10.1051/e3sconf/202454013008

Mexis, I., & Todeschini, G. (2020). Battery energy storage systems in the United Kingdom: A review of current state-of-the-art and future applications. *Energies, 13*(14), 3616. DOI: 10.3390/en13143616

Mia, S., Ahmed, F., Khan, I., Kabir, M. I., Roni, M. H., Cobra, K., ... & Mahmud, S. (2025). Integratıng Renewable Energy Wıth Internet of Thıngs (IoT): Pathways to A Smart Green Planet. *Kufa Journal of Engineering, 16*(1).

Mire, R., Depraz, S., Collacott, B., & Collins, A. (2018, April). Sustainable Development Goals Atlas. In SPE International Conference and Exhibition on Health, Safety, Environment, and Sustainability? (p. D021S014R001). SPE. DOI: 10.2118/190550-MS

Muhammed, G., & Tekbiyik-Ersoy, N. (2020). Development of renewable energy in China, USA, and Brazil: A comparative study on renewable energy policies. *Sustainability (Basel), 12*(21), 9136. DOI: 10.3390/su12219136

Mumin, M. A., Yakubu, I. N., & Adam, I. O. (2025). Integrating Renewable Energy With Internet of Things (IoT) for Smart Logistics. In Developing Dynamic and Sustainable Supply Chains to Achieve Sustainable Development Goals (pp. 155-178). IGI Global Scientific Publishing.

Musau, E. G. (2024). Optimizing transportation and distribution for environmental sustainability: Green logistics. In *Green Supply Chain Management Practice and Principles* (pp. 84–101). IGI Global. DOI: 10.4018/979-8-3693-3486-7.ch005

Nikitakos, N. (2012). Emissions Port. *FME Transactions, 40*, 201–206.

Nzeanorue, C. C., & Okpala, B. C. (2024). Smart Grids and Renewable Energy Integration: Challenges and Solutions. *Traektoriâ Nauki, 10*(9), 3050–3060. DOI: 10.22178/pos.109-30

Oshilalu, A. Z., & Iboh, S. (2025). Integration of renewable energy technologies into transportation infrastructure to minimize carbon footprint and enhance long-term environmental resilience. *Global Journal of Engineering and Technology Advances, 24*(03), 236–251. DOI: 10.30574/gjeta.2025.24.3.0281

Pablo-Romero, M. D. P., Román, R., Sánchez-Braza, A., & Yñiguez, R. (2016). Renewable energy, emissions, and health. *Renewable Energy: Utilisation and System Integration, 173*.

Pan, H., Bayanati, M., Vaseei, M., & Chobar, A. P. (2023). Empowering solar photovoltaic logistic operations through cloud-enabled blockchain technology: A sustainable approach. *Frontiers in Energy Research, 11*, 1293449. DOI: 10.3389/fenrg.2023.1293449

Paschalidou, A., Tsatiris, M., Kitikidou, K., Papadopoulou, C., Paschalidou, A., Tsatiris, M., ... & Papadopoulou, C. (2018). Methods (SWOT Analysis). Using Energy Crops for Biofuels or Food: The Choice, 39-44.

Pu, Y., Liu, H., Wang, J., & Hou, Y. (2023). Collaborative scheduling of port integrated energy and container logistics considering electric and hydrogen-powered transport. *IEEE Transactions on Smart Grid, 14*(6), 4345–4359. DOI: 10.1109/TSG.2023.3266601

Ramasubramanian, B., & Ramakrishna, S. (2023). What's next for the Sustainable Development Goals? Synergy and trade-offs in affordable and clean energy (SDG 7). Sustainable earth reviews, 6(1), 17.

Sadeghi, J., & Haapala, K. R. (2019). Optimizing a sustainable logistics problem in a renewable energy network using a genetic algorithm. *Opsearch, 56*(1), 73–90. DOI: 10.1007/s12597-019-00356-5

Sayeed, M. A., & Manikandan, K. (2024). Smart integration of renewable energy into transportation: Challenges, innovations, and future research directions. [Revue des Energies Renouvelables]. *Journal of Renewable Energies, 27*(2), 319–337. DOI: 10.54966/jreen.v27i2.1300

Sharath Kumar, C. R., & Praveena, K. B. (2023). SWOT analysis. *International Journal of Advanced Research, 11*, 744–748. DOI: 10.21474/IJAR01/17584

Shi, Z., Fan, F., Tai, N., Qing, C., Meng, Y., & Guo, R. (2024). Coordinated Operation of the Multiple Types of Energy Storage Systems in the Green-seaport Energy-logistics Integrated System. *IEEE Transactions on Industry Applications, 60*(3), 4482–4493. DOI: 10.1109/TIA.2024.3359585

Shuaibu, A. S., Mahmoud, A. S., & Sheltami, T. R. (2025). A review of last-mile delivery optimization: Strategies, technologies, drone integration, and future trends. *Drones (Basel), 9*(3), 158. DOI: 10.3390/drones9030158

Steinberg, D. C., Brown, M., Wiser, R., Donohoo-Vallett, P., Gagnon, P., Hamilton, A., & Prasanna, A. (2023). *Evaluating impacts of the inflation reduction act and bipartisan infrastructure law on the US power system (No. NREL/TP-6A20-85242). National Renewable Energy Laboratory*. NREL.

Su, J. P., Wang, C. A., Mo, Y. C., Zeng, Y. X., Chang, W. J., Chen, L. B., & Chuang, C. H. (2017, May). i-Logistics: An intelligent Logistics system based on Internet of things. In *2017 International Conference on Applied System Innovation (ICASI)* (pp. 331-334). IEEE. DOI: 10.1109/ICASI.2017.7988419

Sun, X., Yu, H., & Solvang, W. D. (2022). Towards the smart and sustainable transformation of Reverse Logistics 4.0: A conceptualization and research agenda. *Environmental Science and Pollution Research International, 29*(46), 69275–69293. DOI: 10.1007/s11356-022-22473-3

Takman, J., & Andersson-Sköld, Y. (2021). A framework for barriers, opportunities, and potential solutions for renewable energy diffusion: Exemplified by liquefied biogas for heavy trucks. *Transport Policy, 110*, 150–160. DOI: 10.1016/j.tranpol.2021.05.021

Uckelmann, D. (2008, September). A definition approach to smart logistics. In *International Conference on Next Generation Wired/Wireless Networking* (pp. 273-284). Berlin, Heidelberg: Springer Berlin Heidelberg.

Ugochukwu, N. A., Goyal, S. B., & Arumugam, S. (2022). Blockchain-Based IoT-Enabled System for Secure and Efficient Logistics Management in the Era of IR 4.0. *Journal of Nanomaterials, 2022*(1), 7295395. DOI: 10.1155/2022/7295395

Wen, D., & Gao, W. (2023). Impact of renewable energy policies on solar photovoltaic energy: comparison of China, Germany, Japan, and the United States of America. In *Distributed energy resources: solutions for a low carbon society* (pp. 43–68). Springer International Publishing. DOI: 10.1007/978-3-031-21097-6_3

Yap, W. Y., & Notteboom, T. (2024). Renewable energy options for seaport cargo terminals with application to mega port Singapore. *Maritime Business Review, 9*(4), 294–310. DOI: 10.1108/MABR-12-2023-0083

Zedak, C., Belfqih, A., Boukherouaa, J., & El Mariami, F. (2023). An intelligent energy management system for optimum design and real-time operation. [IJPEDS]. *International Journal of Power Electronics and Drive Systems*, *14*(1), 480–495. DOI: 10.11591/ijpeds.v14.i1.pp480-495

Zhang, Y. (2022, December). Smart logistics warehouse management system based on RFID and Internet of Things technology. In *2022 2nd International Conference on Networking, Communications and Information Technology (NetCIT)* (pp. 489-492). IEEE. DOI: 10.1109/NetCIT57419.2022.00121

Zsifkovits, H., & Woschank, M. (2019). Smart Logistics—Technology Concepts and Potentials. *BHM. Berg- und Hüttenmännische Monatshefte*, *164*(1), 42–45. DOI: 10.1007/s00501-018-0806-9

Chapter 10
Smart Carbon Emission Monitoring and Forecasting for Sustainable Logistics

S. D. Lalitha
https://orcid.org/0000-0003-3712-8254
R.M.K. Engineering College, India

C. Shanmugapriya
https://orcid.org/0009-0000-5514-021X
R.M.K. Engineering College, India

M. S. Sujitha
https://orcid.org/0009-0005-1856-4394
R.M.K. Engineering College, India

K. Sowmiya
R.M.K. Engineering College, India

ABSTRACT

Sustainable logistics demands efficient and scalable systems for monitoring and forecasting carbon emissions. This work presents a Smart Carbon Emission Monitoring and Forecasting System (SCEMFS) for real-time detection of gases like CO_2, CH_4, NH_3, and VOCs using MQ-series sensors (MQ-2, MQ-7, MQ-9, MQ-135) integrated with a microcontroller. The collected data is used to train machine learning models—decision trees, random forests, SVM, and neural networks—for accurate prediction and classification of gas concentrations. Real-time analytics

DOI: 10.4018/979-8-3373-3790-6.ch010

enables early intervention and improved environmental compliance. The system is scalable and suitable for various logistics settings, contributing to reduced carbon footprints and promoting sustainability in supply chain operations.

1. INTRODUCTION

Rapid industrialisation has increased carbon emissions while also fostering economic expansion and urbanisation. Achieving sustainability goals in urban settings requires a balance between economic development and carbon reduction measures (Xu et. al, 2023). Many industries have flourished because of industrialisation, which has also boosted global incomes and population levels (Saba et. al, 2023). The global economy over the last two hundred years has been driven by the extraction of natural resources and the manipulation of biogeochemical cycles (Rees et. al, 1989). The world's population is projected to reach 9.9 billion in 2050, up from 7.8 billion in 2020, driven by growing living standards. As a result, there will be a 70% increase in food consumption and an 80% increase in energy use (Wang, F. et. al, 2021). In addition, when planning the city's future, it's essential to build environmentally friendly facilities and attend to important human development factors like the number of bike lanes, the accessibility of educational institutions, and green infrastructures (Naumann, S et. al, 2021). As a result of these environmentally conscious renovations, city planners are thinking about how to gauge how happy the public is with the Net-Zero Carbon City (NZCC) transition (Szczepańska, A et. al, 2023).

Furthermore, NZCC initiatives are vital for accomplishing the UN's Sustainable Development Goals (SDGs), particularly SDG 11, which aims to make cities safer, more resilient, and more sustainable so that surrounding communities can live more conveniently (Lützkendorf, T et. al, 2019). The future of urban areas depends on achieving near-zero carbon emissions, as these pollutants contribute to climate change and global warming (Hansen, et. al, 2013). On the other hand, the world has already been warned about the negative effects of early industrialisation, with evidence of rising carbon dioxide emissions leading to more frequent and severe weather events (Sarkodie, S.A. et. al, 2020). Reducing emissions of greenhouse gases (GHGs), especially CO2, requires immediate action. If we want to study the spatiotemporal patterns in CO2 and find out how effective emission reduction strategies are, we need to monitor carbon emissions often using trustworthy datasets (An et. al, 2022). In the context of Net-Zero Carbon Cities (NZCCs), Biber and Krogstie (2021) highlighted the significance of developing frameworks for data-driven technologies (Bibri, S.E et. al, 2021). To track the present state of cities and foretell their future developments, data-driven approaches enable digital visualisation tools (Seto K.C. et.al, 2021) Modern cities are becoming smarter and more livable as a result of this development. For policy/decision-makers to be able to monitor and change the city's situation, they must be able to use the right digitalisation technologies, know the city's strengths and weaknesses, conduct thorough data analyses, and then choose the best decision to change the city's state (Stevens, D. et. al, 2007).

A growing number of businesses in the industrial sector are considering using carbon monitoring and mitigation strategies based on the city digital twin approach in order to evade the additional carbon tax (Zhu et. al, 2020). It has been demonstrated that creating NZCCs as the principal goal requires long-term sustainability, and the implementation strategies of digital governance in smart cities have already demonstrated the ability to decrease regional carbon emissions. Building a digital platform for a city allows for more analysis and computations to be used for monitoring and modifying city growth, including flood simulation and the possibility of employing solar energy (Thornbush et. al, 2019). As a bonus, AI, city digital twins (CDTs), Building Information Modelling (BIM), and Geographic Information Systems (GIS) can all help with future NZCC planning by making decisions more accurately.

A state-of-the-art system, the Carbon Footprint Tracking System blends several gas sensors with sophisticated machine learning algorithms to offer real-time analysis, forecast, and monitoring of greenhouse gas emissions. This system allows users to acquire important insights into emissions levels and trends through the exact and continuous tracking of numerous gases. It is designed for industrial, urban, and environmental applications. In order to aid in the prediction of possible dangers and the understanding of long-term environmental impacts, the system makes use of machine learning models such as support vector machines, decision trees, random forests, and neural networks to detect complex patterns, evaluate the importance of features, and spot anomalies in gas concentration data. The system is a great asset for sustainability and proactive environmental management, and it helps with regulatory compliance thanks to its real-time feedback loop that lets you respond instantly to changing emission levels. The adaptability of the system's architecture makes it suitable for a wide range of uses, keeping up with the changing demands of cities and businesses who are trying to lessen their impact on the environment.

2. LITERATURE REVIEW

Li et. al (2019) discussed the urgent and pervasive issue of air pollution and emphasize that while it is a global concern, effective solutions must be tailored to local contexts. They highlight that the sources, severity, and health impacts of pollution vary significantly across regions, necessitating distinct approaches to regulation, technology, and community engagement. The authors advocate for policies that reflect local pollution sources and health burdens, along with innovative, place-specific strategies to reduce emissions and promote sustainable urban growth. The article calls for cooperative action and evidence-based interventions to address air quality on both a global and local scale.

Han et. al (2016) proposed a novel Bayesian Long Short-Term Memory (LSTM) model to assess the impact of air pollution control policies in China. This approach combines LSTM neural networks, which are effective for time-series predictions, with Bayesian inference to enhance model reliability and quantify uncertainties in predictions. Using real-world air quality data, the model evaluates the effectiveness of recent pollution control measures in China, providing insights into regulatory impacts and helping to guide future policy decisions for improved air quality.

Bai et. al (2018) a comprehensive review of air pollution forecasting methods, discussing various models and technologies used to predict air quality. They cover traditional statistical models, machine learning techniques, and hybrid approaches, emphasizing the strengths and limitations of each. The authors also address the challenges in forecasting accuracy and data availability, highlighting how improved models can support public health by offering timely pollution information. Additionally, Hijmansns et. al (2012) introduced the Raster R package, which facilitates geographic data analysis and spatial modelling.

Ding et. al (2019) presented a novel method for identifying writers based on air-handwriting movements captured by inertial sensors. The study leverages deep learning algorithms to analyse unique writing patterns, using sensor data to accurately identify individuals based on their air-handwriting styles. This approach demonstrates promising applications in security and authentication by enabling non-intrusive and contactless writer identification.

Dotse et. al (2018) Silva explores the use of computational intelligence methods to predict PM10 exceedances (particulate matter levels above standard thresholds) in Brunei. They utilize models such as neural networks and support vector machines to forecast daily pollution levels, aiming to provide early warnings for potential air quality issues. The study demonstrates that these techniques offer reliable predictions, which can aid in timely public health responses and pollution management.

M. Jia (2019) provided a comprehensive review of IoT technologies applied in the context of smart buildings. It explores how IoT can revolutionize building management by integrating advanced sensors, cloud computing, and big data analytics. Specifically, the authors emphasize the role of IoT in improving air quality monitoring, energy optimization, and occupant comfort. Real-time data collected through IoT-enabled sensors allows for adaptive building systems that respond dynamically to environmental changes and human activities. The paper concludes by highlighting the potential of IoT in creating energy-efficient and sustainable urban infrastructure while addressing challenges such as data privacy, interoperability, and scalability.

S.Abirami et. al (2020) presented a novel hybrid deep learning framework combining spatial and temporal data for predicting PM2.5 concentrations. The framework utilizes convolutional neural networks (CNNs) to capture spatial patterns and recurrent neural networks (RNNs), specifically Long Short-Term Memory (LSTM) networks, to model temporal dependencies. By leveraging geospatial data and historical air quality trends, this approach enhances forecasting accuracy compared to traditional models. The study includes experiments using real-world datasets, validating the model's effectiveness in providing timely and reliable air pollution predictions. Applications include urban planning, public health risk assessment, and proactive environmental management.

Y.Cheng et. al (2019) investigated strategies for balancing indoor air quality (IAQ) and energy savings in HVAC (heating, ventilation, and air conditioning) systems. The authors focus on optimizing the fresh outdoor air ratio using a stratum ventilation system, which delivers air directly to the breathing zone. The approach minimizes the energy required to cool or heat fresh air while maintaining adequate IAQ. The research includes computational fluid dynamics (CFD) simulations to evaluate airflow patterns and thermal comfort. The results demonstrate that targeted ventilation can achieve significant energy savings without compromising occupant health, offering a sustainable solution for modern buildings.

A. C. Cosma et. al (2019) explored the use of machine learning to predict individual thermal preferences in transient environmental conditions, such as fluctuating temperatures. The authors develop a predictive model that uses environmental data (temperature, humidity, air velocity) and user feedback to adjust HVAC settings dynamically. The study emphasizes the importance of non-invasive data collection, avoiding intrusive sensors or direct user interaction. The proposed system enhances occupant comfort by tailoring thermal conditions to individual needs while improving energy efficiency. The findings suggest that this approach could be integral to future smart building systems, offering personalized climate control in real time.

S.F.J. Raza et. al (2019) introduced an affordable, cloud-integrated air quality monitoring system that combines IoT sensors with machine learning algorithms. The system collects data on pollutants such as PM2.5, CO, and NO2, processes it using machine learning models, and provides real-time insights via a cloud-based dashboard. The authors emphasize the system's low cost and scalability, making it suitable for deployment in resource-constrained environments. Applications include public health monitoring, urban planning, and citizen engagement. The study also highlights the potential of predictive analytics for early warning systems, helping mitigate the effects of air pollution on vulnerable populations.

3. PROPOSED METHODOLOGY

To keep tabs on emissions of greenhouse gases, most current carbon footprint tracking systems use either simple sensor networks or methods that involve periodic sampling. Inadequate real-time monitoring and prediction capabilities are common among these systems due to a lack of sophisticated data analytics. Alarms for high amounts of particular gases are typically based on simple rules or thresholds in more conventional setups, but these methods are inadequate for detecting complicated patterns or accurately predicting trends. Also, they don't cover a wide variety of gases, therefore they're not useful in many different kinds of urban or industrial environments. As a result, current systems can only give you the high-level facts about emissions, which isn't helpful for making informed decisions or preparing for the future of the environment.

By integrating several gas sensors with state-of-the-art machine learning algorithms, the proposed Comprehensive Carbon Footprint Tracking System improves environmental monitoring as shown in Figure 1. This allows for scalable, predictive, and real-time study of emissions. This approach departs from conventional methods by utilising algorithms such as support vector machines, decision trees, random forests, and neural networks to analyse gas concentration data for irregularities, abnormalities, and complicated patterns. A more thorough comprehension of emission patterns and possible dangers is made possible by the system's automated analysis of interactions between various gases and environmental factors, which results in excellent accuracy. It also helps with proactive environmental management and regulatory compliance by allowing early detection of anomalous trends through its predictive capabilities. Offering a strong, future-proof solution to lowering carbon footprints and bolstering sustainability initiatives, the suggested system is scalable, making it suitable for wider uses in metropolitan regions, smart city infrastructures, and industrial settings.

Figure 1. Overview of the Proposed System

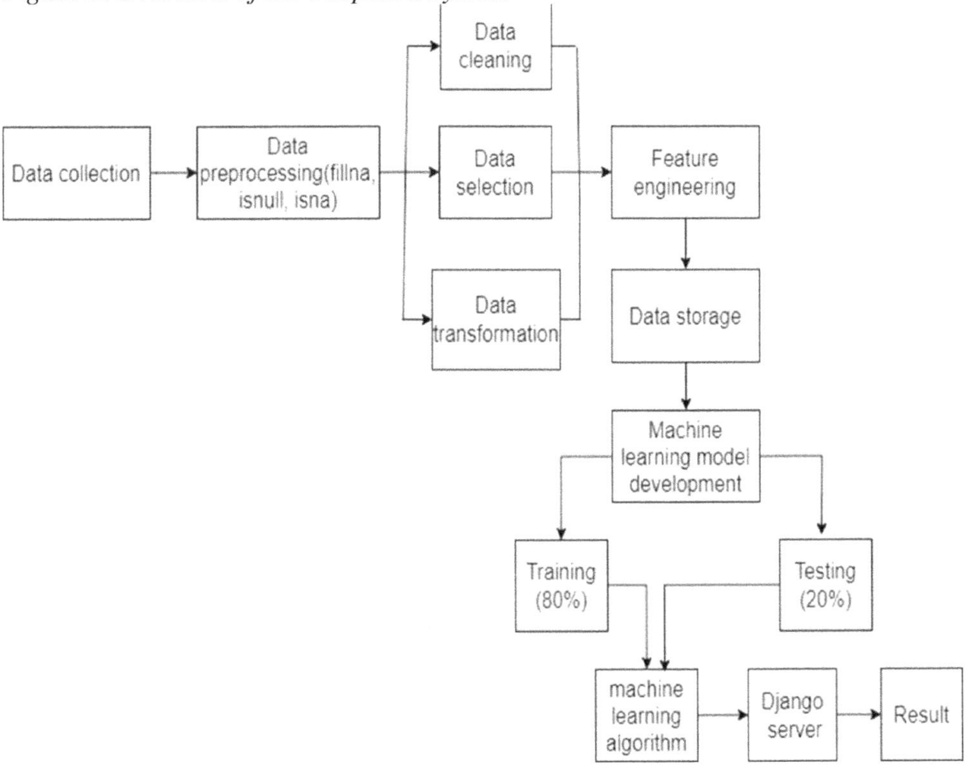

Data Collection

A wireless sensor network that tracks variables including humidity, air quality, and temperature is used to gather data for this project. Some of the sensors, such as the MQ series of micro-gas sensors, are put on or near people in order to gather data from their immediate environments in real-time. A low-power microcontroller, like an Arduino board, is linked to these sensors so that they can be read and processed initially. A central server receives the collected data wirelessly by a Bluetooth or ESP8266 module, guaranteeing a continuous and real-time data flow. Incoming data is stored and managed centrally by the server, which is created using the Django framework. Users can view both current and past data from any location thanks to the system's ability to continuously upload sensor readings to a cloud storage system. While keeping portability and user-friendliness for end-users in mind, this approach guarantees thorough and trustworthy data collecting. The data collection method is further improved with the incorporation of machine learning, which analyses incoming data and identifies air quality patterns or anomalies.

Data Preprocessing

1. Data Acquisition

Air pollution and carbon emissions raw data collection requires information to be gathered from a variety of sources. Particulate matter (PM2.5 and PM10), gases (CO2, CH4, VOCs), and other pollutants (NOx, HBr, etc.) can be detected in real time using sensors. Data collected from weather stations is particularly crucial for understanding patterns of pollution dispersal because it captures environmental variables including temperature, humidity, wind speed, and precipitation. Furthermore, aerial data can be used by remote sensing technologies to track pollution levels, particularly aerosol concentrations, on a broad scale. Uniform sampling intervals, such as minute-by-minute or hourly records, minimise temporal gaps and maintain reliability in data collecting, ensuring accuracy and consistency.

2. Data Cleaning

Reliable analysis relies on removing inconsistencies and errors from the dataset. When dealing with time-series data, there are a few ways to handle missing values: interpolation, guessing missing points from neighbouring values, or, for smaller gaps, statistical methods such median or mean imputation. If you want to tell the difference between real anomalies and sensor mistakes, you need to analyse and validate any outliers that statistical tests like Z-scores or the Interquartile Range (IQR) reveal. Using smoothing techniques like as moving averages or exponential smoothing can help reduce noise and interference-induced variations in sensor readings.

3. Feature Extraction

In order to improve the accuracy of the model, it is vital to derive meaningful variables from the raw data. Fast Fourier Transform (FFT) and statistical properties like standard deviation, periodicity, peaks, and patterns assist identify cyclic behaviours in time-series data. In order to improve pollutant detection in spatial data, especially photographs, it is helpful to extract visual properties such as gas plume intensity, pollution dispersion, or hotspots. Then, using edge detection algorithms, one may outline regions of interest for localisation. Because of their outsized impacts on pollutant distribution patterns, meteorological variables including wind speed and direction must also be considered.

4. Data Augmentation

By increasing variability and enhancing generalisation, extra data samples boost model resilience. In order to capture varied temporal patterns in time-series data, one can alter time windows to generate overlapping intervals and add synthetic noise to replicate sensor variability. Images can be transformed to reveal different pollution patterns using operations like flipping, rotating, and adjusting brightness. Improving the model's adaptability to real-world applications is another benefit of including realistic circumstances like haze or fog in photos.

5. Data Integration

Data integration improves analysis and model performance by bringing together disparate datasets. In multi-sensor fusion, data from several sensors, such as CH4 and CO2 detectors, are aligned and combined into a single dataset using common timestamps to ensure consistency. A complete picture of pollution levels can be obtained by integrating data from both satellite and ground-based sensors through geospatial alignment, which involves mapping satellite observations to their corresponding on-ground locations. A more complete picture of pollution patterns and their effects can be achieved by adding external data, such as socioeconomic variables like industrial activity or population density, which offer contextual insights.

6. Temporal Alignment

Ensuring data synchronization across sources is crucial for consistency and accuracy in analysis. Time synchronization involves matching timestamps from various devices to align data points correctly, with interpolation techniques used to handle differences in sampling rates, such as reconciling datasets recorded at minute and hourly intervals. Time windowing further simplifies analysis by aggregating data into fixed intervals, such as hourly or daily, allowing for structured comparisons. Additionally, calculating rolling averages helps smooth short-term fluctuations, improving the stability and reliability of the dataset for modeling and interpretation.

7.Noise Filtering

To improve data quality and model performance, it is vital to remove undesired fluctuations or distortions in the data. For time-series data, low-pass filters are useful for removing high-frequency noise, while for dynamic systems, Kalman filters are great for reducing noise in real-time. In the context of image data, a median filter is ideal for eliminating "salt-and-pepper" noise while keeping edges and other details intact, and a Gaussian blur improves overall clarity by smoothing pixel-level noise. These methods guarantee that the data used for modelling and analysis is cleaner and more trustworthy.

8. Labelling (for Supervised Learning)

Structured analysis and training of models rely on data samples being assigned categories or targets. Data is categorised into "Good," "Moderate," or "Hazardous" according to air quality classifications, which are based on pollution criteria such as AQI readings. Data points can be labelled as "Normal" or "Anomalous" for anomaly detection purposes. This allows models to be trained to find patterns of unusual pollution. Environmental monitoring systems are made more effective and easier to understand by classifying data according to the sources of emissions, whether they be natural, vehicular, or industrial. This allows for more precise pollution control methods and easier source identification.

9. Dimensionality Reduction

Improving the efficiency of a model relies on minimising the amount of input variables while keeping all relevant data. By reducing correlated variables to a smaller collection of uncorrelated components, techniques like Principal Component Analysis (PCA) aid in linear feature reduction. Autoencoders allow for the efficient compression of high-dimensional data while maintaining important patterns, making them ideal for non-linear feature reduction. By lowering computing requirements without drastically sacrificing information, these methods streamline complicated datasets, make them more interpretable, and speed up model training.

10. Data Transformation

For better model performance and interpretability, it is necessary to convert raw data into an appropriate format for analysis. Using methods like one-hot encoding or label encoding, categorical encoding converts place names and other non-numerical data into numerical representations. Logarithmic scaling improves the performance of statistical and machine learning models by applying log transformations to pollutant concentrations, which helps linearise connections for variables with exponential distributions. This makes patterns more detectable. All while keeping crucial information intact, these preprocessing techniques make the data more usable.

11. Data Standardization

The key to smooth integration and precise analysis is data representation uniformity. To avoid inconsistencies, unit conversion is used to standardise measurements across datasets, like using ppm for gases and $\mu g/m^3$ for particulate matter. Data integrity is further enhanced by ensuring that timestamps, geographical locations, and pollutant names are formatted consistently. After completing these preparation procedures, the dataset is prepared for use in real-time carbon emission tracking, pollution forecasting, anomaly detection, and other similar applications by being clean, consistent, and enhanced with significant features.

Data Cleaning

For this endeavour, data cleaning is an essential step in ensuring the sensor data used to monitor and analyse air quality is accurate and reliable. Errors, inconsistencies, or noise in the raw sensor data is common and might be caused by transmission problems, environmental influence, hardware restrictions, or both. The cleaning procedure employs a number of essential methods to overcome these obstacles. We begin by filling gaps with averages or interpolation, two imputation approaches that are applicable based on the readings' context, when data contains missing values. After that, we look for data points that are extremely out of the ordinary, or "outliers," to eliminate any errors or unusual circumstances that may have caused the sensors to fail. Furthermore, in order to avoid analytical redundancy, duplicate records that may occur during wireless transmission are recognised and deleted.

For the sake of consistency, the data is also normalised, meaning that all temperature measurements are converted to the same unit or that sensor values are scaled to a similar standard. To further enhance data quality and reduce fluctuations, noise reduction techniques like smoothing algorithms or filters are utilised. The final step is to put the cleaned data in a structured style so that the machine learning algorithms can analyse it further. The system's results are guaranteed to be accurate and actionable with this comprehensive data cleansing approach.

Data Transformation

One of the most important parts of this project is data transformation, which entails making the cleansed sensor data more analytically and model-friendly. Standardisation is necessary for uniform processing of raw data acquired from sensors, which might be in various units or formats and includes readings of gas levels, humidity, and temperature.

Data normalisation or standardisation occurs at the first stage of transformation, which is feature scaling. In this way, we can be sure that all the sensor values, irrespective of their initial units or magnitude, are normalised. Some common examples of such transformations include the Celsius and Fahrenheit temperature units, and the parts per million (ppm) scale for gas concentration quantities. After that, if the data contains any labels or classes, category transformation is used. For instance, in order to make them compatible with machine learning models, certain sensor statuses (such as "active" or "inactive") could be encoded into numerical numbers. To further synchronise the sensor data acquired at different intervals, time-series transformation is also executed. The data must be aggregated, interpolated, or resampled in order to guarantee that the timeframes for analysis are consistent. In order to improve the dataset, derived features are subsequently generated. One possible use case is the creation of heat index features by merging data on temperature and humidity. Another is the representation of overall air quality by aggregating readings of gas concentrations. The inputs to the prediction models are enhanced by these engineered properties. The last step in preparing data for use in a machine learning pipeline is to arrange it into a structured dataset, which may take the form of tabular data or data frames. The data is prepared for model training, air quality level predictions, and actionable insight generation by undergoing these changes.

Data Selection

The goal of data selection in this project is to improve the efficiency and performance of the model by selecting the most important features from the data that has been collected. Due to the fact that sensor raw data may contain a multitude of characteristics, not all of which contribute equally to the accuracy of air quality predictions, this step is crucial. Data selection helps minimise dimensionality, enhance computing efficiency, and prevent overfitting by focussing solely on the most informative features. This results in models that are more resilient and easier to comprehend.

Data selection for this project may make use of statistical methods and domain expertise to determine which factors (e.g., levels of particular gases, temperature, humidity, etc.) significantly affect air quality forecasts. One example is the use of correlation analysis to identify features with a strong relationship to air quality measures; this allows one to reject features that aren't important or are duplicated. Also, you may find out which features are most important for accurate predictions by looking at the feature importance rankings produced by machine learning algorithms like random forests or decision trees.

To make the system more effective and responsive, the project prioritises the most impacting variables. This way, the machine learning model can be trained on a dataset that captures the essential aspects affecting air quality. Making good predictions while making the most efficient use of resources and processing power relies heavily on this selection stage.

Feature Engineering

As it pertains to carbon emission tracking via air quality monitoring, feature engineering entails extracting useful information from raw sensor data in order to enhance model performance and deliver practical insights. It is essential to perform data preprocessing first, which includes removing or imputation for missing data and dealing with outliers using methods like clipping or log transformation to lessen skewness. Because air quality changes over time, temporal aspects are essential for capturing trends like industrial activity or heavy traffic by extracting time-of-day, day-of-week, and seasonal features. To better understand emission trends, it is helpful to aggregate sensor data by averaging over regions or by applying rolling statistics such as standard deviation and percentiles. This helps to smooth out swings.

Because air quality can vary greatly depending on location, spatial factors are as significant. To pinpoint locations of heavy pollution, it is helpful to include geographical details like latitude, longitude, and distance from major thoroughfares or factories. Another way to get a better look at emission trends is to divide the country into urban and rural sections.

The model must take into account environmental and meteorological factors that impact air quality, including but not limited to temperature, humidity, wind speed, and precipitation. Since the dispersion and accumulation of pollutants are affected by weather conditions, these characteristics contribute to the understanding of pollution fluctuations. Because of seasonal changes in both temperature and humidity, air quality can alter throughout the year, which means that environmental factors can also play a role.

In order to capture non-linear correlations that could impact emissions, advanced techniques are used, such as the creation of interaction terms between features (such as temperature and wind speed). To capture the inertia in emissions, lag features can be utilised. These features use historical pollutant levels to forecast future values. It is possible to differentiate between the various sectors' contributions to total carbon emissions by using categorical criteria, such as the kind of emission source (e.g., transportation, industrial, residential).

By bringing all features to a consistent scale, transformation techniques such as normalisation or standardisation improve the model by making it more stable and effective. When dealing with data that is not ideal for modelling, such as skewed pollutant distributions, log transformation is frequently used. In order to minimise noise and keep only the most important features when working with a large number of features, dimensionality reduction approaches such as Principal Component Analysis (PCA) might be employed. Carbon intensity calculations and other emission estimates can sometimes be determined from pollutant concentrations recorded by sensors; when combined with data on energy consumption other emission factors, this gives a fuller picture of the carbon footprint in particular areas. By utilising feature selection approaches, one may assess the significance of features for machine learning models. These techniques aid in identifying the most predictive aspects of carbon emissions, which in turn guides future analysis and policy decisions.

Data Storage

Data storage in carbon emission tracking and air quality monitoring is essential for managing and organizing large volumes of data in a way that ensures scalability, accessibility, and efficient processing for analysis and modeling. Relational databases like MySQL or PostgreSQL are often used when the data is structured and has a well-defined schema, such as sensor readings over time with clear relationships between tables. For instance, air quality data might be stored in a table with fields like timestamp, sensor ID, pollutant type, concentration levels, and geographic location, which can be queried and analyzed easily.

NoSQL databases such as MongoDB or Cassandra are suitable for storing unstructured or semi-structured data, making them ideal for handling large volumes of data from diverse sources. These databases are flexible and scalable, capable of storing sensor data where the schema might not be fixed, and they handle high-volume, real-time data streams efficiently. Cloud storage solutions like Amazon S3, Google Cloud Storage, or Azure Blob Storage are highly scalable and cost-effective options for storing vast amounts of raw data in various formats such as CSV, JSON, or Parquet. These platforms are reliable and provide easy integration with data processing and machine learning tools.

For analytical processing, data warehouses such as Google BigQuery, Amazon Redshift, or Snowflake are designed to store large datasets in structured formats. They enable fast querying and analysis, making it easier to aggregate and report air quality data across time and space. A data warehouse can combine data from various sources, like multiple sensors, weather data, or external datasets, which is beneficial for complex analyses. Time series databases, like InfluxDB or TimescaleDB, are particularly useful for storing time-stamped air quality data, as they are optimized for handling time-based data, supporting continuous queries, downsampling, and retention policies.

When handling diverse, large-scale datasets that include both structured and unstructured data, data lakes such as Apache Hadoop or Azure Data Lake offer a centralized repository for storing raw data. Data lakes allow for flexible storage of data, which can be processed later for advanced analytics or machine learning. Edge computing storage is useful for remote or distributed sensor networks, where data can be processed and stored locally on edge devices like Raspberry Pi or IoT sensors. This reduces the need for constant data transmission and allows real-time processing, with periodic syncing to central storage.

The format in which data is stored is important for efficient processing. Common formats include CSV for simple and human-readable data, JSON for semistructured data, and Parquet or ORC for columnar storage, which is more efficient for big data environments. Avro is another binary format often used with Hadoop for serialization, especially for streaming data. Backup and redundancy are crucial to prevent data loss, and cloud services often provide built-in redundancy. Implementing version control and data retention policies ensures that historical data can be accessed when needed, while optimizing storage costs by archiving old data.

Data security is a major consideration in storing air quality data, especially when dealing with public health or emissions-related information. Strong security measures such as encryption, both at rest and in transit, along with access control mechanisms, are essential to protect the data

To start, supervised learning models are commonly used for predicting air quality metrics or carbon emissions based on historical data. Regression models like linear regression or more complex ones like random forests and gradient boosting machines are useful when the goal is to predict continuous outcomes, such as the concentration of pollutants (e.g., CO2, particulate matter) or carbon intensity. These models work by learning the relationship between input features—such as sensor readings, meteorological data, and time—on the training dataset and then making predictions on the test dataset. A key part of this process involves splitting the data into training and testing subsets. The training data is used to train the model, while the testing data is used to evaluate its accuracy and generalization ability.

For more advanced modeling, support vector machines (SVM) can be used, especially for classification tasks like identifying whether a given air quality level exceeds a certain threshold. SVM works by finding the optimal hyperplane that separates different classes in the feature space. In the context of air quality, this could involve classifying data into categories such as "good," "moderate," or "dangerous" air quality based on pollutant levels.

Neural networks, particularly deep learning models, can also be applied if the dataset is large and complex. Models such as feed-forward neural networks or convolutional neural networks (CNNs) can be used to model highly non-linear relationships between input features and air quality or carbon emission predictions. These models require a substantial amount of data for training but can capture complex patterns and interactions between features. When using neural networks, the training dataset is fed through the model iteratively, adjusting weights via backpropagation, while the test dataset is used to evaluate how well the model generalizes to unseen data.

Time series forecasting models like ARIMA (AutoRegressive Integrated Moving Average), LSTM (Long Short-Term Memory), and prophet can be effective when predicting air quality trends over time. These models take into account the temporal dependencies in the data, such as previous sensor readings, weather conditions, and other environmental factors. For instance, using the training data, an ARIMA model can predict future pollutant levels by capturing the timedependent structure in the dataset, and testing it on unseen time periods evaluates its forecasting accuracy.

The model's performance is measured by comparing predicted values (from the test data) to actual observed values using various evaluation metrics. For regression models, mean absolute error (MAE), mean squared error (MSE), and R-squared are commonly used to assess the accuracy of predictions. For classification tasks, accuracy, precision, recall, and F1 score are used to evaluate how well the model categorizes air quality into predefined classes. The choice of evaluation metric depends on the specific problem and the model's purpose.

Once a model is trained and evaluated, further steps like hyperparameter tuning (using techniques like grid search or random search) can be performed on the training dataset to optimize the model's performance. After the final model is selected, it can be deployed for real-time predictions or to inform policy and decision-making processes, continuously receiving new sensor data for ongoing training and refinement.

Simple regression

If you have historical data for a continuous metric, such air quality levels or carbon emissions, you can use linear regression to make predictions about it. To make it work, the data points are fitted to a line in a way that minimises the sum of squared discrepancies between the predicted and actual values. When there is a linear relationship between the independent factors (like pollution or weather data) and the dependent variable (like air quality), this approach is simple and effective.

Decision Trees

For the purpose of this project's air quality level classification and carbon emission category prediction, decision trees are another helpful approach. Based on the feature that yields the optimal split (lowest impurity or largest information gain), a decision tree divides the dataset into smaller subsets. By iteratively building a tree structure, the technique eventually arrives at a prediction. Decision trees are both interpretable and visually appealing, making it easy to see how different input features are used to make predictions.

Random Forest

Using a combination of decision trees, random forests are a kind of ensemble learning. The data and attributes used to train each tree in the forest are selected at random. In regression, the average of all the trees' predictions is used to make the final prediction, while in classification, the majority vote is used. Ideal for complicated, noisy data, random forests greatly improve prediction accuracy by lowering variance and overfitting.

Support Vector Machines

The two main applications of support vector machines (SVMs) are regression and classification. Finding the hyperplane that maximises the margin between classes while best separating data points from distinct classes is how SVM works. Support vector machines (SVMs) attempt to fit a hyperplane or line that minimises the error within a set margin in regression problems. Support vector machines (SVMs) excel in high-dimensional spaces and, with the help of kernel functions, can manage non-linear interactions.

K-neighbors (KNN) method

One easy approach for regression or classification is K-nearest neighbours (KNN). It generates forecasts by averaging the values of the K closest neighbours in the feature space or using the majority class. Utilising KNN, air quality levels can be classified in a carbon emissions tracking project by comparing newly added data points to their nearest neighbours in the dataset. With big datasets, KNN can be computationally expensive despite its simplicity.

Artificial neural networks (ANN)

One family of deep learning algorithms, artificial neural networks (ANNs) excel at processing massive datasets with complicated interactions. Each neurone in an ANN processes input data through a set of weighted connections; ANNs are composed of layers of these interconnected nodes. In order to find out if the neurone will “fire” and transmit information, activation functions are applied to the output. Predicting air quality using a combination of variables such as pollution, weather, and traffic data is a good example of a big dataset with complex patterns that ANNs excel at handling.

Long short-term memory (LSTM)

Recurrent neural networks (RNNs) with long short-term memory (LSTM) capabilities excel in time-series data tasks, such as estimating future levels of air pollution or carbon emissions. Long short-term memory (LSTM) models can handle sequential data with long-term dependencies and trends because they remember information from earlier time steps. Their ability to capture trends and variations in data over extended time periods makes them ideal for long-term air quality forecasting.

K-means clustering

One unsupervised learning approach that can be utilised for clustering data points according to their similarity is K-means clustering. The method is iterative; it first finds the closest cluster centre for each data point in the dataset, and then recalculates the cluster centres using those locations. To classify areas with similar carbon emissions, K-means can be applied to locate pollution hotspots or places with equivalent air quality trends in this project.

Predicting future values from past observations is a common task for time-series forecasting systems such as ARIMA (AutoRegressive Integrated Moving Average). By integrating discrepancies between values (integrated), prior values (autoregressive), and a moving average of the errors, ARIMA models air quality or carbon emissions data. For trend analysis and short-term forecasts, this strategy works well.

In order to better understand and make decisions about air quality and environmental effect, these algorithms are essential to a carbon emissions tracking project. They provide accurate predictions, classifications, and pattern recognition. The data type, the challenge at hand, and the project's unique goals all play a role in determining the method to use.

4. RESULTS AND DISCUSSIONS

Figure 2. Login and Dashboard of the proposed system

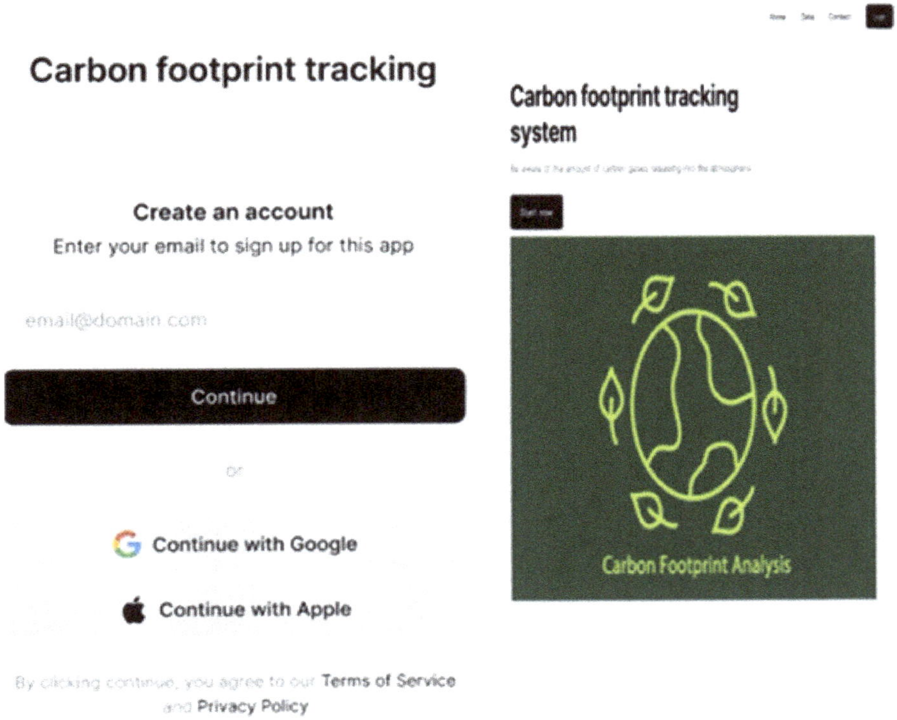

(a) Login page *(b)Dashboard*

The home page provides users with the option to sign up or log in to access personalized features, ensuring secure data management and customization. The design focuses on simplicity, allowing users to easily navigate between sign-up, login, and other app sections for an enhanced user experience as shown in Figure 2. Users can navigate to their desired page and explore the unique features of this system. The system will track in real time the carbon footprint based on multi-gas sensors and applied machine learning algorithms in order to predict gas concentrations, assess anomalous levels, and then decide whether it should perform an action for environmental conditions. This system allows for the scalable nature of deployment within industrial, ur ban, and domestic environments based on the specific monitoring needs of such an environment.

Figure 3. System analyses the carbon footprints

The developed system effectively integrates machine learning models to calculate and visualize the carbon footprints as shown in Figure 3. The system incorporates IoT platforms to enable a robust, real-time monitoring experience, collecting data from distributed sensors and transmitting it to a cloud-based platform where all information can be accessed and managed centrally. This setup allows users to log into a web-based dashboard where they can monitor emissions data continuously, gaining insights through visualizations and metrics tailored to their needs.

Performance analysis

The models evaluated include traditional machine learning algorithms such as Decision Trees, Random Forest, and Logistic Regression, as well as more advanced techniques like Gradient Boosting, Neural Networks, and LSTMs. The results are summarized in the Table 1.

Table 1. Results - The models evaluated include traditional machine learning algorithms such as Decision Trees, Random Forest, and Logistic Regression, as well as more advanced techniques like Gradient Boosting, Neural Networks, and LSTMs

Model	Accuracy	Precision	Recall	F1-Score
Decision Tree	0.85	0.83	0.84	0.83
Random Forest	0.9	0.89	0.88	0.88
Support Vector Machine	0.87	0.86	0.85	0.85
Neural Network	0.92	0.91	0.92	0.91
Logistic Regression	0.8	0.79	0.78	0.78
K-Nearest Neighbors	0.83	0.81	0.82	0.81
Gradient Boosting	0.91	0.9	0.91	0.9
LSTM (Long Short-Term Memory)	0.93	0.92	0.93	0.92

Figure 4. Performance analysis of the proposed system

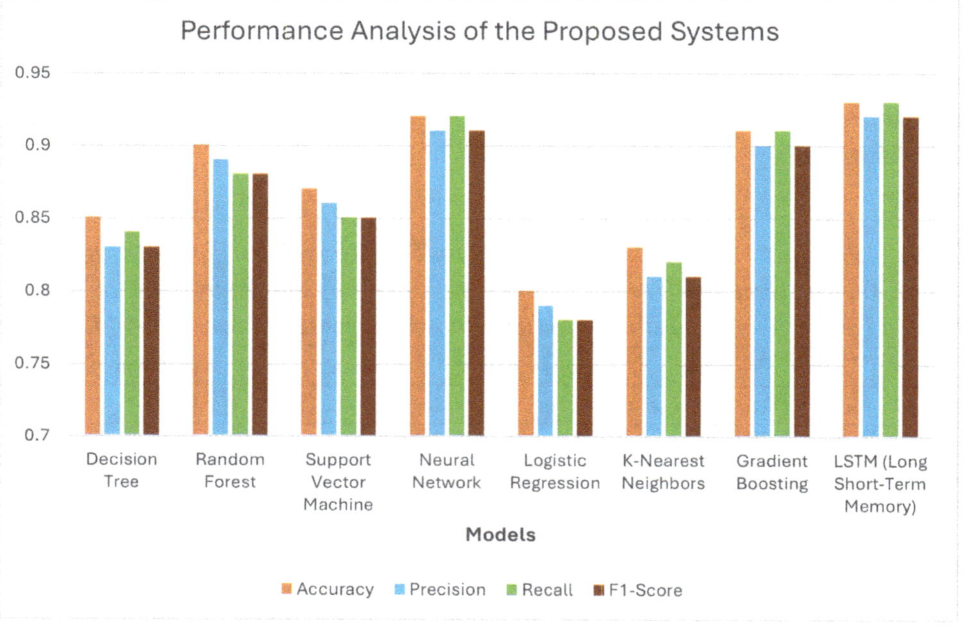

Based on their capacity to capture long-term dependencies, LSTM (Long Short-Term Memory) networks proved to be the most effective for time-series forecasting, as they attained the maximum accuracy of 0.93 in the performance analysis of different models for air pollution and carbon emission prediction as shown in Figure 4. An accuracy of 0.92 was also achieved by Neural Networks, demonstrating their proficiency in handling intricate environmental data. Both Random Forest (0.90) and Gradient Boosting (0.91) showed excellent predictive ability, therefore they are good options for structured datasets. With an accuracy of 0.87, Support Vector Machines (SVM) demonstrated balanced performance, following by Decision Trees with an accuracy of 0.85 and K-Nearest Neighbours with an accuracy of 0.83, all of which gave competitive results. Because of its simplicity, Logistic Regression performed poorly on complicated pollution data, with an accuracy of only 0.80. For both historical and real-time pollution forecasting, deep learning models—and LSTM in particular—proved to be superior to more conventional machine learning approaches. To further increase the accuracy and generalisability of predictions, future developments may incorporate adaptive learning methods, hybrid modelling approaches, and sophisticated feature engineering.

CONCLUSION

This paper presents a Comprehensive Carbon Footprint Tracking System that effectively integrates multiple gas sensors with advanced machine learning algorithms for real-time environmental monitoring. By harnessing the strengths of these technologies, the system is designed to provide accurate and reliable data on various gas emissions, allowing for timely insights that can drive informed decision-making. Its scalability is a key feature, enabling it to be deployed in diverse environments ranging from industrial facilities to urban landscapes, thereby addressing the unique challenges associated with carbon emissions in different contexts. The predictive capabilities of the system are particularly noteworthy, as they empower users to anticipate environmental changes and identify potential hazards before they escalate. This proactive approach is essential in today's world, where rapid industrialization and urbanization are contributing to increasing levels of greenhouse gas emissions. The ability to monitor and predict emissions in real-time equips stakeholders with the information needed to implement effective mitigation strategies and ensure compliance with environmental regulations. In conclusion, the Comprehensive Carbon Footprint Tracking System represents a significant step forward in environmental monitoring technology. By leveraging multiple gas sensors and machine learning, it offers a robust and scalable solution that can meet the evolving needs of industries and urban environments. As research and development efforts continue, this system has the potential to play a vital role in supporting sustainability initiatives and fostering a deeper understanding of our environmental impact.

REFERENCES

Abirami, S., Chitra, P., Madhumitha, R., & Kesavan, S. R. "Hybrid spatio-temporal deep learning framework for particulate matter (PM2.5) concentration forecasting," in *Proc. Int. Conf. Innov. Trends Inf. Technol. (ICITIIT)**, Feb. 2020. DOI: 10.1109/ICITIIT49094.2020.9071548

An, N., Mustafa, F., Bu, L., Xu, M., Wang, Q., Shahzaman, M., Bilal, M., Ullah, S., & Feng, Z. (2022). Monitoring of Atmospheric Carbon Dioxide over Pakistan Using Satellite Dataset. *Remote Sensing (Basel)*, *14*(22), 5882. DOI: 10.3390/rs14225882

Bai, L., Wang, J., Ma, X., & Lu, H. (2018). Air pollution forecasts: An overview. *International Journal of Environmental Research and Public Health*, *15*(4), 780. DOI: 10.3390/ijerph15040780 PMID: 29673227

Bibri, S. E., & Krogstie, J. (2021). A novel model for data-driven smart sustainable cities of the future: A strategic roadmap to transformational change in the era of big data. *Future Cities Environ.*, *7*(1), 1–25. DOI: 10.5334/fce.116

Cheng, Y., Zhang, S., Huan, C., Oladokun, M. O., & Lin, Z. (2019, January). Optimization on fresh outdoor air ratio of air conditioning system with stratum ventilation for both targeted indoor air quality and maximal energy saving. *Building and Environment*, *147*, 11–22. DOI: 10.1016/j.buildenv.2018.10.009

Cosma, A. C., & Simha, R. (2019, January). Machine learning method for real-time non-invasive prediction of individual thermal preference in transient conditions. *Building and Environment*, *148*, 372–383. DOI: 10.1016/j.buildenv.2018.11.017

Ding, Y., & Xue, Y. (2019). A deep learning approach to writer identification using inertial sensor data of air-handwriting. *IEICE Transactions on Information and Systems*, *E102-D*(10), 2059–2063. DOI: 10.1587/transinf.2019EDL8070

Dotse, S.-Q., Petra, M. I., Dagar, L., & De Silva, L. C. (2018, March). Application of computational intelligence techniques to forecast daily PM10 exceedances in Brunei Darussalam. *Atmospheric Pollution Research*, *9*(2), 358–368. DOI: 10.1016/j.apr.2017.11.004

Han, Y., Lam, J. C. K., & Li, V. O. K. "A Bayesian LSTM model to evaluate the effects of air pollution control regulations in China," in Proc. IEEE Big Data Workshop (Big Data), Dec. 2018, pp. 4465–4468. DOI: 10.1109/BigData.2018.8622417

Hansen, J., Kharecha, P., Sato, M., Masson-Delmotte, V., Ackerman, F., Beerling, D. J., Hearty, P. J., Hoegh-Guldberg, O., Hsu, S.-L., Parmesan, C., Rockstrom, J., Rohling, E. J., Sachs, J., Smith, P., Steffen, K., Van Susteren, L., von Schuckmann, K., & Zachos, J. C. (2013). Assessing "dangerous climate change": Required reduction of carbon emissions to protect young people, future generations and nature. *PLoS One*, *8*(12), e81648. DOI: 10.1371/journal.pone.0081648 PMID: 24312568

R. J. Hijmans and J. van Etten, "Raster: Geographic analysis and modeling with raster data," R Package Version 2.0-12, Jan. 12, 2012

Jia, M., Komeily, A., Wang, Y., & Srinivasan, R. S. (2019, May). Adopting Internet of Things for the development of smart buildings: A review of enabling technologies and applications. *Automation in Construction*, *101*, 111–126. DOI: 10.1016/j.autcon.2019.01.023

Li, X., Jin, L., & Kan, H. (2019, June). Air pollution: A global problem needs local fixes. *Nature*, *570*(7762), 437–439. DOI: 10.1038/d41586-019-01960-7 PMID: 31239571

Lützkendorf, T., & Balouktsi, M. On net zero GHG emission targets for climate protection in cities: More questions than answers? In *Proceedings of IOP Conference Series: Earth and Environmental Science*; IOP Publishing: Bristol, UK, 2019; p. 012073. DOI: 10.1088/1755-1315/323/1/012073

Naumann, S., Davis, M., Kaphengst, T., Pieterse, M., & Rayment, M. Design, implementation and cost elements of Green Infrastructure projects; Final report to the European Commission, DG Environment, Contract no. 070307/2010/577182/ETU/F.1, Ecologic institute and GHK Consulting; European Commission: Brussels, Belgium, 2011.

Raza, S. F. J., Cullen, P., & Keeney, P. (2019). Implementation of low-cost, cloud-based air quality monitoring system using IoT sensors and machine learning techniques. *Procedia Computer Science*, *151*, 877–884.

Rees, J. (1989). Natural resources, economy and society. In *Horizons in Human Geography* (pp. 364–394). Palgrave. DOI: 10.1007/978-1-349-19839-9_19

Saba, C. S., & Ngepah, N. (2023). Empirics of convergence in industrialisation and their determinants: Global evidence. *Discover Sustainability*, *4*(1), 25. DOI: 10.1007/s43621-023-00136-8

Sarkodie, S. A., Owusu, P. A., & Leirvik, T. (2020). Global effect of urban sprawl, industrialization, trade and economic development on carbon dioxide emissions. *Environmental Research Letters*, *15*(3), 034049. DOI: 10.1088/1748-9326/ab7640

Schrotter, G., & Hürzeler, C. (2020). The digital twin of the city of Zurich for urban planning. *PFG–J. Photogramm. Remote Sens. Geoinf. Sci.*, *88*, 99–112.

Seto, K. C., Churkina, G., Hsu, A., Keller, M., Newman, P. W., Qin, B., & Ramaswami, A. (2021). From low-to net-zero carbon cities: The next global agenda. *Annual Review of Environment and Resources*, *46*(1), 377–415. DOI: 10.1146/annurev-environ-050120-113117

Stevens, D., Dragicevic, S., & Rothley, K. (2007). iCity: A GIS–CA modelling tool for urban planning and decision making. *Environmental Modelling & Software*, *22*(6), 761–773. DOI: 10.1016/j.envsoft.2006.02.004

Szczepańska, A., Kaźmierczak, R., & Myszkowska, M. (2023). Smart City Solutions from a Societal Perspective—A Case Study. *International Journal of Environmental Research and Public Health*, *20*(6), 5136. DOI: 10.3390/ijerph20065136 PMID: 36982043

Thornbush, M. J., & Golubchikov, O. (2019). *Sustainable Urbanism in Digital Transitions: From Low Carbon to Smart Sustainable Cities*. Springer.

Wang, F., Harindintwali, J. D., Yuan, Z., Wang, M., Wang, F., Li, S., Yin, Z., Huang, L., Fu, Y., & Li, L.. (2021). Technologies and perspectives for achieving carbon neutrality. *Innovation*, *2*, 100180. PMID: 34877561

Xu, D., Abbasi, K. R., Hussain, K., Albaker, A., Almulhim, A. I., & Alvarado, R. (2023). Analyzing the factors contribute to achieving sustainable development goals in Pakistan: A novel policy framework. *Energy Strateg. Rev.*, *45*, 101050. DOI: 10.1016/j.esr.2022.101050

Xu, D., Abbasi, K. R., Hussain, K., Albaker, A., Almulhim, A. I., & Alvarado, R. (2023). Analyzing the factors contribute to achieving sustainable development goals in Pakistan: A novel policy framework. *Energy Strateg. Rev.*, *45*, 101050. DOI: 10.1016/j.esr.2022.101050

Zhu, N., Bu, Y., Jin, M., & Mbroh, N. (2020). Green financial behavior and green development strategy of Chinese power companies in the context of carbon tax. *Journal of Cleaner Production*, *245*, 118908. DOI: 10.1016/j.jclepro.2019.118908

Chapter 11
Stakeholders' Engagement in the Transition to Sustainable Supply Chains

Jihen Msekni
https://orcid.org/0000-0003-2938-766X
FSEG Sfax, Tunisia

Ahmed Turki
https://orcid.org/0000-0003-3706-3024
IHEC Sfax, Tunisia

ABSTRACT

In the Tunisian context, the topic of Corporate Social Responsibility (CSR) starts to gain importance, previously in the post-revolution and currently during the "Covid-19" health crisis, due to social pressures which have become increasingly more demanding and threatening. It is in this context that this research is taking place, which seeks to analyze the stakeholders' engagement process and the actions implemented by companies as part of the transition to sustainable supply chains. The qualitative method was chosen to address this issue, by conducting research on the study of five cases from different sectors (chemical, service, agri-food industries, banking, textile). The results reveal that the impact of the crisis has been minimal, due to stakeholder engagement by diverse companies that are no longer limited to their own performance, but are becoming responsible to their stakeholders.

DOI: 10.4018/979-8-3373-3790-6.ch011

INTRODUCTION

Faced with today's environmental, social, and economic challenges, companies are being urged to radically overhaul the structure and governance of their supply chains. Beyond a simple operational performance approach, the move towards sustainable supply chains requires a systemic transformation that integrates ethical, environmental and societal criteria. In this context, corporate social responsibility (CSR) initiatives play a key role in helping to build an organizational culture based on collective engagement. In the Tunisian context, the topic of Corporate Social Responsibility (CSR) starts to gain importance, previously in the post-revolution and currently during the “Covid-19” health crisis, due to social pressures which have become increasingly more demanding and threatening. According to the African Development Bank Group (2012), the persistent political instability that has persisted since 2011 has obstructed the execution of essential economic reforms, restricting Tunisian companies' ability to adequately incorporate Corporate Social Responsibility (CSR) into their supply chain strategies. Also, companies are not only looking to make a profit where they operate, but also to contribute to the well-being of the society in which they operate (Wirba, 2024). As a result, companies are required to react to these new demands by following an adapted social approach that enables the engagement of stakeholders to ensure their sustainability. In this context, several authors have addressed the subject of CSR and the relationship between companies and stakeholders (Bowen, 1953; Freeman, 1984; Frederick, 1994; Carroll, 1999; Morsing and Schultz, 2006; Bowen et al., 2010). However, few studies have been carried out on this type of issue in the Tunisian context, especially in the context of crisis. One of the major levers of this influence lies in the redefinition of performance criteria. Traditional financial indicators are now being supplemented, or even supplanted, by ESG (environmental, social, and governance) indicators, demanded not only by responsible investors but also by civic-minded consumers. This development is profoundly changing the way supply chains are conceived: it's no longer just a question of producing efficiently but of producing fairly, traceably, and in an ecologically responsible way. A number of initiatives are emerging to overcome this asymmetry. Multi-stakeholder platforms aim to redistribute decision-making power, integrating all stakeholders right from the design stage. Understanding the influence of stakeholders on the sustainability of supply chains requires us to question not only their roles but also the power relationships that structure them.

Stakeholders play a normative and structuring role here by imposing new frames of references that become industry standards. Indeed, Huang et al., (2020) suggest that companies are threatened in the future by the impact of crises on the economy, causing supply chain dysfunction, reduced demand, or even a complete shutdown of operations and/or business. To deal with these threats, companies need to put in place measures to preserve their stability and resilience. In the same context, by referring to Freeman's stakeholder theory (1984), companies are encouraged to revise their priorities, taking into account the needs of stakeholders other than shareholders in achieving organizational objectives, going beyond the traditional vision of maximizing profits. Establishing a favorable relational climate with stakeholders will enable socially responsible companies to effectively identify environmental changes by obtaining the necessary resources and support in order to achieve organizational resilience (Lv et al., 2019).In this sense, Mackey et al. (2007) support the idea that the establishment of reciprocal interactions by companies with the various stakeholders will increase their resilience. Similarly, the research results of Ursic & Cestar (2022) show that the first few months of the Covid-19 crisis brought total social and business upheaval. However, the overall impact was attenuated by those companies that were ready for the crisis even before its onset. The study's findings also indicate that the majority of these companies have developed or already implemented business practices related to CSR and crisis management. Indeed, CSR can be an opportunity for organizations in times of crisis. Indeed, Souto, B. F. F (2009) has listed a series of necessities that could help companies in times of crisis. Among these necessities is the role of stakeholders. In fact, CSR implements an adapted relationship with stakeholders, transforming the situation of risk towards their company into a position of union. In other words, the company can no longer ignore the practices of its stakeholders. It becomes accountable for the global impact of its sourcing, reverse logistics and subcontracting choices. This transforms the logic of supply chain management: ethics, transparency, and circularity become fundamental axes. Stakeholders are no longer peripheral constraints, but catalysts in the systemic transition towards sustainable, fairer, and more resilient supply chains.

The focus of this chapter is on understanding the strategic role of stakeholders in the transition to sustainable supply chains. How do companies engage their stakeholders as part of their CSR strategy? How does this collective participation affect the sustainable restructuring of the supply chain? The intention is to investigate the processes by which CSR initiatives promote not only the strengthening of trust and collaboration between stakeholders, but also the creation of shared value through more resilient and sustainable supply chains. Due to the complexity of the societal issue. A qualitative approach is chosen, which is better suited to the question and aims of the research. Thus, this chapter deals with the presentation of the Tunisian context by analyzing the effect of the Covid-19 health crisis on Tunisian companies and on the economy in general, and deals with the presentation of CSR in the Tunisian context.

1. LITERATURE REVIEW

1.1. Corporate Social Responsibility

CSR has been perceived as something new in recent years. We're becoming to notice more and more of the social, environmental and economic preoccupations from companies, who are increasingly starting to give importance to their environment, and their stakeholders for their well-being, as well for the maintenance of their activity. The publication of Bowen's book "Social Responsibilities of Businessman" (1953) expanded and promoted research and work on the CSR concept, providing a solid basis for managers to refer to in their strategic planning and decision-making. And since that time, debates have emerged to provide further precision and clarification. In this context, Frederick (1994) argues that the emergence of CSR would have resulted from a diverse range of sources, manifested by the necessity for certain influential corporate entities to satisfy the conflicting interests of stakeholders and for the company to conform with social standards in order to be a responsible citizen.

CSR is a vague and highly controversial concept that has been the subject of a great deal of research. Indeed, several definitions have been proposed by different researchers and practitioners, but no consensus has emerged, which has consequently led to confusion over a standard definition. According to Gond and Servayre (2004), the concept of CSR is known by several names, « the company's societal responsiveness », which refers to the way in which it acts in the face of societal pressures (Frederick, 1978), as well as « the company's societal performance », which translates into the maximization of company profits as a result of societal actions, and « the company's societal rectitude », which covers the notion of the justice and righteousness of the actions taken and the decisions made.

CSR is an equivocal concept that is constantly and rapidly evolving, due to its exceptional richness. However, the theories behind the concept still need to be researched further and in greater depth. From a theoretical point of view, five theories can explain the issues raised by CSR: Resource dependence theory, neo-institutional theory, resource-Based View theory, dynamic capabilities theory and stakeholder's theory. According to resource dependency theory of Salancik and Pfeffer (1978), the organization is dependent on its resource providers and aims to reduce uncertainty, by controlling the resources it needs, in order to ensure its sustainability. In fact, through a socially responsible approach, companies can satisfy their providers and thus gain access to the resources they need. As for neo-institutional theory, developed by Meyer and Rowan (1977) and then by DiMaggio and Powell (1983), which postulates that in order to obtain legitimacy, organizations are required to adopt certain practices and procedures defined and institutionalized by the society, regardless of their degree of effectiveness. Thus, those that provide high degrees of legitimacy are more likely to be implemented by companies, such as CSR practices. The adoption of such behavior is explained by pressure from society (coercive isomorphism), imitation (mimetic isomorphism) and professionalization (normative isomorphism). Thus, Barney's (1991) Resource-Based View (RBV) posits that firms must effectively control and manage their resources especially those that are immobile and heterogeneous, such as capabilities, assets, information, know-how, and organizational processes. When these resources are rare, inimitable, valuable, and non-substitutable, they can provide the firm with a sustainable competitive advantage by differentiating it from its rivals. Building on RBV, Teece et al. (1997) developed the theory of dynamic capabilities, which emphasizes the evolving nature of resources. In contrast to the relatively static perspective of RBV, dynamic capabilities account for the firm's ability to adapt, integrate, and reconfigure internal and external competencies in response to environmental changes (Midhani, 2009).

The latest theoretical approach is stakeholders' theory, which examines with the relationship between the company and society. According to Mullenbach (2007), the aim of this theory is essentially to switch from Shareholder's Theory, which considers that managers have an obligation to act exclusively for the interests of shareholders, to Stakeholder's Theory, which aims to take stakeholders' interests into consideration.

1.2. Stakeholder's engagement challenges in the transition to sustainable supply chains

The current state of the global economy cannot be sustained. Indeed, as the planet's resources are not infinite but quantitatively limited, regulatory authorities are forcing manufacturers to take this issue into account. The challenge is not simply to meet regulatory compliance, but to respond with real innovation in ecologically intelligent products and services, developed and marketed responsibly (Grandval & Soparnot, 2005). Thus, sustainable supply chain practices have become increasingly important as global concerns about climate change, resource degradation and social responsibility grow (Gourire et al. 2025). According to Soulier (2022), a sustainable supply chain provides a rational, positive and constructive approach to the resolution of sustainable development issues, in particular through two fundamental leverages: the decarbonization of logistics operations and the implementation of circular economy principles.

Indeed, since globalization has promoted free trade, it has challenged companies to behave in an ethical and socially responsible manner by adopting social responsibility policies and managerial strategies. Thus, the supply chain includes both upstream and downstream players, and can involve the company's suppliers and customers, intermediaries, service providers and even end consumers. In this context, the engagement of these actors in the transition of supply chains will be necessary, due to their power of influence (El aidouni & Abbassi, 2012). Also, Siems et al., (2023) reviewed 78 studies showing that NGOs, regulators, competitors and consumers play a driving or facilitating role in sustainable chains. In this context, Sadeghi et al., (2025) explored the metaverse as a groundbreaking influence on supply chain information systems, especially concerning decision-making processes aimed at sustainable development goals. Therefore, this research investigates in what ways does the metaverse among stakeholders enhance supply chain decision-making processes related to sustainable development goals. In the same context of stakeholders and the supply chain, Inês et al., (2025) explored the concept of eco-innovation (EI) and its contribution to sustainable economic growth by integrating environmental and social factors. The study emphasizes the recognition of essential practices and links that promote EI in supply chains through collaboration among various stakeholders.

Supply chain sustainability (SCS) is currently a crucial strategic issue in the transition to more responsible production models. Faced with the rapid rise of environmental, social and ethical issues, companies are under increasing pressure to re-evaluate their logistics in light of the demands and restrictions coming from their stakeholders. Whether internal stakeholders (such as shareholders, employees and suppliers) or external (such as customers, NGOs, public authorities and local communities). They all have a considerable impact on the implementation of sustainable practices in the supply chain and exert influence via normative, notoriety, economic or social constraints that directly affect companies' organizational, strategic and operational decisions. On the regulatory front, governments and international organizations are progressively tightening their compliance requirements, notably through mechanisms such as the Duty of Vigilance Act and European directives on supply chain transparency. These rules require companies to assess the environmental and social risks associated with their suppliers, set up traceability systems and publish reports on their Corporate Social Responsibility (CSR).

In addition, NGOs and consumers have a role to play in monitoring and opposition. By reporting ethical or environmental breaches, they can considerably affect the reputation of brands. This pressure from society is pushing companies to embrace more responsible practices, often embodied by the award of labels or certifications (ISO 14001, Fair Trade, etc.).

The role of suppliers in this dynamic is also crucial. As key players in the supply chain, they are both carriers and potential recipients of sustainability. According to the work of Iyere and Misopoulos (2022), coordination, information exchange and senior management involvement are essential for effective management of stakeholder's risks and influences. Establishing sustainable collaborations, educational programs or collaborative management tools (such as blockchain for tracking) helps consolidate the resilience of supply chains while ensuring strict adherence to sustainability standards. Nevertheless, it is important to note that the expectations of various stakeholders can sometimes be divergent or even conflicting. Nevertheless, it is important to note that the expectations of various stakeholders can sometimes be divergent, or even conflicting. For example, investors generally tend to aim for short-term profits, whereas sustainability commitments require long-term investments. The qualitative analysis carried out by Uttam et al. (2024) among small and medium-sized Indian companies reveals a variety of reactions to external pressures: some choose active compliance approaches; others favor intermediate solutions or even evasion strategies. This variety highlights the need for multi-stakeholder mediation and consultation mechanisms to manage relations between stakeholders and the supply chain. As Siems et al (2022) point out, proactively engaging stakeholders right from the creation of supply chains helps to increase the company's legitimacy, promote organizational innovation and consolidate collective learning capabilities.

1.2.1. Stakeholders' Engagement: Definition and Typologies

The notion of stakeholders has been the subject of significant development over the last twenty years, mainly in North American management research. This notion has undoubted heuristic value: it enables us to conceive of the environment in a richer and more complex way than in most organizational theories and approaches to strategy, and to emphasize the relativity and porosity of the company's boundaries. Moreover, the environment is no longer conceived in an essentially competitive and commercial mode, and stakeholder theory aims to go beyond a narrowly economic vision of the firm, inviting socio-political variables to be taken into account in strategic analysis. It thus advocates a fairly open vision of the company, whose boundaries with the environment and society are no longer stable and clear-cut, but rather fluid and open to negotiation and redefinition (Cazal, 2011).

The concept of stakeholder is defined as « *an individual or group of individuals who can affect or be affected by the achievement of the organization's objectives* » (Freeman, 1984, p. 46). Thus, the company is confronted with a large number of stakeholders to take into consideration, which constitutes a difficult mission. As a result, several authors have proposed typologies of stakeholders in order to distinguish them according to several criteria, for example, Savage et al (1991) distinguished stakeholders according to their cooperation in the decision-making processes of the company: cooperative, marginal and non-cooperative stakeholders. Clarkson (1995) proposes a typology of primary and secondary stakeholders. Carroll & Näsi (1997) distinguish two types of stakeholders: internal and external. Mitchell et al (1997) identify latent, expectant and definitive stakeholders, based on the possession of one, two or all three attributes: power, legitimacy and urgency. The Stakeholder's Theory aims to demonstrate that the company is required to react positively to society's demands, as expressed by its stakeholders. If it fails to fulfil these demands, stakeholders can exercise pressure on the company and force it to change its behavior.

To this aim, stakeholder engagement is referred to in most definitions as a process that includes individuals or groups who have a direct relationship with the organization in its actions and practices to achieve an objective. Indeed, according to Sloan (2009) stakeholder engagement can be defined as the process of taking into account, participating in, or involving individuals and groups who influence or are influenced by the activities of the organization. Morsing and Schultz (2006), inspired by Grunig and Hunt's (1984) public relations models, assume that there are three types of strategy with stakeholders, according to the way in which companies strategically engage in CSR communication. The « *Stakeholder information strategy* » is grounded on « *one-way communication* », from the organization to its stakeholders. This strategy is intended to provide information to the public rather than trying to persuade them (Morsing and Schultz, 2006, p.326). Regarding the « *Stakeholder response strategy* » it seems to be based on "*asymmetrical two-way*" communication, but it's more of a unilateral strategy, as the company only aims to persuade its stakeholders of its own attractive status (Morsing and Schultz, 2006, p.327). As for the stakeholder « *involvement strategy* », it implies a dialogue based on the persuasion of both parties. This strategy suggests that companies engage in a dialogue with their stakeholders to investigate reciprocally in advantageous actions, on the premise that both parties involved in the dialogue are ready to make changes (Morsing and Schultz, 2006, p.328). Similarly, Bowen et al., (2010) propose three stakeholder engagement strategies: transactional, transitional and transformational.

After presenting the definitions and typologies of stakeholders, we now explore the determinants that influence their engagement and active participation in organizational activities.

1.2.2. Determinants of Stakeholder Engagement

The choice of a stakeholder engagement strategy is influenced by a combination of internal factors (managers' vision, company size, nationality, activity sector) and external factors (stakeholder pressures). The first is associated with managers' vision. The adoption of a social approach may be imposed (for fear of sanctions, loss of market position, loss of reputation, etc.) or voluntary, due to the manager's conviction. In this context, Paine (2001) assumes that many companies engage in responsible initiatives because they just need to effectively inform the larger public about what they're up to in order to generate and keep up favorable stakeholder support. The second is the company's size. Several authors mention the necessity of considering the size of the company during the implementation of a corporate social responsibility approach. In this context, Yu & Choi (2016) argue that company size and employee position can imply different pressures on CSR practices. As larger companies usually confront a higher pressure from stakeholders to adopt CSR and different levels of employee management producing different levels of CSR awareness. According to their analysis, the ability of a larger company to support CSR costs is due to their access to more resources than smaller companies. The third factor relates to the nationality of the company. According to Germain and Gates (2010), European companies aim for greater harmony between the economic and social dimensions of their actions. In another context, Goergen et al., (2019) multinational firms are less embedded in a unique domestic context and may be exposed to a wide variety of pressures from both their home and foreign countries in the places where they operate. Hence, they will adopt a CSR initiative in order to increase their legitimacy and thus their competitiveness compared with domestic companies. The fourth concerns the activity sector. According to some authors, the activity sector has a major role to influence the choice of companies to implement a social initiative (Blombäck & Wigren, 2008; Germain & Gates, 2010; Oueghlissi, 2013). Indeed, companies operating in the industrial sector are highly exposed to the use of chemical ingredients, and are consequently vulnerable to further evaluation and pressure in terms of working conditions and environmental impact (Blombäck & Wigren, 2008). As a result, they will be required to implement a CSR strategy to reduce pressures. The last relates to the degree of threat posed by stakeholders. Stakeholders have influence that they can exert on the company to take their interests into account. In this context, Ahinful et al., (2022) emphasize the ability of stakeholders to affect the company's current and future activities. Firms face a conflictual situation with their stakeholders, who are constantly protesting and demanding the satisfaction of their needs. As a result, they find themselves obliged to adopt a CSR strategy, in order to reduce these pressures. However, the degree of influence varies from one group of stakeholders to others. As a result, the firm may be inclined to privilege a particular group of stakeholders, while ignoring others. In conclusion, the choice of a stakeholder engagement strategy is determined by several factors affecting both the company and its stakeholders. The company is required to choose its engagement strategy as a first step, but also to be able to choose an appropriate CSR communication strategy.

Following the literature review, the next section outlines the research methodology pertinent to the Tunisian context.

2. RESEARCH METHODOLOGY

The current situation in Tunisia seems critical, with the emergence of the Covid-19 health crisis causing economic deterioration across all sectors, companies and even individuals, leaving Tunisia's managers and their stakeholders uncertain and confused about their own future. The aim of this research is to analyze the stakeholders' engagement process and the actions implemented by companies as part of the transition to sustainable supply chains. The qualitative method was chosen to address this issue. By conducting research on multiple cases of Tunisian companies in the context of the COVID-19 pandemic and which have set societal actions as criteria for case selection. Thus, the societal aspect is a complex subject that requires in-depth understanding, so the case study method is the most appropriate. According to Tuli (2010), qualitative methods are more concerned with understanding the meaning of social phenomena. Among the many methods of data collection, the case method is the preferred choice. Indeed, according to Crowe et al. (2011) the case method is a research model widely adopted in various branches, particularly in the social sciences disciplines, as it aims to go deeper into the clarity and comprehensibility of complex subjects in their real-life situation. This approach is based on the study of five cases from different sectors (chemical, service, agri-food industries, banking, textile). The five cases were chosen through a purposive sampling method that focused on the significance of sustainability practices, the variety in size and sector for effective cross-case analysis, and the availability of respondents. The selection of these cases was based on the principle of saturation and replication, ensuring that the data collected was sufficient to cover all aspects of the phenomenon under study. Data saturation was achieved at a clearly defined stage of the study. The selected cases were sufficient to cover both predefined and emergent perspectives relevant to the research objectives. Responses showed redundancy across the five cases, indicating that further cases were unnecessary for uncovering new insights. Two data- collection techniques are used in this research: semi-structured individual interviews and documentary analysis using an interview guide (see table 1) covering the different themes associated with the research (attitude towards CSR, stakeholders, stakeholder engagement strategy before and during the Covid-19 health crisis). The interview guide was intentionally limited to core themes to avoid influencing participants' responses and introducing bias. Additional topics, such as supply chain reconfiguration, resilience, were allowed to emerge naturally during the interviews through an open coding process, ensuring that the analysis captures both predefined and emergent themes while preserving the authenticity of participants' insights.

Table 1. Interview guide

Themes	Questions
Attitude to CSR	When I say CSR, what comes to mind? What do you consider to be the priorities in adopting a CSR approach? What were the reasons that led you to adopt a societal approach? What are the main social practices you've implemented? How do you evaluate your CSR practices?
Stakeholders	Who are the main actors involved in this societal approach? What criteria did you use to select your stakeholders? Why do you engage your stakeholders?
Stakeholder's engagement before and during the "Covid-19" crisis	What are the elements that lead to an engagement process? Based on your experience within the company, are there any risks involved in not engaging? What are the benefits of engagement? What benefits will your stakeholders gain from this engagement? How do you encourage your stakeholders to get engaged? How did you react to the pandemic? What actions were taken during the crisis? How did your stakeholders react? Who were the key actors who helped or prevented you from overcoming the crisis? How did you manage the consequences of the crisis? How would you describe your company's current situation?

Thus, processing the data is a more or less difficult task, given the richness of the data. As a result, we opted for qualitative content analysis, in order to analyze the information collected and produce a well-structured work. The interviews conducted are organized into segments based on the unit of syntactic analysis, with each segment receiving codes that correspond to the themes or categories outlined in the interview guide. This study employs a coding process that merges closed coding, which helps meet the research aims, with open coding, allowing for the exploration of unforeseen issues not initially included in the research aims. To enhance the reliability of the coding process, the coding operation was conducted by another researcher who independently analyzed the data. The results of the coding were then cross-checked and discussed to ensure consistency in code interpretation and to minimize potential bias.

ATLAS.ti software is used to support the content analysis task by reinforcing the organization of the data collected. In total, 21 interviews were carried out, lasting between an hour and an hour and a half. The characteristics of the sample are summarized in the following table.

Table 2. Sample characteristics

Case	Activity sector	Workforce	Interviews	Position
1	Chemical industries	1100	3	Administrative Department Manager, Social Affairs Manager, Environmental Protection Manager.
2	Service: issuing Meal Vouchers & Employee Benefits Solutions	100	6	CEO, Customer Department Manager, Marketing Department Manager, Internal Control and Treasury Department Manager, Development Department Manager, Human Resources Manager.
3	agri-food industries	+500	4	CSR manager, management control manager, energy and HSE manager, human capital development manager.
4	Banking	130	3	CSR Manager, Brand and Communication Manager, Purchasing Manager.
5	textile	+500	5	Marketing manager, product quality and development manager, CSR and HSE manager, regulatory quality and environment manager, HR manager.

After defining the research methodology, data collection, and analysis methods, the next section presents the key findings derived from the five case studies.

3. RESULTS OVERVIEW

For some firms, CSR is considered as an imposed coercive constraint through standards and laws. In order to preserve a certain legitimacy that will guarantee their survival, as explained by neo-institutional theory, or by the theory of resource dependence, companies find themselves obliged to align themselves with the pressures exerted by the environment and specifically by resource owners, to assure their existence. This is the case of the company (case 1), which is trying to build a responsible corporate "image" to align itself with the constraints of the environment and the constraining context in which it operates. Similarly, the survey shows that the company (case 1) is more concerned about the environmental dimension, due to the negative impact of its activity on the environment. It adopts a CSR strategy in order to reduce any form of pressure that might affect its survival. This attitude towards CSR is explained by a fear of social pressure from the environment. However, the results also reveal that some firms (cases 2,3,4.5) give particular attention to CSR by implementing a voluntary corporate social responsibility strategy from the start. This can be explained by their consciousness of the need to adopt such an initiative, which could be beneficial to the company, even if they belong to the service sector. Indeed, according to the interviewees, the social dimension plays an important role in these companies. The company's managers adopt a social initiative as their principal goal. They consider that taking a strong social dimension into account could have a positive impact on the company's activity.

Table 3. Results relating to the theme: attitude towards CSR

Case	*Verbatim*
1	*« We started to take an increasing interest in CSR after the Tunisian revolution, exercising some societal practices in order to mainly involve our external stakeholders who represent a great danger, but which have not been concretized in a well formalized CSR framework. »* ***(Verbatim Social Affairs manager).*** *« Currently, in order to conquer certain markets and maintain a competitive position with other countries as an exporter of chemical fertilizers, we have received funding from a foreign agency, in order to introduce a concrete societal approach within our company »* ***(Verbatim, administrative department manager).***
2	*« For me, CSR means serving and investing in the community, creating a favorable social climate, extrapolating happiness and helping people voluntarily. In fact, it's the company's shared vision, it's in the company's DNA, all the stakeholders are involved, they even have the freedom to express their ideas or proposals, where they will be discussed afterwards in Workshops »* ***(Verbatim, Marketing Department Manager).*** *« Company performance is perceived as plural in the sense that you can't be economically successful without being socially successful in an environmental sense, and when I say socially, I mean internally in a societal, environmental and economic way all at the same time. It's a collective vision, people are more mature than products that co-produce, that give, that's why real performance comes first through social performance, today a manager's vision is to go first through social then economic »* ***(Verbatim, CEO).***
3	*« For us, CSR is a mind-set, responsible behavior that starts with the individual and can then be replicated throughout the company. Without engagement, we won't be able to talk about CSR, even if we have the material resources to implement green investments »* ***(Verbatim, CSR manager).*** *« CSR is a lever for overall performance, a kind of anticipation, the day or the moment will come when CSR will be an obligation, and we're in the process of preparing the ground for it. Our ambition is to build a sustainable CSR management system, and we're looking to anchor the CSR culture in our activities, that's the most important thing for us. We'll be able to concentrate for 9 months on our work, provide the necessary documents and we could easily be awarded the label, but we're looking for our approach to be not only credible, but also operational* ***». (Verbatim, human capital development manager).***
4	*« CSR is a voluntary engagement based on 4 key pillars: economic, social, civic and environmental. We always look to act as a responsible and respectful company. It used to depend on the policy of the group that helped us to get here, but we're continuing and improving our social policy, even after the group's move in 2019. Which proves that we have earned the Label Engaged CSR in 2022 »* ***(Verbatim, CSR Manager).*** *« First of all, it's a question of ethics and morality. We're convinced of the importance of CSR within our bank, and we're always looking to have a positive impact and maintain our image as a socially responsible bank »* ***(Verbatim, Brand and Communication Manager).***
5	*« Our CSR approach is structured, we have to declare our extra-financial balance sheet with the countries where we export, we have appointed CSR officers to occupy positions in the various entities »* ***(Verbatim, Marketing manager).***

All of the companies surveyed demonstrated community motivations in the form of pressure from internal or external stakeholders and in the form of the companies' desire to avoid this pressure and engage their stakeholders. The company (case 1) has been under pressure from its stakeholders and has adopted a CSR approach in response to the pressure from its external stakeholders. However, the other companies adopt a proactive CSR approach in order to avoid potential pressures from stakeholders that could affect their survival. The company's stakeholders are a source of cooperation, if they feel respected and involved, as well as a source of danger and threat to the company, if their needs are not satisfied. Engagement appears to be an essential step for the company to ensure its sustainability over the long term. Indeed, the results reveal that even though the companies (cases 2,3,4,5) were not the target of pressure from their stakeholders, this did not prevent them from engaging them in their CSR strategy. This can be explained by the proactive attitude adopted to avoid any pressures in the future. As for the company (case 1), the results show that it finds itself in a difficult and delicate situation due to the intense pressure exerted by its external stakeholders, especially after the Tunisian revolution, hence the need to react and respond to their needs in order to survive.

Table 4. Results relating to the theme: Stakeholders

Case	*Verbatim*
1	*« We are suffering from external pressure and we regret not having introduced CSR before, we are trying to adjust in order to reduce the risk of pressure. We're most interested in our internal stakeholders, who are the solid core of our company. We were under pressure from our external stakeholders during the 2011 revolution, but our employees were there for us, even sacrificing their lives to defend us from the revolvers at that time »* ***(Verbatim, Environmental Protection Manager).*** *« Pressure from farmers, fishermen, residents of the cities where our factories are located because of the emission of SO2 gas (a polluting and instantaneous gas that attacks the respiratory system), in addition to strikes and claims by neighbors of the industrial zone requesting employment positions in our company as compensation for the negative effects of our company »* ***(Verbatim, Environmental Protection Manager)***
2	*« Our social strategy primarily concerns our employees, who are our first ambassadors, but also our retailers, restaurants, consumers and users of our vouchers. Happy employees mean happy customers, and happy customers mean happy shareholders. We try to make the best decision, which can never be 100% perfect, but rather optimal»* ***(Verbatim, Human Resources Manager)***
3	*« When I invest in social actions for the local community, it offers social stability in the areas where we operate, no problems, it's a win-win situation. First of all, we assume our responsibility towards our environment through actions such as saving energy, water resources and waste, and then we generate money »* ***(Verbatim, Energy and HSE manager).***
4	*« Our goal is to take into account the needs of all our stakeholders. The dialogue with our stakeholders is at the core of our bank's CSR strategy. Our stakeholders are: customers, employees, suppliers, subcontractors, shareholders and civil societies. When our stakeholders are engaged, problems are minimized. »* ***(Verbatim, Brand and Communication Manager)***
5	*« We are looking first and foremost for credibility in our engagement with society, we cannot work with suppliers who are not responsible in terms of CSR, we cannot be a responsible company without having mutual relations based on transparency with our stakeholders, we even engage our external stakeholders in order to attract and influence them towards CSR practices, we are looking to create shared value »* ***(Verbatim, CSR and HSE manager).***

The results of the interviews reveal that as much as stakeholders are engaged, no form of pressure was detected, this was reflected by a proactive stakeholder engagement strategy being adopted. This engagement is based on a communication strategy aimed at engaging stakeholders in a dialogue in order to encourage them to support the social practices implemented. However, the company (case 1) has suffered from pressure from external stakeholders, which has forced it to adjust to their demands in order to find a climate of understanding that will guarantee the survival and continuity of its activity. This company has adopted a reactive engagement strategy in the sense that it has decided to introduce a social initiative that takes into account the expectations of stakeholders in response to the pressures applied to it.

Table 5. Results relating to the theme: Stakeholder engagement strategy pre-crisis Covid-19

Case	*Verbatim*
1	*« After the Tunisian revolution, we decided to engage our external stakeholders, as they represent a great danger to us. Societal practices were organized (trucks to collect the waste generated by our activity, the renovation of schools close to our factories and we even had an order from the government to create a subsidiary company to offer jobs to young people in the region, to minimize the pressure). As for our internal stakeholders, it's a continuation of our practices already in place before (meal vouchers, health insurance, excursions, gifts for their children on the occasion of their success, sports and entertainment clubs), as well as the creation of a trade union and a listening unit whose mission is to discuss and negotiate with internal and external stakeholders»* ***(Verbatim, Administrative Department Manager)***
2	*« Our stakeholder engagement strategy is materialized through social network communication... sensitization days, training, satisfaction inquiries, a CSR charter signed by our suppliers on respect for the environment and the way they relate to their stakeholders. Among the strategic practices adopted by our company are projects for the profit of our affiliated restaurants which aim to collect used frying oils, plastic packaging and metal packaging. We have also initiated a project to raise consciousness of the importance of a balanced diet and a healthy lifestyle among Meal vouchers holders»* ***(Verbatim, Development Department Manager).***
3	*« Our engagement is entirely voluntary, and we are always proactive. Many actions have been implemented for the well-being of our stakeholders, such as the purchase of missing materials for nearby schools, and the organization of excursions, seminars and sensitization days every year. Even our purchases are responsible and are carried out with suppliers by drawing up a supplier charter »* ***(Verbatim, Management control Manager).***
4	*« Our CSR approach is based on dialogue and communication with our stakeholders through annual CSR reports published on site and on social networks, seminars and public awareness days. Our CSR approach is voluntary, we respond to our customers' needs in an ethical manner, and we have equipped certain schools to combat school drop-out. We also try to limit our environmental impact by transforming all our cards into ecological ones. »* ***(Verbatim, Purchasing Manager)***
5	*« This is the first 100% recyclable product, which is highly innovative in the world of textiles. After the product has reached the end of its life, the customer brings it back to the point of sale, where it can be washed 50 times, and we give them a voucher or a bonus as a way of raising consumer awareness of the need to protect the environment, and then we recycle it»* ***(Verbatim, regulatory quality and environment manager).***

The results show that even in times of crisis, all the companies studied did not cease their social initiatives. The impact of the crisis was minimal, due to the engagement of stakeholders in the company's corporate social responsibility. In other terms, stakeholders were cooperative and grateful for the company's engagement during the crisis. This was the case for companies (2,3,4,5) which insisted on communication with stakeholders. For these companies, stakeholder engagement went beyond the simple dissemination of information, and involved communication and dialogue. The results also reveal that the company (case 2) succeeded in profiting from this crisis by transforming their products towards digitalization in order to minimize the risks of contamination due to the circulation of their meal vouchers from hand to hand. Even the supply disruptions were an opportunity for companies to switch to local suppliers and further develop their staff's machine maintenance skills. These unpredictable changes confirmed their ability to adapt to unexpected circumstances, and further encouraged employees' ability to overcome and withstand crises.

Table 6. Results relating to the theme: Stakeholder engagement strategy post-crisis Covid-19

Case	*Verbatim*
1	*« During the Covid-19 crisis, the support of our internal stakeholders was apparent, we communicated well with them, because of the critical situation that our company has been facing since the Tunisian revolution, our employees were understanding and cooperative, they continued to work as if nothing serious had happened. Of course, we offered all sanitary measures (masks, antiseptic gels, etc.) and even dedicated a driver for their displacement. For our external stakeholders, we offered financial subsidies, an ambulance for the hospital nearby our worksite, oxygen concentrators, masks, etc.»* ***(Verbatim, Administrative Department Manager).*** *« We have offered donations for hospitals, oxygen concentrators for our employees and for hospitals, and food for homes close to our company. We already had a crisis unit and an infirmary established when the company was created. We were truly exemplary in crisis management, as the nature of our business requires us to take these protective measures even before the pandemic, due to our high exposure to chemicals. For this reason, the damage caused by the pandemic was minimal »* ***(Verbatim, Environmental Protection Manager)***
2	*«The crisis has affected us positively, and our employees have moved to teleworking. A lot of hard work with a cohesive team has helped us to achieve a feat in this period! we have moved to innovation, in fact, the circulation of money or vouchers from hand to hand allows the virus to be transmitted, which is why we have moved to dematerialization of our vouchers, creating a card that will ensure greater flexibility by recharging it remotely »* ***(Verbatim, Internal Control and Treasury Department Manager)***
3	*« we did encounter some difficulties, especially with the closure of the borders, when our foreign suppliers were banned from arriving, which impacted our business for a period. Purchases during the Covid period were disrupted by foreign supplies »* ***(Verbatim, management control manager).*** *« The crisis was an opportunity for us to duplicate our local purchases, we really felt the need for our local suppliers and we encouraged this approach by minimizing our dependence on foreign suppliers, in addition to the contribution of our human capital was shown at this time, we trusted our internal skills to solve technical problems»* ***(Verbatim, human capital development manager).***
4	*« During the Covid-19 pandemic, we implemented a crisis unit that ensures the continuity of the company's activities and communication for all stakeholders. We covered medical expenses, tests, antiseptic gels and masks for our employees, and donated medical equipment to hospitals. We also discussed deadlines with our customers, and were flexible and understanding with our customers and suppliers. This period of crisis has shown us that our stakeholders are reassured. They are very pleased and grateful for our efforts, and as a responsible bank, we can only feel proud and confident»* ***(Verbatim, CSR Manager)***
5	*« We decided to start manufacturing masks to meet the requirements of the pandemic and to face up to this crisis, we were agile and we proved to be very adaptable»* ***(Verbatim, HR manager).***

The following discussion analyzes these findings in relation to existing literature.

4. DISCUSSION

The decision to engage in a CSR approach with various stakeholders depends on a number of factors, mainly the nature of the activity in which companies operate. Indeed, as Blombäck & Wigren (2008) assert, companies operating in the industrial sector are heavily dependent on chemical ingredients, and are likely to come under additional pressure in terms of working conditions and environmental impact. As a result, companies operating in the industrial sector, and more specifically in the chemical industry, are concerned about their impact on society, and are constantly trying to apply socially responsible practices to reduce stakeholder pressure, so that they are seen as socially legitimate. However, this is not always the case for all companies adopting CSR practices. The results show that a company's social initiative can be determined by the manager's experience and vision of CSR. This has enabled him to acquire the social skills needed to implement a social strategy. Such a vision demonstrates a high degree of openness and awareness of CSR, which will be passed on to his employees. These results are in line with those of Paine (2001) who asserts that many companies engage in social initiatives because their managers feel it is morally "*the right thing to do*".

Similarly, the company's nationality may explain its high level of social engagement. It should be remembered that the bank (case 4) was previously a French group subsidiary, which is particularly interested in and committed to the social dimension. This finding confirms that of Germain and Gates (2010), who assume that European companies are placing greater importance on societal concerns. Stakeholders are therefore a pillar of the CSR initiative, and their engagement remains a necessity. It therefore comes down to the proactive attitude and conviction of companies to engage their stakeholders. These findings concur with those of Johnson (1971), who states that a socially responsible manager is one who has a utility function of the second kind, so that he is interested not only in his own well-being but also that of other members of the company and his fellow citizens. Thus, companies sometimes find themselves obliged to engage their stakeholders, as a result of pressure exerted by them. This comes down to their "passive" attitude as designated by (Saulquin & Schier, 2007) in the sense that "the company waits until the pressure from stakeholders is strong to carry out social actions. The results also showed that all surveyed companies agree on the seriousness and influence of stakeholders on their business. They therefore try to find a climate of agreement with them by favoring certain stakeholders over others according to their degree of threat. These results coincide with the assertions of Mitchell et al. (1997) which assume that the company is required to better identify and prioritize its stakeholders according to the possession of one, two or three attributes: power, legitimacy and urgency, so that leaders can achieve certain goals.

In the same context, the size factor plays a significant role in involving and reconciling the interests of stakeholders. As part of our research, the small and medium-sized enterprises surveyed are better able to manage their stakeholders, in relation to the large undertaking investigated (case 1). During the Covid-19 health crisis, stakeholder pressures were minimal or absent. Similarly, stakeholder support was evidence of early engagement by companies before the crisis. This is consistent with Huang et al., (2020) which states that when faced with unexpected environmental turbulence, support from key stakeholders provides critical resources for the company to adapt to and overcome the crisis quickly. The sector of activity therefore plays an important role in managing health crises, especially for industrial companies, which require health care even before a pandemic occurs. Thus, the survey shows that the crisis can be a source of opportunity for some companies, in the sense that the company can achieve organizational resilience through a coherent and communicated CSR vision and mutual commitment built on values of support, trust, recognition and sharing with stakeholders. In this regard, Saulquin & Schier (2007) argue that if CSR is perceived as a lever of internal dynamics its role is summarized in innovation, in the sense that the company seeks opportunities to find new products and services and have innovative approaches.

The COVID-19 crisis was therefore an opportunity for the companies surveyed to assess the effectiveness of their CSR approach. The results show that the internal and external stakeholders of the companies surveyed demonstrated their unconditional support during the COVID-19 crisis. The companies were able to overcome the crisis by adapting their supply chain due to the cooperation of their stakeholders, who proved to be extremely agile and flexible. Therefore, the active engagement of stakeholders enables the supply chain to be reconfigured, transforming the threat of crisis into an opportunity, which in turn creates a resilient and sustainable supply chain. Building on the insights from the discussion, the conclusion summarizes the main findings, highlights the study's limitations, and directions for future research.

CONCLUSION

The company becomes a responsible coordination center, integrating the CSR vision into its relationships with its stakeholders. The sustainable supply chain emerges from this dynamic, and the impact is reinforced in times of crisis through the trust and cooperation of stakeholders. Firms evaluate vulnerabilities and pinpoint critical disruptions. CSR-focused engagement guarantees that responses take social and environmental effects into account, emphasizing the well-being of suppliers and employees alongside the continuity of operations. They adapt their current processes and frameworks to address emerging challenges. CSR initiatives inform these adaptation choices, encouraging sustainable sourcing, ethical labor practices, and effective environmental management, while input from stakeholders ensures alignment with societal expectations. Then, they restructure their supply chain designs to bolster resilience and promote long-term sustainability. This involves reorganizing supplier networks, incorporating new technologies, and adopting circular economy practices. CSR engagement offers a basis for assessing strategic decisions in relation to social and environmental goals. And the last phase solidifies changes, integrating sustainability and resilience into routine operations. Ongoing monitoring, driven by CSR, guarantees that the adjusted supply chain stays in sync with stakeholders, responds to new risks, and provides both economic and social benefits.

Through the results analyzed and discussed, we propose this model which highlights the central role of the company engaged in a corporate social responsibility (CSR) approach, over and above its economic performance alone. Driven by a vision based on transparency and credibility, such a company integrates the expectations and interests of its stakeholders into the sustainable structuring of its supply chain. Internal stakeholders (employees, managers), external stakeholders (customers, suppliers), and societal stakeholders (local authorities, institutions) are no longer observers but active actors in the transition. The company engages its suppliers through responsible charters and audits, mobilizes its customers around environmental projects, and develops a constructive dialogue with its ecosystem to build a more ethical, resilient, and eco-designed supply chain. This dynamic encourages the creation of shared value, generating economic, social, and environmental benefits for all those involved. In this way, the company no longer seeks solely to maximize its profits, but to strengthen its positive impact on society and the environment. This engagement proves to be a lever for resilience in times of crisis: the stakeholders, mobilized and confident, cooperate more effectively and support the company, ensuring the continuity of the chain and its ability to adapt to disruptions. This model highlights a logic of virtuous interdependence, where shared engagement forms the core of a sustainable and resilient supply chain.

Figure 1. Stakeholders' engagement in the transition to sustainable supply chain

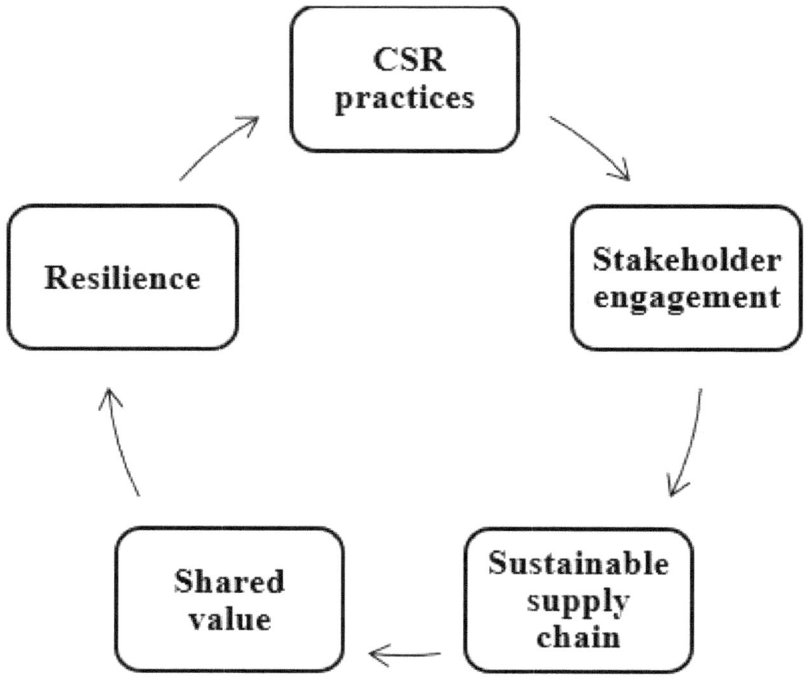

(model proposed by us).

THEORETICAL AND MANAGERIAL IMPLICATIONS

The research additionally emphasizes important contributions that enhance both theoretical insights and practical implementation within the field. The theoretical contribution of this research is based on the analysis of the link between CSR and stakeholders through the study of five Tunisian companies. Drawing on stakeholder theories, RBV, dynamic capabilities, and situational crisis communication theory, this study enriches the existing literature by shedding contextual light on the dynamics of CSR in Tunisia and its role in supply chain transition.

This research enhances the comprehension of sustainable supply chain management by illustrating the dynamic interplay among resilience, shared value creation, and stakeholder engagement in the aftermath of crises. The model developed inductively further advances theoretical frameworks by merging these elements into a cohesive structure that connects empirical findings from Tunisia with broader conceptual insights on sustainability-focused supply chain design. It refines established theories by indicating that sustainability in emerging markets is influenced not only by external institutional pressures but also by the internal learning and adaptive capabilities that firms develop through experiences of crisis. Thus, the managerial contribution of this research lies in understanding the behavior of company managers in choosing their CSR approach and explaining the environmental, social, and economic results of their companies, since a successful strategy of social engagement is one determined in part by the manager's attitude to CSR. Also, the results highlight the importance for Tunisian companies to take a proactive stance on sustainability and resilience. Managers should incorporate social and environmental performance indicators into their supply chain oversight systems, enhance cooperation with local suppliers, and establish enduring partnerships centered around shared value creation.

LIMITATIONS AND DIRECTIONS FOR FUTURE RESEARCH

This study has its own set of limitations. The first relates to the data collection stage. The case study method requires both passive and active observation. However, company instructions refuse the use of this method, which could have further enriched the data collected. Thus, the complexity of the subject justifies the use of a qualitative approach only. The above-mentioned limitations may open up new directions for further research: the use of observation allows the researcher to better understand the phenomenon in its natural environment. Thus, the identification of new determinants seems more relevant to better identify the attitude adopted for engagement in a CSR approach. The adoption of a quantitative approach as a complement to the qualitative one also enables the results to be generalized. Indeed, the model developed in this study could be quantitatively assessed to determine its relevance and effectiveness across different sectors and regions. Performing comparative studies between Tunisia and other developing countries could provide a deeper understanding of how institutional and cultural factors influence sustainable supply chain strategies. Furthermore, longitudinal studies would be advantageous in exploring the long-term evolution of resilience and stakeholders' engagement practices beyond crisis situations.

REFERENCES

Ahinful, G. S., Tauringana, V., Essuman, D., Boakye, J. D., & Sha'ven, W. B. (2022). Stakeholders pressure, SMEs characteristics and environmental management in Ghana. *Journal of Small Business and Entrepreneurship, 34*(3), 241–268. DOI: 10.1080/08276331.2018.1545890

Barney, J. (1991). Firm resources and sustained competitive advantage. *Journal of Management, 17*(1), 99–120. DOI: 10.1177/014920639101700108

Blombäck, A., & Wigren, C. (2008), Are not multinationals also community actors?: Challenging the notion of small versus large firm corporate social responsibility. In *CRRC 2008 Conference Proceedings: The corporate responsibility research conference, Belfast, Northern Ireland, 7-9 September, 2008.*

Bowen, F., Newenham-Kahindi, A., & Herremans, I. (2010). When suits meet roots: The antecedents and consequences of community engagement strategy. *Journal of Business Ethics, 95*(2), 297–318. DOI: 10.1007/s10551-009-0360-1

Bowen, H. R. (1953). *Social Responsibilities of the Businessman*. Harper & Brothers.

Carroll, A. B. (1999). Corporate social responsibility: Evolution of a definitional construct. *Business & Society, 38*(3), 268–295. DOI: 10.1177/000765039903800303

Carroll, A. B., & Näsi, J. (1997), Understanding stakeholder thinking: Themes from a Finnish conference. *Business Ethics: A European Review*, vol. *6, n°*1, p. 46-51.

Cazal, D. (2011). RSE et théorie des parties prenantes: les impasses du contrat. *Revue de la régulation. Capitalisme, institutions, pouvoirs*, (9).

Clarkson, M. E. (1995). A stakeholder framework for analyzing and evaluating corporate social performance. *Academy of Management Review, 20*(1), 92–117. DOI: 10.2307/258888

Crowe, S., Cresswell, K., Robertson, A., Huby, G., Avery, A., & Sheikh, A. (2011). The case study approach| BMC Medical Research Methodology| Full Text.

DiMaggio, P. J., & Powell, W. W. (1983). The iron cage revisited: Institutional isomorphism and collective rationality in organizational fields. *American Sociological Review, 48*(2), 147–160. DOI: 10.2307/2095101

EL AIDOUNI, M., & ABBASSI, A. (2012). La RSE dans le management de la chaine logistique. *Revue Marocaine de Recherche en Management et Marketing*, (6-7).

Frederick, W. C. (1978). From CSR, to CSR: The maturing ofbusiness-and-society thought. Working paper, Katz.

Frederick, W. C. (1994). From CSR1 to CSR2: The maturing of business-and-society thought. *Business & Society*, *33*(2), 150–164. DOI: 10.1177/000765039403300202

Freeman, R. E. (1984). *Strategic management: A stakeholder approach*. Pitman.

Germain, C., & Gates, S. (2010). L'engagement des parties prenantes internes dans les démarches de Responsabilité Globale (RG): L'intégration des indicateurs de RG dans les outils de pilotage du contrôle de gestion. *Management Avenir*, (3), 223–237. DOI: 10.3917/mav.033.0223

Goergen, M., Chahine, S., Wood, G., & Brewster, C. (2019). The relationship between public listing, context, multi-nationality and internal CSR. *Journal of Corporate Finance*, *57*, 122–141. DOI: 10.1016/j.jcorpfin.2017.11.008

Gond, J. P., & Mullenbach-Servayre, A. (2004). Les fondements théoriques de la responsabilité sociétale de l'entreprise. *La Revue des Sciences de Gestion: Direction et Gestion*, (205), 93.

Gourire, M., & Rouggani, K., & AMINE, N. B. (2025). Dévoiler les pratiques de la chaîne d'approvisionnement durable et leur impact sur la performance logistique des entreprises: An exploratory analysis of companies listed on the Moroccan stock exchange. *RMd• Economics, Management &. Social Sciences*, *2*(1), e202510–e202510.

Grandval, S., & Soparnot, R. (2005). Le développement durable comme stratégie de rupture: Une approche par la chaîne de valeur inter-sectorielle. *Management & Avenir*, *5*(3), 7–26. DOI: 10.3917/mav.005.0007

Huang, W., Chen, S., & Nguyen, L. T. (2020). Corporate social responsibility and organizational resilience to COVID-19 crisis: An empirical study of Chinese firms. *Sustainability (Basel)*, *12*(21), 8970. DOI: 10.3390/su12218970

Inês, A., Diniz, A., & Moreira, A. C. (2025). Driving eco-innovation in supply chains through multi-stakeholder collaboration: A review and research agenda. *Journal of Open Innovation*, *11*(1), 100472. DOI: 10.1016/j.joitmc.2025.100472

Iyere, M., & Misopoulos, F. (2022). The degree of stakeholder influences and risks in sustainable supply chains: A systematic literature review. *International Journal of Contemporary Management*, *58*(2), 9–26. DOI: 10.2478/ijcm-2022-0004

Johnson, H. L. (1971). *Business in contemporary society: Framework and issues*. Wadsworth Publishing Company.

Lv, W., Wei, Y., Li, X., & Lin, L. (2019). What dimension of CSR matters to organizational resilience? Evidence from China. *Sustainability (Basel)*, *11*(6), 1561. DOI: 10.3390/su11061561

Mackey, A., Mackey, T. B., & Barney, J. B. (2007). Corporate social responsibility and firm performance: Investor preferences and corporate strategies. *Academy of Management Review*, *32*(3), 817–835. DOI: 10.5465/amr.2007.25275676

Meyer, J. W., & Rowan, B. (1977). Institutionalized organizations: Formal structure as myth and ceremony. *American Journal of Sociology*, *83*(2), 340–363. DOI: 10.1086/226550

Midhani, P. (2009). Resource Based View of Competitive Advantages: Importance, Issues and Implications. *Indian Management Research Journal*, *1*(2), 1–15.

Mitchell, R. K., Agle, B. R., & Wood, D. J. (1997). Toward a theory of stakeholder identification and salience: Defining the principle of who and what really counts. *Academy of Management Review*, *22*(4), 853–886. DOI: 10.2307/259247

Morsing, M., & Schultz, M. (2006). Corporate social responsibility communication: Stakeholder information, response and involvement strategies. *Business Ethics (Oxford, England)*, *15*(4), 323–338. DOI: 10.1111/j.1467-8608.2006.00460.x

Mullenbach-Servayre, A. (2007). L'apport de la théorie des parties prenantes à la modélisation de la responsabilité sociétale des entreprises. La revue des Sciences de Gestion, (1), 109-120.

Oueghlissi, R. (2013). La RSE et les PME. *Revue française de gestion, 236*(7), 163-180.

Paine, L. S. (2001). *Value Shift: Why Companies Must Merge Social and Financial Imperatives to Achieve Superior Performance*. McGraw-Hill.

Sadeghi, K., Ojha, D., Kaur, P., Mahto, R. V., & Dhir, A. (2025). Metaverse technology in sustainable supply chain management: Experimental findings. *Decision Support Systems*, *191*, 114423. DOI: 10.1016/j.dss.2025.114423

Salancik, G. R., & Pfeffer, J. (1978). A social information processing approach to job attitudes and task design. *Administrative Science Quarterly*, *23*(2), 224–253. DOI: 10.2307/2392563

Saulquin, J. Y., & Schier, G. (2007), Des perceptions managérials aux pratiques RSE: une étude exploratoire. *Gestion 2000*, vol. *24, n°*6.

Savage, G. T., Nix, T. W., Whitehead, C. J., & Blair, J. D. (1991). Strategies for assessing and managing organizational stakeholders. *The Academy of Management Perspectives*, *5*(2), 61–75. DOI: 10.5465/ame.1991.4274682

Siems, E., Seuring, S., & Schilling, L. (2023). Stakeholder roles in sustainable supply chain management: A literature review. *Journal of Business Economics*, *93*(4), 747–775. DOI: 10.1007/s11573-022-01117-5

Sloan, P. (2009). Redefining stakeholder engagement: From control to collaboration. *Journal of Corporate Citizenship*, *2009*(36), 25–40. DOI: 10.9774/GLEAF.4700.2009.wi.00005

Soulier, J. M. (2022). Introduction. La révolution supply chain, origines et définition. *Hors collection*, 9-30.

Souto, B. F. F. (2009). Crisis and corporate social responsibility: Threat or opportunity? *International Journal of Economic Sciences and Applied Research*, *2*(1), 36–50.

Teece, D. J., Pisano, G., & Shuen, A. (1997). Dynamic capabilities and strategic management. *Strategic Management Journal*, *18*(7), 509–533. DOI: 10.1002/(SICI)1097-0266(199708)18:7<509::AID-SMJ882>3.0.CO;2-Z

The African Development Bank (AfDB) Group. (2012). Economic and Social Challenges Beyond the Revolution.

Tuli, F. (2010). The basis of distinction between qualitative and quantitative research in social science: Reflection on ontological, epistemological and methodological perspectives. *Ethiopian Journal of Education and Sciences*, *6*(1). Advance online publication. DOI: 10.4314/ejesc.v6i1.65384

Ursic, D., & Cestar, A. S. (2022). Crisis management and CSR in Slovenian companies: The impact of the COVID-19 pandemic. *Sustainability (Basel)*, *14*(5), 2690. DOI: 10.3390/su14052690

Uttam, N., Dutta, P., & Singh, A. (2024). Influence of stakeholders on supply chain social sustainability: New insights from small suppliers in the Indian manufacturing sector. *Journal of Cleaner Production*, *444*, 141015. DOI: 10.1016/j.jclepro.2024.141015

Wirba, A. V. (2024). Corporate social responsibility (CSR): The role of government in promoting CSR. *Journal of the Knowledge Economy*, *15*(2), 7428–7454. DOI: 10.1007/s13132-023-01185-0

Yu, Y., & Choi, Y. (2016). Stakeholder pressure and CSR adoption: The mediating role of organizational culture for Chinese companies. *The Social Science Journal*, *53*(2), 226–235. DOI: 10.1016/j.soscij.2014.07.006

Chapter 12
Transforming Supply Chains Through the Implementation of the Circular Economy and Sustainable Closed-Loop Models

Jose De Jesus Reyes-Sánchez
https://orcid.org/0000-0002-3866-2363
TECNM Instituto Tecnologico Superior de Fresnillo, Mexico

ABSTRACT

The study analyzes how the convergence between circular economy and closed-loop models reconfigures supply chains still anchored in linear patterns. Based on literature review, cross-industry case studies, and examination of emerging policies, technological, financial, and cultural barriers that keep global circularity below ten percent are documented, as well as incentives that favor rapid extraction over resource regeneration. The findings show that the combined adoption of digital twins, smart contracts, and multi-capital metrics increases visibility into material flows, facilitates coordination between actors, and reduces logistics costs by up to 23%, while decreasing the carbon footprint and strengthening resilience to disruptions. It is concluded that effective circularity demands collaborative governance frameworks, green investment and dynamic capacity building that allow organizations to perceive changes, mobilize knowledge and reconfigure assets with agility, turning the regeneration of resources into a central vector of competitiveness and social legitimacy.

DOI: 10.4018/979-8-3373-3790-6.ch012

INTRODUCTION

The ability to sustain human progress without devouring the biophysical foundations of the planet has gone from being an ethical aspiration to constituting a strategic imperative. Under this premise, the circular economy is emerging as a transformative narrative that reconfigures competitiveness by placing the regeneration of resources at the center of business value; However, the global circularity rate remains anchored in single-digit values, evidence of the technical, financial and cultural inertias that continue to relegate the closing of loops to a marginal participation of industrial activity. Pan-European studies show that the lack of incentives for material recovery, coupled with accounting indicators that privilege immediate productivity, limits the flow of capital towards infrastructure reuse and refurbishment, perpetuating the dissonance between discourse and practice (Miatto et al., 2025; Corvellec et al., 2022; Zink & Geyer, 2017).

The latest research agrees that the circular transition cannot be restricted to the end of the useful life of products or focus exclusively on waste management; but which requires redesigning the offer from its conception and reconfiguring supply chains through reduction, reuse and revaluation loops. From the theory of resources and dynamic capabilities, it has been shown that perceiving changes, absorbing knowledge and reconfiguring assets constitutes an indispensable strategic tripod to abandon the extract-produce-discard paradigm and embrace closed-loop configurations. This tripod is legitimized when it is articulated with integrative visions of sustainability that transcend the dichotomy of economy versus ecology and propose robust metrics to capture value throughout the life cycle (De Angelis et al., 2023).

Digitization increases the viability of such proposals by providing granular visibility into the functional degradation of assets and the provenance of materials. The emergence of digital twins, supported by prescriptive analytics and edge computing, synchronizes physical flows with contextual data in near-zero latency and anticipates optimal return windows that reduce logistical inefficiencies and free up capital trapped in recovery inventories. Portfolios of asset research models in the aeronautics, electronics, and textile sectors show decreases in lifecycle costs when twins incorporate predictive maintenance algorithms and integrated carbon certification (Mügge et al., 2024).

However, financial viability will depend on measurement frameworks that combine economic, environmental and social indicators. The proliferation of heterogeneous metrics has diluted the credibility of sustainability reporting and makes it difficult to compare across industries; Thus, a research agenda is invoked that prioritizes hybrid boards capable of simultaneously weighing retained economic value, carbon intensity, effective duration of service, and distributive justice. Causal models based on system dynamics show that the integration of these variables reduces efficiency rebound effects and accelerates convergence towards net-zero targets (Schultz et al., 2024; Nishitani et al., 2022).

A necessary condition for this convergence to take hold is contract innovation, where self-executing smart contracts, anchored on state-of-the-art blockchains, redistribute responsibilities for product returns, reduce litigation, and ensure the integrity of critical data without human mediation. To this end, effective circularity will also depend on the relational density between organizations. The research that systematizes experiences in eco-industrial parks identifies twenty opportunities and twenty-three challenges to implement synergies of waste heat exchange, by-products and logistics services, highlighting the urgency of neutral data platforms and regulatory harmonization in the field of extended producer responsibility. Such findings emphasize that the challenge is both technological and cultural: it requires replacing competitive paradigms with regenerative synergies that distribute risks and benefits equitably (Centobelli et al., 2021; Prajapati et al., 2022).

Territorial inequality introduces additional nuances in emerging economies, highly digitized supply chains coexist with informal segments that handle significant volumes of post-consumer materials without the support of formal standards, for this, pilot programs that integrate the popular economy with low-cost traceability platforms show increases in the recovery of critical metals and substantial improvements in collectors' incomes urban, articulating social inclusion with environmental profitability and avoiding corporate profitability (Schlüter et al., 2023; Sellitto et al., 2025).

Green financing stands as a transversal catalyst, in this sense, the standardization of performance metrics has attracted capital to reverse logistics and modernization projects, generating a secondary market for circularity-linked bonds. To this end, recent issuances structured by multilateral banks channel resources into automatic sorting infrastructures and remanufacturing plants with authenticity holograms; Returns indexed to verifiable carbon reduction have shown competitive risk-adjusted returns against traditional corporate bonds, confirming that profitability and sustainability can coexist under extreme transparency regimes (Zisopoulos et al., 2023).

The complexity of these systems calls for collaborative governance frameworks, where international organizations and industry coalitions have begun to formulate common taxonomies and regulatory sandboxes to accelerate the adoption of circular solutions. To this end, initial evidence indicates that regulatory sandboxes halve licensing times for advanced modernization technologies and reduce the exposure of SMEs to compliance litigation, stimulating innovation in contexts of high uncertainty (Prajapati et al., 2022).

The convergence between data science, collaborative agreements and multi-capital metrics opens, in short, a plausible route for circularity to leave its status as a pending project and become the backbone of competitive, inclusive and environmentally friendly supply chains. This work is situated at such a critical intersection and proposes an integrative framework that articulates dynamic capacities, last-mile digitalization and distributed governance as indissoluble vectors to close the gap between ambition and execution and contribute, with scientific rigor, to the construction of a shared planetary well-being (Sellitto et al., 2025).

STATEMENT OF THE PROBLEM

Despite the proliferation of corporate discourses affirming circularity, the global circular materials metrics reported by the Circular Economy Platform have remained stagnant at around eight percent over the past decade, evidencing a significant gap between ambition and execution. Production systems are still trapped in a linear inertia of extraction and disposal, supported by short-term financial criteria and fragmented regulatory frameworks that penalize the return of materials to the industry. This dichotomy raises questions about the factors that hinder the scalability of closed-loop models and questions the ability of companies to align economic-environmental incentives with the integrity of their value chains (ElBannan and Löffler, 2024; Ordóñez-Borrallo et al., 2024; Shimauchi et al., 2025).

The literature identifies technological obstacles, infrastructure limitations, and governance gaps that prevent articulating responsibilities between producers, distributors, and consumers. Added to this is the absence of business models that capture value in the post-consumer phases without deteriorating the user experience. In this scenario, it is necessary to deepen systemic strategies that integrate principles of eco-design, smart reverse logistics, and green financing to overcome the paradox of declared circularity and effective circularity (Alaassar et al., 2020).

BOTTOM

Historical Evolution of Reverse Logistics

Reverse logistics has gone from being a defensive practice aimed at minimising disposal costs to becoming a strategic axis of circular competitiveness. During the so-called corrective stage, which is between 1995 and 2005, after-sales returns were managed as exceptional events, with linear flows and without traceability; However, the early addition of two-dimensional optical readers paved the way for the first generation of reverse inventories, reducing the recovery cycle time. Between 2006 and 2015, the logic of retained value emerged: high-tech manufacturers demonstrated that the refurbishment of motherboards could generate gross margins comparable to the sale of new inputs; This led to academic forums and industry consortia dedicated to standardizing environmental and financial benefit metrics. The emergence of Industry 4.0 marked, as of 2016, the predictive stage: IoT sensors, prescriptive analytics, and digital twins orchestrated circuits that anticipate failures, schedule collection, and even sell recovered material in the future, which triggered loop closure rates to values (De Angelis et al., 2023; Schultz et al., 2024).

International Regulation and Extended Producer Responsibility

Regulatory convergence is today the dominant driver of circularity. The European Union is moving towards the Ecodesign for Sustainable Products Regulation, which will require durability, repairability and minimum recycled content to be declared in the digital passport, penalising imports that do not comply with circularity parameters with border tariffs. At the same time, the Global Plastics Treaty, negotiated within the UNEA, contemplates levies on unrecovered kilograms in all phases of the life cycle, which pushes global producers to internalize reverse logistics before 2030. Comparative studies in SMEs in the plastics sector in India reveal that the obligation to manage post-consumer material incentivized the adoption of multi-criteria decision-making methods that prioritize PCR resins without compromising mechanical strength. Latin America, for its part, is promoting progressive collection rates for containers and packaging, while North America combines tax incentives and state return frameworks that already register greater reductions in household appliances (Centobelli et al., 2021).

Innovations in Eco-Design for Circularity

Ecodesign has moved from macro approaches, focused on minimizing mass, to microstrategies that integrate virtual twins of materials and molecular degradation algorithms. Interoperable CAD platforms calculate, in real-time, the circularity footprint of each component: a weighted combination of recyclability, reparability, viable chemical recycling, and integrated energy. Emerging concepts, such as design for decoupling, use 3D-printed micrology grids, whose density varies according to the load path, reducing the overall mass without sacrificing rigidity. The bio-reactive design explores lignocellulosic resins reinforced with fungal fibers that, after multiple cycles of use, biodegrade in controlled industrial environments. Blockchain-based digital passports anchor the chemical signature and maintenance history, increasing transparency for second-life buyers (Jensen et al., 2024; Zhang & Seuring, 2024).

Development of Digital Platforms for Industrial Symbiosis

Electronic by-product exchanges have evolved into ecosystems that use knowledge graphs and machine learning to identify material, thermal, and water synergies between heterogeneous plants. In practice, waste from a bioethanol distillery serves as a protein substrate for inland aquaculture, while waste heat from a refinery feeds district heating networks. The success of these smart markets lies in smart contracts that automatically settle carbon credits once the flow of matter has been verified; The material recovery rate reaches higher levels than above and emissions are significantly reduced compared to linear arrays. Trends point to the emergence of *digital twins* of industrial parks capable of simulating the environmental and economic impacts of each exchange in seconds, allowing regions to adjust their symbiosis strategies before investing in physical infrastructure (Makropoulos et al., 2024).

Product-as-a-Service Business Models

Servitization is gaining ground thanks to digital twins that monitor performance and optimize residual service life. A connected device transmits vibration, temperature and consumption data; Upon detecting deviations, the algorithm schedules preventive maintenance and renews a lease that incentivizes repair rather than replacement. This logic reduces the total cost of ownership received and raises the real rate of return to the factory. In aeronautics, hourly power schemes link manufacturer revenue to effective flight time, displacing operational risk, while in fashion, cloud cabinets satisfy prosumers eager for rotation without increasing primary output. The common denominator is a cultural shift: value is associated with continued performance and avoided impact, not ownership (Karkaria et al., 2024).

Emerging Circularity Metrics and Standards

The quantification of circularity is progressing rapidly. The Multi-Capital Dynamic Index, currently tested in sectors with high use of rare earths, combines mass, energy, toxicity and geopolitical criticality, offering a 4D portrait of the efficiency of each loop. Analytics-fed plant twins reveal circularity bottlenecks invisible to traditional flow indicators and suggest concrete interventions, from supplier changes to process parameter readjustments. On a macro scale, ISO 59020 standardizes the reporting of retained value and extended useful life, while the updated GRI-306 modules require the disclosure of material and energy losses, promoting transparency in the face of ESG investors (Kanda et al., 2021).

The Energy Transition and its Influence on Circular Chains

The energy vector of circularity is shifting towards renewable energies and zero-emission fuels. Thus, urban collection fleets with swappable batteries enable noise-free and pollutant-free nighttime operations, while fuel cell trucks are reserved for regional travel, where simulation scenarios show reductions in the carbon intensity of reverse logistics and payback times of less than six years in regions with low solar costs, . Likewise, hybridization with PV and wind microgrids allows flexible management of sorting loads, and renewable energy certificates generated by logistics centers finance the expansion of charging infrastructure (Rout et al., 2022).

Dynamics of Secondary Commodity Markets

The price volatility of post-consumer polymers and critical metals has driven fintech platforms that trade forward contracts based on purity and carbon footprint. These solutions, assisted by deep learning, predict weekly dispersals and automatically adjust premiums and discounts, fostering diversified portfolios between urban waste and primary mining. The future European Critical Commodities Act includes virtual strategic reserves fueled by urban mining, positioning e-waste as a liquid financial asset. This creates opportunities for *specialty traders* to arbitrage between traditional supply chains and secondary markets based on certified quality (Karkaria et al., 2024).

Corporate Zero Waste Strategies in Advanced Manufacturing

The intersection between additive manufacturing and green lean manufacturing fosters high-strength hollow architectures, eliminating supports and reducing final mass. Line twins, trained with federated data from global plants, detect microvariations in metal powder and adjust mixtures in real time, closing the internal material loop. Companies that integrate statistical process control and reinforcement learning algorithms report reductions in loss and return on investment below 18 months. At the same time, zero-waste-to-landfill certification is consolidated as a sign of circular maturity and reputational advantage (Ordóñez-Borrallo et al., 2024; Shimauchi et al., 2025).

Evolution of Organizational Culture Towards Circular Responsibility

Organizations that embrace narratives of regenerative prosperity link managerial bonuses to circularity indicators and deploy sentiment analysis to detect internal resistances where cross-cutting training in systems thinking and data literacy fosters process innovation and decreases voluntary turnover. A study in the construction industry found that multi-stakeholder collaboration, rather than regulation, explains the adoption of circular practices in emerging economies. To this end, the creation of internal communities of practice and the gamification of circularity challenges reinforce the collective ownership of environmental objectives.

THEORETICAL FRAMEWORK

Dynamic Capabilities in the Circular Economy

The adoption of the circular economy has forced organizations to abandon legacy linear routines and develop detection, seizure and reconfiguration processes that allow waste to be transformed into stable streams of value without deteriorating efficiency. The most recent research demonstrates that these three processes constitute the core of a regenerative competitive advantage by enabling early detection of waste resource opportunities, strategic mobilization of complementary investments, and reconfiguration of assets to process material returns in real-time (Psarommatis and May, 2024).

The result is a paradigm shift: circularity evolves from peripheral pilot projects to a structural attribute that defines the entire business model. Technological trends amplify, above all, the detection phase. The mass deployment of IoT sensors, digital twins, and predictive analytics has increased the speed at which companies identify incompatibilities between waste streams and potential recovery routes, while reducing virgin raw material consumption and scope emissions by an average. This agility is supported by micro-fundamentals that turn ambient intelligence into a daily routine (De Sousa, 2024; Fletcher & Evans, 2025).

In the seizure phase, the literature highlights the importance of blockchain traceability platforms and smart contracts that automatically orchestrate remanufacturing orders when digital twins detect critical failure thresholds. By integrating distributed smart contract logic with cloud lifecycle analysis, enterprises reduce downtime and increase perceived reliability on second-cycle assets. At the same time, rapid prototyping laboratories, financed through open innovation funds, make it possible to validate the technical-economic feasibility of new reuse routes in weeks, not months, minimizing the risk aversion associated with green investments (Psarommatis & May, 2024).

The reconfiguration phase requires redesigning structures, processes, and mindsets. Comparative studies in manufacturing clusters in Central Europe show that firms with higher absorptive capacity reconfigure their chains more quickly, thanks to the effective transfer of knowledge between engineering, sourcing and reverse logistics. Similarly, trust-based governance mitigates tensions between short-term efficiency and exploring regenerative innovations. In used clothing chains, the coordination of NGOs, retailers, and online platforms increases social and economic margins through transparent value-sharing schemes (Sánchez-García et al., 2024).

The additional benefits are manifold. First, organizations that master all three capabilities reduce their operational risk to logistical disruptions: time-series-based simulations of port incidents reveal that they restore normal flow faster than their linear peers. Second, the net savings in material, energy, and waste management costs frees up resources to fund automated disassembly technologies, additive printing with recycled material, and AI-based vision-based sorting systems. Third, the creation of advanced remanufacturing cells has increased the average salaries of skilled operators in mature industrial regions, boosting talent retention and strengthening social cohesion (Bressanelli et al., 2024).

Near-term projections point to an intimate integration between dynamic capabilities and augmented no-code analytics platforms. By allowing plant engineers and logistics coordinators to establish their own traceability routines, organizations have reduced the time to implement new circular solutions and increased the success rate by scaling pilots to the corporate level. This citizen-developer paradigm democratizes innovation, strengthens the culture of continuous learning, and distributes circular responsibility throughout the hierarchy (Psarommatis & May, 2024; Sánchez-García et al., 2024).

Dynamic capabilities provide the operational and cognitive scaffolding indispensable for the circular economy to evolve from a normative aspiration to a dominant business strategy. By synchronizing detection, seizure, and reconfiguration with advanced analytics, collaborative governance, and green financial models, organizations not only improve their environmental metrics, but build competitive advantages that are difficult to imitate and resilient in the face of systemic volatility (Psarommatis and May, 2024; Sánchez-García et al., 2024).

Multi-Capital Accounting and Lifecycle Framework

The integration of multi-capital accounting system approaches has shifted the logic of incremental efficiency towards systemic regeneration paradigms, creating tools such as the Material Circularity Indicator, the Circular Transition Indicator or the Circularity Index combine physical flows with natural, human and social capital metrics. generating holistic dashboards that inform eco-design, responsible sourcing, and life extension decisions. The literature shows that the integration of these indicators into corporate dashboards redefines profitability criteria by explicitly internalizing ecological costs (Calzolari et al., 2022; Segara et al., 2024).

Multi-Capital Accounting and Lifecycle Framework

The evolution of corporate Life Cycle Assessment (LCA) has gone from fragmentary carbon footprint estimates to integrated systems that track, weigh and monetize the effects of production processes on various capitals, shifting the logic of incremental efficiency towards paradigms of systemic regeneration. Consolidated tools such as the Material Circularity Indicator, Circular Transition Indicators, and Circularity Index combine physical flows with natural, human, and social capital metrics, generating holistic dashboards that facilitate eco-design, responsible sourcing, and life-extension decisions. The digitization of LCA through *digital twins* and IoT sensors updates emissions inventories in real time, allowing managers to adjust the material mix when impacts approach planetary thresholds; this responsiveness has reduced carbon intensity per ton processed in the chemical sector (Petri et al., 2025).

The social dimension of the life cycle, traditionally relegated, is being incorporated through frameworks such as STI-Social, which quantify decent work, the inclusion of vulnerable populations and job security, linking circularity with social justice objectives. In addition, the publication of integrated reports that combine financial statements with natural capital balance sheets improves communication with institutional investors, generating improved valuation premiums compared to competitors using one-dimensional metrics. These narratives broaden corporate language: impacts on biodiversity, community cohesion or workplace well-being are internalized into the profitability equation, reconfiguring capital allocation criteria (Calzolari et al., 2022).

The joint effect of these practices is visible in the financial instruments market. More than 60% of green bond issuances in 2024 referred to multi-capital metric packages to justify the allocation of funds to material recirculation projects, ensuring traceability of environmental and social benefits. This has created a virtuous cycle: multi-capital accounting strengthens the credibility of issuers, reduces financing costs, and accelerates the adoption of advanced recovery technologies (Segara et al., 2024; Astudillo et al., 2024; Calzolari et al., 2022).

Institutional Pressures and Drivers of Adoption

The expansion of circularity in companies is catalyzed by a network of coercive, normative, and mimetic pressures that reorder market incentives. The European Corporate Sustainability Reporting Directive (CSRD) mandates the disclosure of granular circularity data from FY2025, which has increased investment in reverse infrastructure, especially when the indicators are linked to senior management remuneration schemes. This effect extends beyond Europe: Asian and US conglomerates with operations in the EU have replicated circularity requirements in their supply contracts to avoid operational and reputational penalties (Segara et al., 2024; Astudillo et al., 2024).

Regulatory convergence is reinforced through the International Sustainability Standards Board and the Global Reporting Initiative, which harmonize frameworks to prevent the proliferation of incompatible metrics, reduce administrative burden, and facilitate external verification. At the same time, competitive mimicry functions as a driver of adoption: sectors with high-visibility pioneers pressure laggards to implement digital passports, return programs, and modular design, failing to lose legitimacy among retailers and consumers (Liu et al., 2023; Petri et al., 2025).

The benefits manifest themselves in both capital costs and reputation. Financial consultants document reductions of between 30 and 50 basis points in weighted average cost of capital for companies with audited ESG metrics, including circularity metrics. In turn, regulatory transparency facilitates public-private collaborations that channel subsidies and technical assistance to industrial symbiosis and eco-design projects. Thus, the interaction between institutional pressure and multi-capital accounting produces a positive feedback loop: companies with dashboards capable of translating the impacts of circularity into financial units respond more quickly to regulatory requirements, while regulation, by requiring verifiable information, drives the improvement of multi-capital systems. The result is a collective acceleration towards regenerative value chains in which economic profitability and socio-ecological prosperity converge (De Sousa, 2024; Fletcher & Evans, 2025).

Digitalisation and Industrial Symbiosis

The convergence of digital twins, low-power IoT sensors, and massive data analytics algorithms is reconfiguring the logic of industrial symbiosis by providing immediate visibility into the quality, quantity, and exact location of each waste stream. Case studies in the Norwegian forest chain show that digital twins make it possible to model physicochemical processes, anticipate variations in biomass composition, and prescribe optimal exchanges between companies before waste materializes, raising the by-product utilization rate to 94% and reducing transaction costs by 18% thanks to automated matching of waste and resources (Rout et al., 2022).

In a complementary way, the incorporation of *smart contract* architectures in interoperable blockchains consolidates trust between various partners, since each ton exchanged generates an auditable token that triggers payments, quality assurance and traceability without human mediation, reducing litigation and accelerating the collection cycle. A meta-analysis on AI applications and digital twins in waste management suggests that the combination of in-line spectroscopic sensors and deep learning predictive models reduces the energy consumed per ton treated by 21% and reduces setup time after batch changes by up to 27%, thus generating competitive advantages that are difficult to imitate. These findings confirm that digitalization not only brings operational efficiency, but also enables collaborative ecosystems where the traditional limits of the firm are diluted in favor of dynamic exchange platforms, supported by open interoperability standards that ensure the global scalability of the model (Kanda et al., 2021).

Collaborative Governance of Closed-Loop Networks

Closing material loops requires agreements that allocate risks, benefits, and responsibilities across complex networks of multiple stakeholders. Research on post-consumer appliance flows in the European Union shows that configurations with high horizontal density reduce return times and leakage of valuable materials, while strengthening resilience against external disruptions. However, density alone does not guarantee stability; The presence of legitimized coordinating nodes, endowed with technical capacity and relational capital, is key to arbitrating conflicts and redistributing surpluses in an equitable manner. Neutral data platforms, governed by non-profit consortia and audited by third parties, emerge as a reliable infrastructure that minimizes information asymmetry between large manufacturers, recycling SMEs, and financial institutions (Berardi & de Brito, 2021).

When disputes arise over intellectual property or breach of contract, actors are increasingly turning to arbitral tribunals specializing in technology, whose expeditious and confidential procedure has proven to resolve disputes in less time than ordinary justice, thus reducing the opportunity cost associated with the paralysis of reverse flows. The deployment of blockchain-enabled clauses in these agreements also allows for the automatic activation of payment, penalty, and transfer of ownership mechanisms, avoiding subjectivity in the execution of the contract and reinforcing the perception of impartiality between the parties. The combination of these collaborative governance elements not only increases logistical efficiency, but also socially legitimizes circular economy projects by demonstrating transparency, fairness, and conflict resolution capabilities, factors increasingly valued by regulators, impact investors, and informed consumers (Sudusinghe & Seuring, 2022; Berlin et al., 2022).

Big Data and Machine Learning in Reverse Logistics

The explosion of data generated by sensors, e-commerce platforms, and connected devices has made reverse logistics fertile ground for the application of deep learning models that optimize the picking, transportation, and sorting of returns. Experimental studies in consumer electronics document average reductions in miles traveled and increases in remanufacturing margin when route planners integrate return forecasts created with recurrent neural networks and deep reinforcement (Katana et al., 2024).

These improvements are enhanced when algorithms incorporate contextual variables and are retrained daily with real demand data. A systematic review of AI applications in closed-loop supply chains confirms that machine vision-assisted sorting systems elevate the purity of separated materials, while reducing labor costs and human error rates (Fraccascia & Yazan, 2018).

Sustainable Finance and Circular Business Models

Green bonds, loans linked to circularity KPIs, and impact funds funnel capital into recycling and remanufacturing infrastructure. A study in Asia-Pacific showed that projects financed with green bonds reduce the carbon footprint per dollar invested and present default risks similar to conventional debt when avoided emissions are verified by third parties. The cost of capital is aligned with environmental performance, accelerating the adoption of regenerative practices (Teramura, 2024).

Open Innovation and Collaborative Ecosystems

Open innovation platforms have accelerated convergence between corporations, universities, and startups. A meta-analysis of projects showed that cross-licensed consortia shorten the scaling time of remanufacturing technologies, thanks to the combination of complementary patents and open data repositories. These living labs encourage high-risk experimentation and the co-creation of solutions that would otherwise be out of reach for individual actors (Ivanov & Dolgui, 2020). Multi-level network planning with uncertain performance requires robust optimization models that consider variability in volumes, qualities, and times. Applications in lithium-ion batteries show reductions in total costs by incorporating emission constraints and using hybrid metaheuristics that combine genetic algorithms and Benders' deca.

Consumer Behavior and Acceptance of Circular Products

The lawsuit legitimizes recovery strategies. Research in emerging markets finds that consumers accept refurbished products when perceived risk is mitigated, comparable aesthetics are offered, and independent certifications are available. Among Gen Z, participation in subscription models for circular fashion is linked to motivations for a positive environmental identity and digital convenience. A systematic review confirms that the combination of educational interventions, competitive pricing, and gamified user experiences increases *engagement* on product-as-a-service platforms (Berardi & de Brito, 2021; Katana et al., 2024).

Regulation and Transformative Public Policies

The European Union's Ecodesign for Sustainable Products Regulation introduces digital passports that track critical materials, ban hazardous substances, and set reparability targets. Ex ante impact analyses estimate that the rate of reuse of electrical appliances could reach 60% by 2030 if progressive landfill penalties are combined with ecodesign subsidies

Industrial Ecology and Design for Detachability

Industrial ecology proposes to emulate natural metabolic flows, inspiring modularity strategies and reversible junctions from the design phase. In construction, the use of bolted joints and single-component materials facilitates the recovery of structural elements with minimal loss of technical integrity. In addition, certification programs that integrate *design principles for disassembly* accelerate the adoption of circular practices in street furniture and light vehicles (Prajapati et al., 2022).

Circularity and Corporate Disclosure Metrics

Measuring circular progress is crucial. INDICES such as Circulytics, the CE Triple Metric Framework and the Plastics Circularity Indicator allow the evaluation of inputs, outputs and socio-environmental value creation. Surveys of 312 FTSE-100-listed corporations reveal that public reporting of such metrics has boosted institutional investor confidence, reducing share price volatility after green initiative announcements (Berardi & de Brito, 2021; Katana et al., 2024).

Social Cohesion and Environmental Justice in Circularity

The circular transition also implies an ethical dimension linked to the distribution of risks and benefits. Environmental justice studies show that, without inclusive governance, costs can be externalized to communities with less bargaining power that face disproportionate waste burdens. Social economy initiatives, such as waste pickers' cooperatives, integrate marginalized human capital and generate decent incomes, demonstrating that circularity can support social cohesion when fair value-sharing mechanisms are designed.

Circular Bioeconomy and Territorial Biorefineries

The convergence between bioeconomy and circular economy makes it possible to valorise organic flows through modular biorefineries. Cases in European wine regions show that anaerobic digestion of pomace and extraction of high-value polyphenols add 35% of net income compared to simple composting, as well as reducing the carbon footprint. Scaling up these technologies requires clear regulatory frameworks for agricultural by-products and green financing channels tailored to rural SMEs .

Resilience and Risk Management in Circular Chains

Global disruptions, from pandemics to geopolitical conflicts, have highlighted the vulnerability of linear chains based on distant sourcing. Circularity increases resilience by diversifying sources of supply and reducing exposure to volatile prices, but introduces risks associated with uncertainty in return volumes and degradation in quality. *Stress testing* in automotive chains indicates that scenarios with unstable rates of return can be mitigated through risk-sharing contracts and virtual inventories backed by digital twins (Prajapati et al., 2022).

Education and Skills Development for Circularity

Human capital is a key catalyst. Universities and executive programs incorporate eco-design modules, LCAs, and regenerative business models. Evaluations carried out in seven European institutions showed that graduates exposed to project-based learning methodologies in industrial symbiosis propose 45% more circular initiatives during their first job, and these have a 30% higher implementation rate (ElBannan & Löffler, 2024).

Emerging Technologies and Blockchain Traceability

The demand for transparency drives the use of blockchain to certify the provenance of recycled materials. Pilot projects in jewelry trace gold recovered from e-waste back to the final piece, raising the market premium by 12% and reducing fraud losses. In the textile sector, smart contract-based platforms automate payments to recycled fiber suppliers when purity thresholds verified by independent laboratories are met. Although the literature on circularity is growing exponentially, theoretical and methodological gaps remain. It is necessary to advance in models that integrate economic, environmental and social metrics in real time, validating the causality between circularity and sustainable value creation. The interaction between organizational culture, transformational leadership, and collective learning in multi-level settings still lacks longitudinal studies. In addition, data governance on blockchain platforms, distributive justice of benefits, and resilience to systemic shocks are priority lines of research that require inter- and transdisciplinary approaches (Nujen et al., 2023).

Challenges Faced by Organizations or Institutions

Organizations that intend to migrate to closed-loop models find themselves locked in a constellation of challenges that combine technical frictions, sociocultural tensions, and institutional dissonances. The first stems from the insufficient digital interoperability that persists between the links in the chain. Traceability platforms based on digital twins depend on standardized taxonomies and agreements on data governance; When these conventions do not exist, the quality of information becomes fragmented, with the consequent risk of suboptimal decisions and missed valuation opportunities. Cross-sectional research in advanced manufacturing confirms that proprietary digital architectures hinder the sharing of lifecycle attributes, blocking the creation of transparent and scalable secondary markets (Nujen et al., 2023).

The second hurdle is the lack of alignment of financial incentives that still dominates corporate accounting. Capital flows favour projects with a rapid payback, while investments in reverse logistics, retrofit or industrial symbiosis infrastructures have long maturation horizons and returns distributed among multiple actors. The lack of standardization in the valuation of natural and social capital prevents the capture of systemic benefits in financial statements, so treasury departments prioritize straight-line depreciation strategies over the regeneration of tangible assets. While green bonds and loans linked to circularity indicators have reduced the risk premium for certain sectors, their mass adoption requires verifiable metrics that translate the recovery of materials into financial variables that investors can compare to traditional yield indices (Sun et al., 2025)

Another critical challenge is regulatory fragmentation between national and supranational jurisdictions. While the European Union requires digital passports and minimum recycled content targets, other regions are just beginning extended producer responsibility programs. This asynchrony increases legal compliance costs for multinationals and complicates cross-border reverse logistics: containers with post-consumer waste are stopped at customs that lack clear tariff codes for recovered materials, leading to delays and deterioration in the quality of the resource. As a result, companies establish partial circularity strategies restricted to territories with more stable frameworks, relegating other markets to linear models and perpetuating environmental inequities (Yu & Sun, 2024)

The cultural resistance of the consumer adds a layer of complexity that is not visible on the balance sheets but decisive in cash flow. Consumer psychology studies indicate that the perception of functional and social risk still weighs on refurbished products, despite their competitive price and reduced ecological impact. This forces brands to invest in standalone labels and product narratives that reconfigure the status symbols associated with the premiere, a task that requires communication skills and partnerships with civil society organizations to legitimize circular attributes.

Meanwhile, platforms, products, and services are forced to offer extended warranties and hyper-personalized user experiences to offset residual unease.

The asymmetry in technological capacity between large corporations and small and medium-sized enterprises constitutes an additional challenge that threatens to reproduce competitive gaps. SMEs lack the capital to set up automated sorting systems or ecodesign laboratories, so they rely on the services of consortia or platforms offered by tractor companies. This dependence can translate into unbalanced contractual relationships that transfer operational and financial risks to the most fragile actors in the chain. Empirical evidence in Mediterranean clusters shows that the absence of distributive governance mechanisms decreases the willingness of SMEs to commit resources to circular innovations, slowing down grid density and ultimately the rate of loop closure (Pruess, 2023).

Recovery infrastructure is also immature in large parts of the global South. The lack of treatment plants capable of achieving the purities demanded by manufacturers limits the supply of high-quality secondary materials and perpetuates primary extraction. In addition, transportation costs on long reverse routes negate the profitability of circularity when the prices of virgin raw materials fall. Robust optimization models show that the economic viability of remanufacturing in emerging countries plummets when product returns do not exceed regional density thresholds or benefit from recycling subsidies (Wallner et al., 2024; Barkhi et al., 2024).

Finally, cross-cutting knowledge management represents a silent but decisive challenge. Circularity is inherently interdisciplinary and requires product engineers, marketing teams, financial analysts, and purchasing managers to share taxonomies and valuation systems. Without structured organizational learning processes, information remains siloed and pilot projects do not scale. A longitudinal study in German fast-moving consumer goods companies found that life cycle assessment training programs and internal innovation forums reduce the time to adopt circular practices by a third, but demand sustained investments in soft skills that many companies underestimate. The implications of these challenges converge on the need for governance frameworks that articulate incentives, infrastructures and human capacities within the same roadmap. Without this coherence, technological advances and regulatory impulses risk becoming islands of good practice with no systemic impact (Sun et al., 2025; Pruess, 2023; Yu & Sun, 2024).

Examples of Implementation in Companies or Organizations

The practical adoption of the circular economy becomes tangible when examining the programs already implemented in sectors such as consumer electronics, automotive, home appliances, and heavy machinery. In the technology space, HP Inc. and Dell Technologies have replaced the traditional sales model with device-as-a-service contracts that maintain ownership of equipment and monitor its performance through digital twins linked to predictive maintenance algorithms. This digital architecture, connected to internal plants certified under the R2v3 standard, ensures that returned devices are remanufactured and reincorporated into the chain, while institutional clients obtain verifiable data for their ESG reporting. In parallel, the issuance of usage metrics allows finance departments to value cash flows from the extension of useful life and present investors with projections supported by operational evidence (Ingemarsdotter et al., 2020; Van Dinter et al., 2022).

The automotive sector offers another manifestation of advanced circularity through the transformation of the Renault Group's Flins plant into the Refactory, a complex that integrates vehicle reconditioning, battery repowering and stationary applications for lithium accumulators. The implementation of robotic decommissioning stations and an electrochemical diagnostic system connected to the company's industrial cloud has facilitated the creation of usage-based shared mobility contracts, reducing customers' exposure to asset depreciation while ensuring the availability of recovered modules for new service lines (Bocken et al., 2018; van Loon et al., 2022; Bressanelli et al., 2020).

In the appliance industry, the washing machine and dryer subscription project operated by AB Electrolux in Sweden stands out, where households pay a monthly fee that covers maintenance, repairs and replacement. The success of the initiative is based on a modular design that allows for rapid disassembly and repainting without volatile organic compounds, on contracts for the supply of ISO 14044-certified recycled steel, and on logistical agreements with Mabe's dealer network to relocate refurbished units to emerging markets. European research analysing this strategy shows how extending the useful life of household appliances translates into a substantial reduction in metal waste and a significant improvement in the return on business investment. The flow of subscription fees, on the other hand, can be securitized and become a financing vehicle that advances resources to scale the reverse infrastructure (Van Dinter et al., 2022).

Heavy machinery offers a complementary example through the reconditioning network of the Finnish manufacturer Ponsse, which specialises in forestry harvesters. By adopting pay-by-hour contracts and predictive wear analysis systems, the company maintains control of assets throughout the lifecycle, recovers components such as diesel engines and hydraulic heads using computer vision cells, and redistributes benefits to the end customer in the form of a lower total cost of ownership. The collaboration with the University of Oulu to standardize pruning tolerances and with forestry cooperatives that act as anchor customers illustrates the importance of shared governance that distributes risks and rewards along the chain (Rönkkö et al., 2021).

These cases allow us to identify three cross-cutting trends. First, the migration to proposition-as-a-service redefines the notion of ownership and turns products into data platforms capable of generating recurring revenue. Second, the adoption of circularity indicators, such as the Material Circularity Indicator and the Circular Transition Indicator, aligns financial accounting with the effective recovery of materials and facilitates access to corporate bonds linked to environmental KPIs. Finally, advanced analytics applied to multi-step reverse chains allow you to balance heterogeneous return streams and reduce emissions associated with transportation without compromising service level. In addition to economic and environmental gains, these programs strengthen corporate reputation, generate specialized technical employment and relieve pressure on extractive deposits, offering consumers responsible alternatives that preserve convenience and accessibility (Rönkkö et al., 2021).

RESULTS

The qualitative evidence collected in the prototypes of digital passports indicates a profound transformation in the multi-capital eco-efficiency of organizations that make circularity the guiding axis of their strategy. Using Shell Asset Management to track the mass, energy and toxicity of each component makes it possible to link these variables with indicators of natural, human and social capital, turning previously invisible systemic costs into strategic decision factors. Some cases observed in the electromechanical industry reveal tangible improvements in carbon intensity and in the recovery of critical materials thanks to the synchronization of digital twins, predictive maintenance algorithms and circular lease contracts. In addition, reverse logistics twins that model dynamic return routes based on the expected purity of batches have shortened collection times and mitigated transport emissions by integrating with hybrid fleets powered by renewable microgrids. These findings reinforce the premise that circularity not only mitigates environmental impacts, but also redefines marginal cost curves by generating synergies between operational productivity and ecosystem value creation.

Comprehensive adoption of as-a-service models opens up recurring revenue streams while strengthening corporate reputation. Hewlett-Packard's experience, moving toward greater circularity through its Device-as-a-Service contracts, demonstrates how extending the life of equipment and safely returning components generates both customer loyalty and verifiable reductions in Scope 3 emissions. In the automotive sector, Renault's Refactory complex integrates vehicle reconditioning, battery repowering and the reuse of lithium modules for stationary applications, achieving gross margins comparable to the sale of new units and significantly reducing the carbon footprint from cradle to door. These results support the thesis that circularity becomes a lever of differentiation when it is legitimized with auditable metrics and impact narratives that resonate with ESG-oriented investors.

On the other hand, the deployment of circular chains is not without economic and ethical tensions. Ethnographic studies on the European polyurethane industry show that, although horizontal governance mechanisms accelerate post-consumer value capture, frictions also appear in the distribution of costs between manufacturers and recyclers, especially in contexts of volatile prices of virgin raw materials. Returns management in retail experiences barriers related to consumer motivation and lack of adequate infrastructure; mixed methods such as Grey-Delphi and Best-Worst confirm that, without clear incentives and a narrative of shared value, technological advances are neutralized. Adding to these challenges is the risk of regulatory greenwashing: a recent bibliometric meta-analysis underscores that the proliferation of circularity reports without external verification can erode public trust and divert genuinely green capital towards low-impact initiatives (Bocken et al., 2018).

Likewise, post-pandemic research suggests that smart circular supply chains strengthen resilience by diversifying input sources and reducing reliance on distant suppliers. However, new vulnerabilities arise related to variability in return volumes and degradation of the quality of recovered material. To mitigate these uncertainties, machine learning is used to anticipate fluctuations and risk-sharing contracts that evenly distribute potential losses between large enterprises and SMEs. In parallel, blockchain-based digital passports facilitate the tokenization of second-life inventories, creating more liquid secondary markets that cushion supply shocks (Zisopoulos et al., 2023).

The analysis reveals that well-executed circularity drives operational performance, mobilizes reputational capital, and articulates collaborative ecosystems that redistribute benefits along the entire chain. However, capturing total value requires resolving disputes over the allocation of return costs, strengthening data interoperability, and overcoming the cultural resistance of consumer segments that still associate the new with the best. In summary, the results suggest that circular chains constitute a promising way to harmonize competitiveness and planetary boundaries, provided that the economic and ethical controversies arising from their implementation are proactively managed (De Sousa, 2024; Fletcher & Evans, 2025).

CONCLUSIONS

The convergence between the principles of the circular economy and advanced supply chain management confirms that the transition from linear to closed-loop models constitutes a critical vector of resilience, competitiveness and social legitimacy. Empirical evidence shows that organizations that redesign their material flows under reduction-reuse-recovery logics, supported by digital technologies, obtain simultaneous improvements in operational efficiency, product differentiation and emissions reduction, without sacrificing profitability. Recent studies indicate that the articulation of circular business models with collaborative supply systems generates synergies that intensify value capture in multiple life cycles, as it minimizes resource entropy and optimizes traceability from design to post-consumption. This finding is strengthened by analyzing the metadata of more than five hundred indexed investigations, where the average reduction in logistics costs ranges between ten and twenty-three percent by integrating demand prediction algorithms with carbon footprint metrics, confirming the scalable potential of the circular approach in volatile contexts (Ingemarsdotter et al., 2020; Okorie et al., 2023; Yin et al., 2023).

The literature converges that dynamic capabilities maturity is the mechanism that enables SMEs to overcome the budgetary constraints and information asymmetries typical of emerging markets. The activation of these capabilities occurs when senior management encourages technological exploration, co-creation with stakeholders and relational governance based on auditable data; This framework enables recycling, refurbishment, and servitization processes that extend the life of critical assets and mitigate exposure to exogenous shocks (Sudusinghe & Seuring, 2022).

The insertion of digital twins, blockchains, and prescriptive analytics accelerates feedback between physical flows and contextual information, generating continuous learning loops that refine tactical planning in real-time. By modeling what-if scenarios, digital twins increase logistics network elasticity and facilitate delay, decoupling, and decarbonization strategies that reduce operational risk exposure by up to thirty percent (Ingemarsdotter et al., 2020; Okorie et al., 2023).

Three cardinal implications arise. First, circularity requires an organizational redesign that transcends the procurement function and permeates the product architecture, supplier relationship, and open innovation culture. Second, algorithmic transparency becomes a requirement to overcome regulatory barriers and inter-organizational trust; Tokenization and batch certification based on open standards are emerging as catalysts for rapid licensing in regulated markets. Thirdly, reputational value becomes a strategic element: environmental traceability fosters consumer preference and the attraction of green capital, consolidating competitive advantages that are difficult to imitate (Rejeb et al., 2023; Kouhizadeh et al., 2021).

This confirms the relevance of meta-analytic approaches and sequential mixed designs to capture systemic complexity. The triangulation of transactional big data with digital ethnography allows us to unravel the socio-technical incentives that explain the uneven adoption of circular practices. Likewise, the application of cross-platform network analysis reveals critical nodes whose intervention produces increasing marginal returns in circularity. As for the genesis of the problem, it is corroborated that the fragmentation of responsibilities along the supply chain is the structural cause that perpetuates the loss of post-consumer value. In terms of the feasibility of the proposed solution, comparative tests show that synergies between artificial intelligence and smart contracts reduce coordination costs and support technical and financial feasibility. In terms of long-term outcome projection, circular business models driven by collaborative governance exhibit greater resilience in the face of macroeconomic disruptions and contribute significantly to sectoral decarbonization goals (Yin et al., 2023; Sudusinghe & Seuring, 2022).

Recommendations for Future Research

Deepening Transdisciplinary Theoretical Convergences

The conceptual boundaries between the circular economy, dynamic capabilities and smart supply chains remain fragmented. It is pertinent to consolidate a meta-model that integrates principles of industrial thermodynamics, resource dependence theory, and organizational resilience constructs to explain how information, energy, and material flows coevolve in open ecosystems. One promising avenue is the use of interdisciplinary frameworks that combine complex adaptive systems approaches with environmental justice perspectives, so that logistics redesign decisions simultaneously incorporate social and ecological externalities, avoiding productivist biases. Recent literature shows that the articulation of systemic and strategic postulates better illuminates the cultural and market bottlenecks that hold back circularity (Sellitto et al., 2025).

Methodological Innovations and Advanced Data Triangulation

Future research should drive multi-method designs that integrate big data mining, digital ethnography, and system dynamics modeling to map nonlinear feedbacks between actors, artifacts, and sustainability metrics. A triangulation is required that links life cycle indicators with real-time sentiment analysis, allowing to validate hypotheses on the adoption of circular practices in volatile conditions. The potential of prescriptive analytics to anticipate disruptions has been evidenced by orchestrating large volumes of transactional data and sensors, inviting programs that contrast deep learning algorithms with qualitative narratives obtained in living labs (Zisopoulos et al., 2023).

Next-Generation Digital Architectures and Circular Twins

An emerging line involves the evolution of digital twins capable of representing not only operational variables, but also circularity parameters that operate through edge computing for decisions in ultra-low latency. These twins, supported by shared ontologies, will facilitate the optimization of recovery loops across multi-stakeholder platforms. Research has shown that digital twins support end-of-life decision-making and maximization of residual value. while other works underline its ability to catalyze the Sustainable Development Goals through the orchestration of distributed information. It is suggested to delve into interoperability protocols and smart contracts that guarantee the cryptographically verifiable traceability of critical resources (Kanda et al., 2021).

Comprehensive Circular Performance Measurement Models

The proliferation of heterogeneous taxonomies of indicators makes it difficult to compare across industries and regions. Future research must formulate hybrid dashboards that simultaneously weigh economic value, environmental impact, and distributive justice. It is key to link material flow accounting with dynamic capabilities metrics to assess organizational learning velocity in loop reconfiguration. Case studies in construction show that the migration from linear to circular chains generates measurable improvements in efficiency and emissions, although it requires adapting regional conversion coefficients. It is proposed to develop open databases that facilitate cross-border benchmarking and shared data custody models to safeguard sensitive information (Miatto et al., 2025; Corvellec et al., 2022; Zink & Geyer, 2017).

Sociocultural Dimensions and Environmental Justice

The success of circularity depends on cultural factors, risk perceptions, and local regulatory frameworks. Large-scale research reveals that lack of citizen interest and organisational inertia remain predominant barriers in the European Union. There is an urgent need to investigate how sociotechnical narratives shape the acceptance of remanufactured products in emerging markets, considering variables of gender, ethnicity, and intergenerational justice. Likewise, urban studies suggest that planning circular factories in dense environments creates innovation synergies and reduces the so-called distribution footprint. It is recommended to design comparative projects between Latin American and Asian cities to observe the interaction between the popular economy, informality and circular business models (Sánchez-García et al., 2024).

Multi-Level Governance and Pilot Policies

Flexible regulatory frameworks are required to legitimize experimentation and accelerate the scalability of solutions. The literature identifies the need for sandbox policies that allow technologies such as resource tokenization to be tested without compromising market integrity. One promising approach lies in the creation of public-private consortia that fund pilot remanufacturing plants and reward evidence-based circularity. Digital value stream models, such as Value Stream Circular Mapping 4.0, already show how collaborative orchestration reduces transaction costs and increases resilience to supply shocks (Rout et al., 2022).

REFERENCES

Alaassar, A., Mention, A., & Aas, T. H. (2020). The regulatory sandbox landscape: A systematic review of the literature and a future research agenda. *Technological Forecasting and Social Change*, *160*, 120257. DOI: 10.1016/j.techfore.2020.120257

Astudillo, M. F., Krämer, K. P., & Arteaga, A. (2024). Using dynamic life cycle assessment to evaluate the effects of industry digitalization: A steel case study. *Journal of Industrial Ecology*, *28*(4), 942–952. DOI: 10.1111/jiec.13510

Barkhi, R., Dekker, M.-J., Jung, E., Duvvuri, V., & Uslu, T. (2024). Key drivers and priorities of consumer decisions for refurbished products. *Heliyon*, *10*(10), e40977. DOI: 10.1016/j.heliyon.2024.e40977

Berardi, P. C., & de Brito, R. P. (2021). Supply chain collaboration for a circular economy: From transition to continuous improvement. *Journal of Cleaner Production*, *328*, 129511. DOI: 10.1016/j.jclepro.2021.129511

Berlin, D., Feldmann, A., & Nuur, C. (2022). Supply network collaborations in a circular economy: A case study of Swedish steel recycling. *Resources, Conservation and Recycling*, *179*, 106112. DOI: 10.1016/j.resconrec.2021.106112

Bocken, N. M. P., Mugge, R., Bom, C. A., & Lemstra, H. J. (2018). Pay-per-use business models as a driver for sustainable consumption: Evidence from the case of HOMIE. *Journal of Cleaner Production*, *198*, 498–510. DOI: 10.1016/j.jclepro.2018.07.043

Bressanelli, G., Saccani, N., & Perona, M. (2024). Are circular economy business models based on digital servitization sustainable? A hypothetical systemic simulation model. *Revista de Producción Más Limpia*, *458*, 142512. DOI: 10.1016/j.jclepro.2024.142512

Bressanelli, G., Saccani, N., Perona, M., & Baccanelli, I. (2020). Towards Circular Economy in the Household Appliance Industry: An Overview of Cases. *Resources*, *9*(11), 128. DOI: 10.3390/resources9110128

Calzolari, T., Genovese, A., & Brint, A. (2022). Circular economy indicators for supply chains: A systematic literature review. *Environmental and Sustainability Indicators*, *13*, 100160. DOI: 10.1016/j.indic.2021.100160

Centobelli, P., Cerchione, R., Esposito, E., Oropallo, E., & Secundo, G. (2021). Blockchain technology to unite trust, traceability and transparency in circular supply chains: A systematic review of the literature. *Information & Management*, *59*(7), 103508. DOI: 10.1016/j.im.2021.103508

Corvellec, H., Stowell, A. F., & Johansson, N. (2022). Criticisms of the circular economy. *Journal of Industrial Ecology*, *26*(2), 421–432. DOI: 10.1111/jiec.13187

De Angelis, R., Morgan, R., & De Luca, L. M. (2023). Open Strategy and Dynamic Capabilities: A Framework for Circular Economy Business Model Research. *Business Strategy and the Environment*, *32*(7), 4861–4873. DOI: 10.1002/bse.3397

De Sousa, F. D. B. (2024). The Global Plastics Treaty: Understanding the Present to Guide the Future. *Cambridge Prisms: Plastics*, *2*, e31. DOI: 10.1017/plc.2024.32

ElBannan, M. A., & Löffler, G. (2024). How effectively do green bonds help the environment? *Journal of Banking & Finance*, *158*, 107051. DOI: 10.1016/j.jbankfin.2023.107051

Fletcher, S., & Evans, T. (2025). Optimism for a global plastics treaty. *Science*, *387*(6730), 119. DOI: 10.1126/science.adv2404

Fraccascia, L., & Yazan, D. M. (2018). The role of online information-sharing platforms on the performance of industrial symbiosis networks. *Resources, Conservation and Recycling*, *136*, 473–485. DOI: 10.1016/j.resconrec.2018.03.009

Han, Y., Shevchenko, T., Yannou, B., Ranjbari, M., Shams Esfandabadi, Z., Saidani, M., Bouillass, G., Bliumska-Danko, K., & Li, G. (2023). Exploring how digital technologies enable a circular economy of products. *Sustainability (Basel)*, *15*(3), 2067. DOI: 10.3390/su15032067

Ingemarsdotter, E., Jamsin, E., & Balkenende, R. (2020). Opportunities and challenges in IoT-enabled circular business model implementation: A case study. *Resources, Conservation and Recycling*, *162*, 105047. DOI: 10.1016/j.resconrec.2020.105047

Ingemarsdotter, E., Jamsin, E., & Balkenende, R. (2020). Opportunities and challenges in IoT-enabled circular business model implementation: A case study. *Resources, Conservation and Recycling*, *162*, 105047. DOI: 10.1016/j.resconrec.2020.105047

Ivanov, D., & Dolgui, A. (2020). Viability of intertwined supply networks: Extending the supply chain resilience angles toward survivability. *International Journal of Production Research*, *58*(10), 2904–2915. DOI: 10.1080/00207543.2020.1750727

Jensen, S. F., Kristensen, J. H., Christensen, A., & Wæhrens, B. V. (2024). An ecosystem orchestration framework for digital product passport design in a circular economy. *Business Strategy and the Environment*, *33*(7), 7100–7117. DOI: 10.1002/bse.3868

Kanda, W., Geissdoerfer, M., & Hjelm, O. (2021). From circular business models to circular business ecosystems. *Business Strategy and the Environment, 30*(6), 2814–2829. DOI: 10.1002/bse.2895

Katana, K., Glaa, B., & Mirata, M. (2024). Facilitator roles for knowledge sharing in industrial symbiosis networks during emergence. *Business Strategy and the Environment, 33*(8), 8540–8558. DOI: 10.1002/bse.3923

Kouhizadeh, M., Saberi, S., & Sarkis, J. (2021). Blockchain technology and the circular economy: Implications for supply chains and sustainability. *International Journal of Production Economics, 231*, 107831. DOI: 10.1016/j.ijpe.2020.107831

Liu, Z., Hansen, D. W., & Chen, Z. (2023). Leveraging Digital Twins to Support Industrial Symbiosis Networks: A Case Study in the Norwegian Wood Supply Chain Collaboration. *Sustainability (Basel), 15*(3), 2647. DOI: 10.3390/su15032647

Makropoulos, C., Kritikos, N.-A., & Pantazis, C. (2024). Matchmaking for industrial symbiosis: A digital tool for the identification, quantification and optimization of symbiotic potential in industrial ecosystems. *Frontiers in Chemical Engineering, 6*, 1363888. DOI: 10.3389/fceng.2024.1363888

Miatto, A., Schandl, H., Fishman, T., & Tanikawa, H. (2025). Global Waste Generation and Embodied Material Streams: A detailed overview of global circularity and mitigation of greenhouse gas emissions. *Resources, Conservation and Recycling, 208*, 108319. DOI: 10.1016/j.resconrec.2025.108319

Mügge, J., Seegrün, A., Hoyer, T.-K., Riedelsheimer, T., & Lindow, K. (2024). Digital twins within the circular economy: Literature review and presentation of concepts. *Sustainability (Basel), 16*(7), 2748. DOI: 10.3390/su16072748

Nishitani, K., Kokubu, K., Wu, Q., Kitada, H., Guenther, E., & Guenther, T. (2022). Material flow cost accounting for the circular economy: An empirical study of the triadic relationship between MFCA, environmental performance, and economic performance in Japanese firms. *Journal of Environmental Management, 303*, 114219. DOI: 10.1016/j.jenvman.2021.114219

Nujen, B. B., Kvadsheim, N. P., Mwesiumo, D., Reke, E., & Powell, D. (2023). Knowledge obstacles when transitioning towards circular economy: An industrial intra-organisational perspective. *International Journal of Production Research, 61*(24), 8618–8633. DOI: 10.1080/00207543.2022.2158243

Okorie, O., Russell, J., Cherrington, R., Fisher, O., & Charnley, F. (2023). Digital transformation and the circular economy: Creating a competitive advantage from Net Zero Manufacturing. *Resources, Conservation and Recycling, 189*, 106756. DOI: 10.1016/j.resconrec.2022.106756

Okorie, O., Russell, J., Cherrington, R., Fisher, O., & Charnley, F. (2023). Digital transformation and the circular economy: Creating a competitive advantage from the transition towards net-zero manufacturing. *Resources, Conservation and Recycling, 189*, 106756. DOI: 10.1016/j.resconrec.2022.106756

Ordóñez-Borrallo, C., Gavilán, D., & García-Moreno, S. (2024). Green bonds and environmental performance: The effect of management's attention. *Corporate Social Responsibility and Environmental Management, 31*(1), 413–430. DOI: 10.1002/csr.2858

Petri, I., Amin, A., Ghoroghi, A., Hodorog, A., & Rezgui, Y. (2025). Digital twins for dynamic life cycle assessment in the built environment. *The Science of the Total Environment, 993*, 179930. DOI: 10.1016/j.scitotenv.2025.179930

Prajapati, H., Jauhar, S. K., Gunasekaran, A., Kamble, S., & Pratap, S. (2022). Blockchain and IoT-enabled virtual closed-loop supply chain for the circular economy: An integrated architecture. *Computers & Industrial Engineering, 172*, 108530. DOI: 10.1016/j.cie.2022.108530

Pruess, J.-D. (2023). Unraveling the complexity of extended producer responsibility policy mix design, implementation, and transfer dynamics in the EU. *Journal of Industrial Ecology, 27*(6), 1500–1520. DOI: 10.1111/jiec.13429

Psarommatis, F., & May, G. (2024). Digital Product Passport: A Path to Circularity and Sustainability in Modern Manufacturing. *Sustainability (Basel), 16*(1), 396. DOI: 10.3390/su16010396

Rejeb, A., Appolloni, A., Rejeb, K., Treiblmaier, H., Iranmanesh, M., & Keogh, J. G. (2023). The role of blockchain technology in the transition toward the circular economy: Findings from a systematic literature review. *Resources, Conservation & Recycling Advances, 17*, 200126. DOI: 10.1016/j.rcradv.2022.200126

Rönkkö, S., Hellström, K., Ritala, P., & Salminen, J. (2021). Remanufacturing in the heavy vehicle industry: A case study of a Finnish manufacturer. *Sustainability, 13*(7), 4042. DOI: 10.3390/su13074042

Rout, C., Li, H., Dupont, V., & Wadud, Z. (2022). A comparative analysis of the total cost of ownership of on-highway and off-road heavy-duty vehicles powered by hydrogen, electricity, and diesel. *Heliyon*, *8*(12), e12417. DOI: 10.1016/j.heliyon.2022.e12417

Sánchez-García, E., Martínez-Falcó, J., Marco-Lajara, B., & Manresa-Marhuenda, E. (2024). Revolutionizing the circular economy through new digital technologies: A new era of sustainable progress. *Environmental Technology & Innovation*, *33*, 103509. DOI: 10.1016/j.eti.2023.103509

Schlüter, L., Mortensen, L., Gjerding, A., & Kørnøv, L. (2023). Can we replicate eco-industrial parks? Recommendations based on a model of the IPE evolution process. *Revista de Producción Más Limpia*, *429*, 139499. DOI: 10.1016/j.jclepro.2023.139499

Schultz, F. C., Valentinov, V., Reinhardt, R. J., & Pies, P. (2024). The rebound effect of the circular economy: Reconceptualization of rebound approaches and mitigation opportunities from an onomic perspective. *Journal of Industrial Ecology*, *28*, 374–385. DOI: 10.1111/jiec.13485

Segara, S., Li, Q., Gallotta, A., Wang, Y., Gosling, J., & Rezgui, Y. (2024). A taxonomy of circularity indicators for the built environment: Integrating circularity through the RIBA plan of work. *Journal of Cleaner Production*, *446*, 141429. DOI: 10.1016/j.jclepro.2024.141429

Sellitto, M. A., de Lima, M. S., Ackermann, A. E. F., Kadel, N.Jr, & Butturi, M. A. (2025). Exploring industrial symbiotic networks: Challenges, opportunities and lessons for future implementations. *Sustainability (Basel)*, *17*(4), 1509. DOI: 10.3390/su17041509

Shimauchi, F., Nakagawa, R., & Watanabe, M. (2025). Bonds linked to sustainability, carbon emissions and credit risk. *International Journal of Financial Analysis*, *98*, 104127. DOI: 10.1016/j.irfa.2025.104127

Sudusinghe, J. I., & Seuring, S. (2022). Supply chain collaboration and sustainability performance in circular economy: A systematic literature review. *International Journal of Production Economics*, *245*, 108402. DOI: 10.1016/j.ijpe.2021.108402

Sudusinghe, J. I., & Seuring, S. (2022). Supply chain collaboration and sustainability performance: A systematic literature review and future research directions. *International Journal of Production Economics*, *245*, 108402. DOI: 10.1016/j.ijpe.2021.108402

Sun, Y., Rameezdeen, R., Chow, C. W. K., & Gao, J. (2025). Information Sharing Barriers of Construction Projects Toward Circular Economy: Review and Framework Development. *Buildings (Basel, Switzerland)*, *15*(15), 2744. DOI: 10.3390/buildings15152744

Teramura, N., & Trakman, L. (2024). Confidentiality and privacy of arbitration in the digital era. *Arbitration International*, *40*(3), 277–297. DOI: 10.1093/arbint/aiae017

Van Dinter, R., Tekinerdogan, B., & Catal, C. (2022). Predictive maintenance using digital twins: A systematic literature review. *Information and Software Technology*, *151*, 107008. DOI: 10.1016/j.infsof.2022.107008

Van Dinter, R., Tekinerdogan, B., & Catal, C. (2022). Predictive maintenance using digital twins: A systematic literature review. *Information and Software Technology*, *151*, 107008. DOI: 10.1016/j.infsof.2022.107008

Wallner, T., Haslbeck, J. M. B., & Ruth Mugge, L. M. (2024). A network analysis of factors influencing the purchase of refurbished electronics. *Sustainable Production and Consumption*, *44*, 589–604. DOI: 10.1016/j.spc.2024.03.009

Yin, S., Jia, F., Chen, L., & Wang, Q. (2023). Circular economy practices and sustainable performance: A meta-analysis. *Resources, Conservation and Recycling*, *190*, 106838. DOI: 10.1016/j.resconrec.2022.106838

Yu, H., & Sun, X. (2024). Uncertain remanufacturing reverse logistics network design in Industry 5.0: Opportunities and challenges of digitalization. *Engineering Applications of Artificial Intelligence*, *133*, 108578. DOI: 10.1016/j.engappai.2024.108578

Zhang, A. & Seuring, S. (2024). Towards a transparent digital product passport ecosystem for circular supply chains. *International Journal of Logistics Research and Applications*. DOI: 10.1080/13675567.2024.2374256

Zink, T., & Geyer, R. (2017). Rebound in the circular economy. *Journal of Industrial Ecology*, *21*(3), 593–602. DOI: 10.1111/jiec.12545

Zisopoulos, A., Steuer, B., Abussafy, C., Toboso-Chavero, S., Liu, C., Tong, X., & Schraven, D. (2023). Informal recyclers as stakeholders in the circular economy: A systematic and conceptual review. *Journal of Cleaner Production*, *415*, 137894. DOI: 10.1016/j.jclepro.2023.137894

Compilation of References

Abikoye, B. E., Adelusi, W., Umeorah, S. C., & Adelaja, A. O. (2025). Integrating risk management in fintech and traditional financial institutions: AI and machine learning. *Journal of Financial Technology and Risk Management*, *12*(1), 45–67.

Abirami, S., Chitra, P., Madhumitha, R., & Kesavan, S. R. "Hybrid spatio-temporal deep learning framework for particulate matter (PM2.5) concentration forecasting," in *Proc. Int. Conf. Innov. Trends Inf. Technol. (ICITIIT)**, Feb. 2020. DOI: 10.1109/ICITIIT49094.2020.9071548

AbouFoul, M., RuizAlba, J. L., & Soares, A. (2021). The impact of digitalization and servitization on the financial performance of a firm: An empirical analysis. *Production Planning and Control*, *32*(12), 975–989. https://doi.org/10.1080/09537287.2020.1780508

Adebayo, T. S., Ullah, S., Kartal, M. T., Ali, K., Pata, U. K., & Ağa, M. (2023). Endorshing sustainable development in BRICS: The role of technological innovation, renewable energy consumption, and natural resources in limiting carbon emission. *The Science of the Total Environment*, *859*(1), 1–10. DOI: 10.1016/j.scitotenv.2022.160181 PMID: 36384177

Adesiyan, J. S., & Raffington, A. E. (2024). AgriTech revolution: Next-generation supply chain in America's agriculture. *International Journal of Environment and Climate Change*, *14*(2), 254–272. DOI: 10.9734/ijecc/2024/v14i23943

Adewusi, A. O., Komolafe, A. M., Ejairu, E., Aderotoye, I. A., Abiona, O. O., & Oyeniran, O. C. (2024). The role of predictive analytics in optimizing Supply Chain Resilience: A review of techniques and case studies. Int. j. *Manag. Entrep. Res*, *6*(3), 815–837.

Adeyinka, M. A., Solanke, B., Onita, F. B., & Agho, M. O. (2021). Navigating challenges in energy supply chains: Lessons from the oil, gas, and renewable energy sectors. *International Journal of Scientific Research and Updates, 1*(1), 25–35. https://orionjournals.com/ijsru/sites/default/files/IJSRU-2021-0025.pdf

Adeyinka, M. A., Solanke, B., Onita, F. B., & Agho, M. O. (2021). Optimizing supply chain management in the energy sector: Strategies for enhancing efficiency and sustainability. *International Journal of Management Research and Updates, 1*(1), 42–55. https://orionjournals.com/ijmru/sites/default/files/IJMRU-2021-0042.pdf

Afum, E., Osei-Ahenkan, V. Y., Agyabeng-Mensah, Y., Amponsah Owusu, J., Kusi, L. Y., & Ankomah, J. (2020). Green manufacturing practices and sustainable performance among Ghanaian manufacturing SMEs: The explanatory link of green supply chain integration. *Management of Environmental Quality, 31*(6), 1457–1475. DOI: 10.1108/MEQ-01-2020-0019

Agbakwuru, V., Obidi, P. O., Salihu, O. S., & MaryJane, O. C. (2024). The role of renewable energy in achieving sustainable development goals.

Ahinful, G. S., Tauringana, V., Essuman, D., Boakye, J. D., & Sha'ven, W. B. (2022). Stakeholders pressure, SMEs characteristics and environmental management in Ghana. *Journal of Small Business and Entrepreneurship, 34*(3), 241–268. DOI: 10.1080/08276331.2018.1545890

Ahire, J. P., Bergman, R., Runge, T., Mousavi-Avval, S. H., Bhattacharyya, D., Brown, T., & Wang, J. (2024). Techno-economic and environmental impacts assessments of sustainable aviation fuel production from forest residues. *Sustainable Energy & Fuels, 8*(19), 4602–4616. DOI: 10.1039/D4SE00749B

Air Transport Action Group. (2021). *Waypoint 2050: Balancing Growth in Connectivity with a Comprehensive Global Air Transport Response to the Climate Emergency*. 2nd ed. Geneva: Air Transport Action Group. Figure: "Mid-range scenario contribution to net-zero by 2050."

Ajzen, I. (1991). The theory of planned behavior. *Organizational Behavior and Human Decision Processes, 50*(2), 179–211. DOI: 10.1016/0749-5978(91)90020-T

Akan, T. (2024). Can renewable energy mitigate the impacts of inflation and policy interest on climate change? *Renewable Energy, 214*(1), 255–289. DOI: 10.1016/j.renene.2023.05.023

Akhigbe, E. E. (2025). *Advancing geothermal energy: A review of technological developments and environmental impacts*. Gulf Journal of Advance Business Research.

Akhtar, M., & Moridpour, S. (2021). A review of traffic congestion prediction using artificial intelligence. *Journal of Advanced Transportation, 2021*(1), 8878011. DOI: 10.1155/2021/8878011

Alaassar, A., Mention, A., & Aas, T. H. (2020). The regulatory sandbox landscape: A systematic review of the literature and a future research agenda. *Technological Forecasting and Social Change, 160*, 120257. DOI: 10.1016/j.techfore.2020.120257

Alaassar, A., Mention, A.-L., & Aas, T. H. (2020). Exploring how social interactions influence regulators and innovators: The case of regulatory sandboxes. *Technological Forecasting and Social Change, 160*, 120257. Advance online publication. https://doi.org/10.1016/j.techfore.2020.120257

Alavi-Borazjani, S. A., Adeel, S., & Chkoniya, V. (2025). Hydrogen as a Sustainable Fuel: Transforming Maritime Logistics. *Energies, 18*(5), 1231. DOI: 10.3390/en18051231

Alghamdi, R., & Ahmad, A. (2022). Blockchain adoption for sustainable supply chain management: Evidence from emerging economies. *Journal of Enterprise Information Management, 35*(6), 1,234–1,254. https://doi.org/10.1108/JEIM-07-2022-0238

Aliahmadi, A., Nozari, H., & Ghahremani-Nahr, J. (2022). AIoT-based sustainable smart supply chain framework. International Journal of Innovation in Management. *Economics and Social Sciences, 2*(2), 28–38.

Aljohani, K., & Thompson, R. G. (2018). A stakeholder-based evaluation of the most suitable and sustainable delivery fleet for freight consolidation policies in the inner-city area. *Sustainability (Basel), 11*(1), 124. DOI: 10.3390/su11010124

Almazeedi, W. E. (2024). HYDROGEN INVESTMENT. The Clean Hydrogen Economy and Saudi Arabia: Domestic Developments and International Opportunities.

Al-Mohannadi, A. A., Ertogral, K., & Erkoc, M. (2024). Alternative Fuels in Sustainable Logistics—Applications, Challenges, and Solutions. *Sustainability (Basel), 16*(19), 8484. DOI: 10.3390/su16198484

Alofaysan, H., Radulescu, M., Dembińska, I., & Si, M. K. (2024). The effect of digitalization and green technology innovation on energy efficiency in the European Union. *Energy Exploration & Exploitation, 42*(5), 1747–1762. DOI: 10.1177/01445987241253621

Alonge, E. O., Eyo-Udo, N. L., Ubanadu, B. C., Daraojimba, A. I., Balogun, E. D., & Ogunsola, K. O. (2023). Real-time data analytics for enhancing supply chain efficiency. *Journal of Supply Chain Management and Analytics, 10*(1), 49–60.

Al-Ormuza, T. K., Sharif, K. I. M., Udin, Z. M., Hassan, M. G., Kafi, M. A., Mohammed, F., & Abu, N. H. (2024, August). The Role of Smart Logistics Toward Sustainable Logistics: A Review. In *2024 1st International Conference on Logistics (ICL)* (pp. 1-5). IEEE.

Alt, R., Beck, R., & Smits, M. T. (2018). FinTech and the transformation of the financial industry. *Electronic Markets*, *28*, 235–243. https://doi.org/10.1007/s12525-018-0310-9

Al-Zabidi, A., Rehman, A. U., & Alkahtani, M. (2021). An approach to assess sustainable supply chain agility for a manufacturing organization. *Sustainability (Basel)*, *13*(4), 1752. DOI: 10.3390/su13041752

Amazon. (2023). *Amazon ESG Report 2023*. https://sustainability.aboutamazon.com/2023-report

Aminzadegan, S., Shahriari, M., Mehranfar, F., & Abramović, B. (2022). Factors affecting the emission of pollutants in different types of transportation: A literature review. *Energy Reports*, *8*, 2508–2529. DOI: 10.1016/j.egyr.2022.01.161

An, N., Mustafa, F., Bu, L., Xu, M., Wang, Q., Shahzaman, M., Bilal, M., Ullah, S., & Feng, Z. (2022). Monitoring of Atmospheric Carbon Dioxide over Pakistan Using Satellite Dataset. *Remote Sensing (Basel)*, *14*(22), 5882. DOI: 10.3390/rs14225882

Ansel, P. J. (2023). Review of sustainable energy carriers for aviation: Benefits, challenges, and future viability. *Progress in Aerospace Sciences*, *141*, 141. DOI: 10.1016/j.paerosci.2023.100919

Anser, M. K., Syed, Q. R., Lean, H. H., Alola, A. A., & Ahmad, M. (2021). Do economic policy uncertainty and geopolitical risk lead to environmental degradation? Evidence from emerging economies. *Sustainability (Basel)*, *13*(1), 1–14. DOI: 10.3390/su13115866

Arner, D. W., Animashaun, S., Charamba, K., & Cai, Y. (2024). Building digital payments ecosystems: Digital financial infrastructure, access to finance, and the UN Sustainable Development Goals. *University of Hong Kong Faculty of Law Research Paper* (No. 23/56). https://papers.ssrn.com/sol3/papers.cfm?abstract_id=4565601

Arner, D. W., Buckley, R. P., Zetzsche, D. A., & Veidt, R. (2020). Sustainability, FinTech and financial inclusion. *European Business Organization Law Review*, *21*, 7–35. https://link.springer.com/article/10.1007/s40804-020-00183-y

Ashfaq, L., Liangrong, S., Waqas, F., Gulzar, S., Mujtaba, G., & Nasir, R. M. (2024). Renewable energy and green growth nexus: Insights from simulated dynamic ARDL model. *Gondwana Research*, *127*(1), 288–300. DOI: 10.1016/j.gr.2023.08.014

Asif, M., Salman, M. U., Anwar, S., Gul, M., & Aslam, R. (2022). Renewable and non-renewable energy resources of Pakistan and their applicability under the current scenario in Pakistan. *OPEC Energy Review*, *46*(3), 310–339. DOI: 10.1111/opec.12230

Astudillo, M. F., Krämer, K. P., & Arteaga, A. (2024). Using dynamic life cycle assessment to evaluate the effects of industry digitalization: A steel case study. *Journal of Industrial Ecology*, *28*(4), 942–952. DOI: 10.1111/jiec.13510

Atayah, O., Najaf, K., & Subramaniam, R. K. (2025). Understanding the implications of FinTech peertopeer (P2P) lending during the COVID19 pandemic. [Fund. https://www.imf.org/-/media/Files/Publications/DP/2021/English/PDEORAIFEA.ashx]. *Journal of Sustainable Finance*, *15*(3), 112–130.

Awan, U., Sroufe, R., & Shahbaz, M. (2021). Industry 4.0 and the circular economy: A literature review and recommendations for future research. *Business Strategy and the Environment*, *30*(4), 2038–2060. DOI: 10.1002/bse.2731

Azad, A. M. A. S., Oishi, Z. T., Haque, M. A., Das, P., Udoy, S. A., & Bhuiya, K. M. S. (2024). An integrated framework for assessing renewable-energy supply chains using multicriteria decision-making: A study on Bangladesh. *Clean Energy*, *8*(3), 1–15. DOI: 10.1093/ce/zkae019

Azad, A. M. A. S., Oishi, Z. T., Haque, M. A., Das, P., Udoy, S. A., & Bhuiya, K. M. S. (2024). Integration of renewable energy in industrial operations: Experiences from Canada, USA, and Africa. *GSC Advanced Research and Reviews*, *18*(1), 213–221. Advance online publication. DOI: 10.30574/gscarr.2024.18.1.0014

Bag, S., Pretorius, J. H. C., Gupta, S., & Dwivedi, Y. K. (2021). Role of technological dimensions of green supply chain management practices on firm performance. *Journal of Enterprise Information Management*, *34*(1), 220–245. DOI: 10.1108/JEIM-10-2019-0324

Bai, L., Wang, J., Ma, X., & Lu, H. (2018). Air pollution forecasts: An overview. *International Journal of Environmental Research and Public Health*, *15*(4), 780. DOI: 10.3390/ijerph15040780 PMID: 29673227

Ball, P. J. (2021). A review of geothermal technologies and their role in reducing greenhouse gas emissions in the USA. *Journal of Energy Resources Technology*, *143*(1), 010903. DOI: 10.1115/1.4048187

Barbieri, E., & Capoani, L. (2025). Renewable Energy, Resilience, Digitalization, and Industrial Policies in Seaborne Transport. *Energies*, *18*(5), 1089. DOI: 10.3390/en18051089

Barkhi, R., Dekker, M.-J., Jung, E., Duvvuri, V., & Uslu, T. (2024). Key drivers and priorities of consumer decisions for refurbished products. *Heliyon, 10*(10), e40977. DOI: 10.1016/j.heliyon.2024.e40977

Barney, J. (1991). Firm resources and sustained competitive advantage. *Journal of Management, 17*(1), 99–120. DOI: 10.1177/014920639101700108

Batinge, B., Musango, J. K., & Brent, A. C. (2017). Leapfrogging to renewable energy: The opportunity for unmet electricity markets. *South African Journal of Industrial Engineering, 28*(4), 32–49. DOI: 10.7166/28-4-1702

Bazarbash, M. (2019). *FinTech in financial inclusion: Machine learning applications in assessing credit risk* (IMF Working Paper No. 2019/109). International Monetary Fund. https://doi.org/10.5089/9781498314428.001

Bazarhanova, D., Maier, C., & Laumer, S. (2022). Beyond efficiency: Rethinking FinTech performance metrics. *Journal of Information Systems, 36*(2), 147–166.

Behera, P., Haldar, A., & Sethi, N. (2023). Achieving carbon neutrality target in the emerging economies: Role of renewable energy and green technology. *Gondwana Research, 121*(1), 16–32. DOI: 10.1016/j.gr.2023.03.028

Ben Rejeb, N., Zailani, S., & Ali, M. (2021). Green lending practices and SME financing in emerging markets: Evidence from Tunisia. *Journal of Business and Industrial Marketing, 36*(14), 2, 300–302, 318. https://doi.org/10.1108/JBIM-12-2021-0587

Ben-Daya, M., Hassini, E., & Bahroun, Z. (2019). Internet of Things and supply chain management: A literature review. *International Journal of Production Research, 57*(15–16), 4719–4742. DOI: 10.1080/00207543.2017.1402140

Berardi, P. C., & de Brito, R. P. (2021). Supply chain collaboration for a circular economy: From transition to continuous improvement. *Journal of Cleaner Production, 328*, 129511. DOI: 10.1016/j.jclepro.2021.129511

Bergman, N., & Foxon, T. J. (2023). Drivers and effects of digitalization on energy demand in low-carbon scenarios. *Climate Policy, 23*(3), 329–342. DOI: 10.1080/14693062.2022.2145260

Berlin, D., Feldmann, A., & Nuur, C. (2022). Supply network collaborations in a circular economy: A case study of Swedish steel recycling. *Resources, Conservation and Recycling, 179*, 106112. DOI: 10.1016/j.resconrec.2021.106112

Bhattacharjee, S., Riame, L., & Kalita, G. (2025). Utilization of Artificial Intelligence and the Internet of Things in the Renewable Industry. In *Artificial Intelligence and IoT for Cyber Security Solutions in Smart Cities* (pp. 1–15). Chapman and Hall/CRC.

Bhatti, Z. A., Raza, M., Iqbal, S., Mehmood, A., & Mehmood, A. (2023). Sustainable supply chain finance through digital platforms: A pathway to green entrepreneurship. *Annals of Operations Research, 331*(1), 285–319. Advance online publication. DOI: 10.1007/s10479-022-04623-5

Bhimani, A., Mention, A.-L., & Barlatier, P. J. (2021). Artificial intelligence in operations management: Applications, challenges and opportunities. *Production Planning and Control, 32*(12), 975–990. DOI: 10.1080/09537287.2020.1810759

Bibri, S. E., & Krogstie, J. (2021). A novel model for data-driven smart sustainable cities of the future: A strategic roadmap to transformational change in the era of big data. *Future Cities Environ., 7*(1), 1–25. DOI: 10.5334/fce.116

Bingxin, W., Qamri, G. M., Hui, G., Ameer, W., & Majeed, M. A. (2025). From digitalization to renewable energy: How the tech-energy connection drives the green energy in belt and road countries. *Energy Economics, 144*(1), 1–12. DOI: 10.1016/j.eneco.2025.108324

Birkel, H. S., & Hartmann, E. (2019). Impact of IoT challenges and risks for SCM. *Supply Chain Management, 24*(1), 39–61. DOI: 10.1108/SCM-03-2018-0142

Blombäck, A., & Wigren, C. (2008), Are not multinationals also community actors?: Challenging the notion of small versus large firm corporate social responsibility. In *CRRC 2008 Conference Proceedings: The corporate responsibility research conference, Belfast, Northern Ireland, 7-9 September, 2008.*

Bocken, N. M. P., Mugge, R., Bom, C. A., & Lemstra, H. J. (2018). Pay-per-use business models as a driver for sustainable consumption: Evidence from the case of HOMIE. *Journal of Cleaner Production, 198*, 498–510. DOI: 10.1016/j.jclepro.2018.07.043

Boiteux, M., & Baumstark, L. (2001). France. Commissariat général du plan. Transports: choix des investissements et coût des nuisances (p. 325). La documentation française.

Boldyreiva, L., & Gryshko, V. (2020). European experience of smart-technology implementation in logistic process management. *Економіка та держава*, (4), 39-43.

Bondi, A. B. (2000). *Characteristics of scalability and their impact on performance.* In *Proceedings of the Workshop on Software and Performance (WOSP2000)* (pp. 195–203). ACM. https://dl.acm.org/doi/10.1145/350391.350432

Boston Consulting Group & QED Investors. (2024, June 26). *Global FinTech 2024: Prudence, profits, and growth.* Boston Consulting Group. www.bcg.com/publications/2024/global-fintech-prudence-profits-and-growth?utm_source=chatgpt.com

Boukherouaa, E. B., Shabsigh, M. G., AlAjmi, K., Deodoro, J., Farias, A., Iskender, E. S., & Ravikumar, R. (2021). *Powering the digital economy: Opportunities and risks of artificial intelligence in finance*. International Monetary Fund. https://www.imf.org/-/media/Files/Publications/DP/2021/English/PDEORAIFEA.ashx

Bowen, F., Newenham-Kahindi, A., & Herremans, I. (2010). When suits meet roots: The antecedents and consequences of community engagement strategy. *Journal of Business Ethics*, *95*(2), 297–318. DOI: 10.1007/s10551-009-0360-1

Bowen, H. R. (1953). *Social Responsibilities of the Businessman*. Harper & Brothers.

Braun, M., Grimme, W., & Oesingmann, K. (2024). Pathway to net zero: Reviewing sustainable aviation fuels, environmental impacts and pricing. *Journal of Air Transport Management*, *117*, 117. DOI: 10.1016/j.jairtraman.2024.102580

Braun, V., & Clarke, V. (2006). Using thematic analysis in psychology. *Qualitative Research in Psychology*, *3*(2), 77–101. https://doi.org/10.1191/1478088706qp063oa

Bressanelli, G., Saccani, N., & Perona, M. (2024). Are circular economy business models based on digital servitization sustainable? A hypothetical systemic simulation model. *Revista de Producción Más Limpia*, *458*, 142512. DOI: 10.1016/j.jclepro.2024.142512

Bressanelli, G., Saccani, N., Perona, M., & Baccanelli, I. (2020). Towards Circular Economy in the Household Appliance Industry: An Overview of Cases. *Resources*, *9*(11), 128. DOI: 10.3390/resources9110128

Brown, R., & Mason, C. (2017). Looking inside the spiky bits: A critical review and conceptualisation of entrepreneurial ecosystems. *Small Business Economics*, *49*(1), 11–30. https://doi.org/10.1007/s11187-017-9865-7

Brown, W., Johnson, O., & Wilson, G. (2024). Influence of E-commerce technologies on supply chain management in retail. In *Preprints*.

Bullerdiek, N., Neuling, U., & Kaltschmitt, M. (2021). A GHG reduction obligation for sustainable aviation fuels (SAF) in the EU and in Germany. *Journal of Air Transport Management*, *92*(May), 102020. DOI: 10.1016/j.jairtraman.2021.102020

Bureau of Energy Efficiency. (2022). *Annual report 2021–2022*. https://beeindia.gov.in/

Bureau of Energy Efficiency. (2022). *Energy efficiency labelling programmes.* https://beeindia.gov.in/

Buse, J. (2024). *Sustainable Aviation Fuels, transitioning towards green aviation.* CRC Press.

Button, K. (2002). Empty cores in airlines markets, *5th Hamburg* **Aviation** *Conference.* Hamburg.

Calzolari, T., Genovese, A., & Brint, A. (2022). Circular economy indicators for supply chains: A systematic literature review. *Environmental and Sustainability Indicators, 13,* 100160. DOI: 10.1016/j.indic.2021.100160

Cameron, P. D. (2006). The consumer and the internal market in energy: Who benefits. *European Law Review.*

Cao, Z., & Shi, X. (2021). A systematic literature review of entrepreneurial ecosystems in advanced and emerging economies. *Small Business Economics, 57*(1), 75–110. https://doi.org/10.1007/s1118702000326y

Carroll, A. B., & Näsi, J. (1997), Understanding stakeholder thinking: Themes from a Finnish conference. *Business Ethics: A European Review,* vol. *6, n°*1, p. 46-51.

Carroll, A. B. (1999). Corporate social responsibility: Evolution of a definitional construct. *Business & Society, 38*(3), 268–295. DOI: 10.1177/000765039903800303

Carvalho, P., & Marques, R. C. (2014). Computing economies of vertical integration, economies of scope and economies of scale using partial frontier nonparametric methods. *European Journal of Operational Research, 234*(1), 292–307. DOI: 10.1016/j.ejor.2013.09.022

Cazal, D. (2011). RSE et théorie des parties prenantes: les impasses du contrat. *Revue de la régulation. Capitalisme, institutions, pouvoirs,* (9).

Centobelli, P., Cerchione, R., Esposito, E., Oropallo, E., & Secundo, G. (2021). Blockchain technology to unite trust, traceability and transparency in circular supply chains: A systematic review of the literature. *Information & Management, 59*(7), 103508. DOI: 10.1016/j.im.2021.103508

Chen, F., Xu, J., Zhang, W., Wei, B., & Li, T. (2023). Capital formation and natural resources extraction: A source of sustainable development or a curse for the economy? *Resources Policy, 85*(1), 103964. DOI: 10.1016/j.resourpol.2023.103964

Cheng, Y., Zhang, S., Huan, C., Oladokun, M. O., & Lin, Z. (2019, January). Optimization on fresh outdoor air ratio of air conditioning system with stratum ventilation for both targeted indoor air quality and maximal energy saving. *Building and Environment*, *147*, 11–22. DOI: 10.1016/j.buildenv.2018.10.009

Chen, N., Wang, S., & He, Y. (2025). Integrating Hydrogen Fuel Cell Trucks into Freight Transport: Scenarios for Costs, Emissions, and Policy in China. *Energy*, •••, 138609. DOI: 10.1016/j.energy.2025.138609

Chen, R. (2024). Sustainable supply chain management as a strategic enterprise innovation. Advances in Economics. *Management and Political Sciences*, *85*(1), 24–29.

Chen, S. L., Chen, Y. Y., & Hsu, C. (2014). A new approach to integrate internet-of-things and software-as-a-service model for logistic systems: A case study. *Sensors (Basel)*, *14*(4), 6144–6164. DOI: 10.3390/s140406144

Chen, W., Alharthi, M., Zhang, J., & Khan, I. (2024). The need for energy efficiency and economic prosperity in a sustainable environment. *Gondwana Research*, *127*(1), 22–35. DOI: 10.1016/j.gr.2023.03.025

Chen, X., Wang, C., & Li, S. (2023). The impact of supply chain finance on corporate social responsibility and creating shared value: A case from the emerging economy. *Supply Chain Management*, *28*(2), 324–346. https://doi.org/10.1108/SCM-10-2021-0478

Chen, Y., Hu, S., Zheng, Y., Xie, S., Yang, Q., Wang, Y., & Hu, Q. (2024). Coordinated optimization of logistics scheduling and electricity dispatch for electric logistics vehicles considering uncertain electricity prices and renewable generation. *Applied Energy*, *364*, 123147. DOI: 10.1016/j.apenergy.2024.123147

Cho, G. S. (2018). A study on establishment of smart logistics center based on logistics 4.0. *Journal of Multimedia Information System*, *5*(4), 265–272.

Choi, T. Y., Dooley, K. J., & Rungtusanatham, M. (2001). Supply networks and complex adaptive systems: Control versus emergence. *Journal of Operations Management*, *19*(3), 351–366. DOI: 10.1016/S0272-6963(00)00068-1

Choudhury, S., Verma, S., & Chaudhuri, R. (2022). Exploring the role of digital technologies in building sustainable supply chains: A case of Amazon. *Journal of Cleaner Production*, *356*, 131817. DOI: 10.1016/j.jclepro.2022.131817

Christopher, M. (1992). *Logistics and Supply Chain Management*. Pitman Publishing.

Christopher, M., & Peck, H. (2004). Building the resilient supply chain. *International Journal of Logistics Management*, *15*(2), 1–14. DOI: 10.1108/09574090410700275

Chu, S., & Majumdar, A. (2012). Opportunities and challenges for a sustainable energy future. nature, 488(7411), 294-303.

Clarkson, M. E. (1995). A stakeholder framework for analyzing and evaluating corporate social performance. *Academy of Management Review*, *20*(1), 92–117. DOI: 10.2307/258888

CLASP. (2019). *Opportunities for international collaboration on energy efficiency labels*. https://www.clasp.ngo/

Cortés-Borda, D., Polanco, J. A., & Escobar-Sierra, M. (2022). Social perception assessment of hydropower sustainability: A stepwise logistic regression modeling. *Environmental Science & Policy*, *134*, 108–118. DOI: 10.1016/j.envsci.2022.03.026

Corvellec, H., Stowell, A. F., & Johansson, N. (2022). Criticisms of the circular economy. *Journal of Industrial Ecology*, *26*(2), 421–432. DOI: 10.1111/jiec.13187

Cosma, A. C., & Simha, R. (2019, January). Machine learning method for real-time non-invasive prediction of individual thermal preference in transient conditions. *Building and Environment*, *148*, 372–383. DOI: 10.1016/j.buildenv.2018.11.017

Coulombel, N., Dablanc, L., Gardrat, M., & Koning, M. (2018). The environmental social cost of urban road freight: Evidence from the Paris region. *Transportation Research Part D, Transport and Environment*, *63*, 514–532. DOI: 10.1016/j.trd.2018.06.002

Creti, A., & Villeneuve, B. (2004). *Long-term contracts and take-or-pay clauses in natural gas markets*. Energy Studies Review.

Crocker, K. J., & Asten, S. E. M. (1985). Efficient adaptation in long-term contracts: Take-or-pay provisions for natural gas. *The American Economic Review*.

Crowe, S., Cresswell, K., Robertson, A., Huby, G., Avery, A., & Sheikh, A. (2011). The case study approach| BMC Medical Research Methodology| Full Text.

Dalton, J. (2018). SWOT analysis (Strengths, weaknesses, opportunities, threats). In *Great Big Agile: An OS for Agile Leaders* (pp. 249–252). Apress.

Damianov, D., & Demirova, S. (2018, June). Principles of designing automated logistics systems-hybrid component of cyber-physical systems. In *2018 International Conference on High Technology for Sustainable Development (HiTech)* (pp. 1-4). IEEE. DOI: 10.1109/HiTech.2018.8566533

Davis, F. D. (1989). Perceived usefulness, perceived ease of use, and user acceptance of information technology. *Management Information Systems Quarterly*, *13*(3), 319–340. DOI: 10.2307/249008

De Angelis, R., Morgan, R., & De Luca, L. M. (2023). Open Strategy and Dynamic Capabilities: A Framework for Circular Economy Business Model Research. *Business Strategy and the Environment*, *32*(7), 4861–4873. DOI: 10.1002/bse.3397

De Sousa, F. D. B. (2024). The Global Plastics Treaty: Understanding the Present to Guide the Future. *Cambridge Prisms: Plastics*, *2*, e31. DOI: 10.1017/plc.2024.32

de Vass, T., Shee, H., & Miah, S. (2021). IoT in supply chain management: Opportunities and challenges for businesses in early industry 4.0 context. Oper. Supply Chain Manag. Int. J., 148–161.

De Vass, T., Shee, H., & Miah, S. J. (2018). Internet of Things for improving supply chain performance: A Qualitative study of Australian retailers. ACIS2018 Proceedings.

Deka, A., Ozdeser, H., Cavusoglu, B., Seraj, M., & Tursoy, T. (2023). Exchange rate stability in the emerging economies: Does renewable energy play a role—a panel data analysis. *Environmental Science and Pollution Research International*, *30*(9), 23668–23677. DOI: 10.1007/s11356-022-23911-y PMID: 36329243

Dembek, K., Singh, P., & Bhakoo, V. (2016). Literature review of shared value: A theoretical concept or a management buzzword? *Journal of Business Ethics*, *137*(2), 231–267. 10.1007/s10551-015-2554-z

Deng, H., Wang, R., Miao, K., Zhao, Y., Sun, J., & Du, J. (2018, December). Design and implementation of ship gps monitoring and dispatching system for intelligent transportation system. In *Proceedings of the International Conference on Information Technology and Electrical Engineering 2018* (pp. 1-5). DOI: 10.1145/3148453.3306261

Denoncourt, J. (2020). Companies and UN 2030 sustainable development goal 9 industry, innovation and infrastructure. Journal of Corporate law studies, 20(1), 199-235.

DHL. (2022). *Logistics Trend Radar 6th Edition*. https://www.dhl.com/global-en/home/insights-and-innovation/thought-leadership/trend-reports/logistics-trend-radar.html

Diao, X., Gao, A., Jin, X., & Chen, H. (2022). A layer-based relaxation approach for service network design. *Sustainability (Basel)*, *14*(20), 13143. DOI: 10.3390/su142013143

DiMaggio, P. J., & Powell, W. W. (1983). The iron cage revisited: Institutional isomorphism and collective rationality in organizational fields. *American Sociological Review*, *48*(2), 147–160. https://doi.org/10.2307/2095101

Ding, Y., Jin, M., Li, S., & Feng, D. (2021). Smart logistics based on the internet of things technology: An overview. *International Journal of Logistics*, *24*(4), 323–345. DOI: 10.1080/13675567.2020.1757053

Ding, Y., & Xue, Y. (2019). A deep learning approach to writer identification using inertial sensor data of air-handwriting. *IEICE Transactions on Information and Systems*, *E102-D*(10), 2059–2063. DOI: 10.1587/transinf.2019EDL8070

Dong, J., Luo, J., Zhang, W., & Zhang, H. (2024). How does the digital economy improve renewable energy development across the globe?-from a product perspective. [Online]. DOI: 10.2139/ssrn.4806484

Dötsch, J., & Schmid, N. (2021). The effectiveness of energy efficiency labels in influencing consumer choice: A meta-analysis. *Journal of Cleaner Production*, *279*, 123456.

Dotse, S.-Q., Petra, M. I., Dagar, L., & De Silva, L. C. (2018, March). Application of computational intelligence techniques to forecast daily PM10 exceedances in Brunei Darussalam. *Atmospheric Pollution Research*, *9*(2), 358–368. DOI: 10.1016/j.apr.2017.11.004

Dridi, N. (2021). Financing sustainable development: The case of renewable energy in Tunisia. *International Journal of Energy and Environmental Issues*, *43*, 504–521. https://doi.org/10.1504/IJGEI.2021.118939

Dunbar, K. (2024). *FinTech for environmental sustainability: Promises and pitfalls*. Preprint., DOI: 10.1016/j.xxx

Eggleston, H. S., Gaudioso, D., Gorissen, N., Joumard, R., Rijkeboer, R. C., Samaras, Z., & Zierock, K. H. (1993). *CORINAIR working group on emission factors for calculating 1990 emissions from road traffic*. Office for Official Publications of the European Communities.

Egwuonwu, A., Mordi, C., Egwuonwu, A., & Uadiale, O. (2022). The influence of blockchains and Internet of Things on global value chain. *Strategic Change*, *31*(1), 45–55. https://doi.org/10.1002/jsc.2484

EL AIDOUNI, M., & ABBASSI, A. (2012). La RSE dans le management de la chaine logistique. *Revue Marocaine de Recherche en Management et Marketing*, (6-7).

Elavarasan, R. M., Pugazhendhi, R., Jamal, T., Dyduch, J., Arif, M. T., Kumar, N. M., & Nadarajah, M. (2021). Envisioning the UN Sustainable Development Goals (SDGs) through the lens of energy sustainability (SDG 7) in the post-COVID-19 world. *Applied Energy*, *292*, 116665. DOI: 10.1016/j.apenergy.2021.116665

ElBannan, M. A., & Löffler, G. (2024). How effectively do green bonds help the environment? *Journal of Banking & Finance, 158,* 107051. DOI: 10.1016/j.jbankfin.2023.107051

Elkington, J. (1997). *Cannibals with forks: The triple bottom line of 21st century business.* Capstone Publishing.

Elouaourti, Z. (2024). Financial technologies for all MENA citizens: Tackling digital financial inclusion. *ScienceDirect.* https://www.sciencedirect.com/article/pii/S1757780224000519

ElWassal, K. (2022). Crowdfunding renewable energy investments: Investor perceptions and decisionmaking factors in an emerging market. *Technological Forecasting and Social Change, 182,* 121686. Advance online publication. https://doi.org/10.1016/j.techfore.2022.121686

Environmental Protection Agency. (2020). *About ENERGY STAR.* https://www.energystar.gov/

Environmental Protection Agency. (2020). *ENERGY STAR overview.* https://www.energystar.gov/

Environmental Protection Agency. (2022). Inventory of U.S. Greenhouse Gas Emissions and Sinks: 1990–2020. Tables A-98 and ES-6, last updated July 13.

EPA. (2020). *ENERGY STAR Overview.* https://www.energystar.gov/

Esposito, L. (2023). Renewable energy consumption and per capita income: An empirical analysis in Finland. *Renewable Energy, 209*(1), 558–568. DOI: 10.1016/j.renene.2023.03.139

European Commission. (2021). *Consumer research on the effectiveness of the EU energy label.* https://ec.europa.eu/

European Commission. (2021). *EU Energy Label.* https://ec.europa.eu/energy/topics/energy-efficiency/energy-labelling_en

European Commission. (2021). Proposal for a Regulation of the European Parliament and of the Council on Ensuring a Level Playing Field for Sustainable Air Transport: COM(2021) 561 Final. (2023). European Commission.

European Commission. (2021). *The new energy label explained.* https://ec.europa.eu/info/energy-climate-change-environment/standards-tools-and-labels

Eyitoyo, A. K. (2023). A Comparative Study of Alternative Energy Usage in the Global Logistics Sector (Master's thesis, Texas A&M University-Kingsville).

Faiz, A., Weaver, C. S., & Walsh, M. P. (1996). *Air pollution from motor vehicles: standards and technologies for controlling emissions*. World Bank Publications.

Firouraghi, N., Kiani, B., Jafari, H. T., Learnihan, V., Salinas-Perez, J. A., Raeesi, A., Furst, M., Salvador-Carulla, L., & Bagheri, N. (2022). The role of geographic information system and global positioning system in dementia care and research: A scoping review. *International Journal of Health Geographics*, *21*(1), 8. DOI: 10.1186/s12942-022-00308-1 PMID: 35927728

Fletcher, S., & Evans, T. (2025). Optimism for a global plastics treaty. *Science*, *387*(6730), 119. DOI: 10.1126/science.adv2404

Fontes, C. H. D. O., & Freires, F. G. M. (2018). Sustainable and renewable energy supply chain: A system dynamics overview. *Renewable & Sustainable Energy Reviews*, *82*, 247–259. DOI: 10.1016/j.rser.2017.09.033

Fraccascia, L., & Yazan, D. M. (2018). The role of online information-sharing platforms on the performance of industrial symbiosis networks. *Resources, Conservation and Recycling*, *136*, 473–485. DOI: 10.1016/j.resconrec.2018.03.009

Frederick, W. C. (1978). From CSR, to CSR: The maturing ofbusiness-and-society thought. Working paper, Katz.

Frederick, W. C. (1994). From CSR1 to CSR2: The maturing of business-and-society thought. *Business & Society*, *33*(2), 150–164. DOI: 10.1177/000765039403300202

Freeman, R. E. (1984). Strategic management: A stakeholder approach. Pitman. 10.1017/CBO9781139192675

Freeman, R. E. (1984). *Strategic management: A stakeholder approach*. Pitman.

Frioui, E., & Ben Sedrine Doghri, S. (2025). Tunisian fintech's ecosystem: To what extent can Tunisian fintechs be an engine for shared value creation? In Digital technologies for sustainability and quality control (p. 36). IGI Global. DOI: 10.4018/979-8-3693-4373-9.ch006

GAO. (2023). Report to Congressional Committees. Sustainable Aviation Fuel, Agencies should track progress toward ambitious federal goals.

Gartner, W. B. (1985). A conceptual framework for describing the phenomenon of new venture creation. *Academy of Management Review*, *10*(4), 696–706. https://doi.org/10.2307/258039

Gebresenbet, G., Bosona, T., Olsson, S. O., & Garcia, D. (2018). Smart system for the optimization of logistics performance of the pruning biomass value chain. *Applied Sciences (Basel, Switzerland)*, *8*(7), 1162. DOI: 10.3390/app8071162

Geels, F. W. (2002). Technological transitions as evolutionary reconfiguration processes: A multi-level perspective and a case-study. *Research Policy*, *31*(8/9), 1257–1274. DOI: 10.1016/S0048-7333(02)00062-8

Genovese, A., Acquaye, A. A., Figueroa, A., & Koh, S. C. L. (2017). Sustainable supply chain management and the transition towards a circular economy: Evidence and some applications. *Omega*, *66*, 344–357. DOI: 10.1016/j.omega.2015.05.015

Germain, C., & Gates, S. (2010). L'engagement des parties prenantes internes dans les démarches de Responsabilité Globale (RG): L'intégration des indicateurs de RG dans les outils de pilotage du contrôle de gestion. *Management Avenir*, (3), 223–237. DOI: 10.3917/mav.033.0223

Getahun, S., Kefale, H., & Gelaye, Y. (2024). Application of precision agriculture technologies for sustainable crop production and environmental sustainability: A systematic review. *TheScientificWorldJournal*, *2024*(1), 2126734. DOI: 10.1155/2024/2126734 PMID: 39421732

Ghahremani Nahr, J., Nozari, H., & Sadeghi, M. E. (2021). Green supply chain based on artificial intelligence of things (AIoT). International Journal of Innovation in Management. *Economics and Social Sciences*, *1*(2), 56–63.

Ghio, N., & Guerini, M., & RossiLamastra, C. (2019). The creation of hightech ventures in entrepreneurial ecosystems: Exploring the interactions among university knowledge, cooperative banks, and individual attitudes. *Small Business Economics*, *52*(2), 523–543. https://doi.org/10.1007/s11187-017-9958-3

Ghosh, D. (2021). Role of artificial intelligence in green logistics and sustainable supply chain management. *Transportation Research Part E, Logistics and Transportation Review*, *148*, 102265. DOI: 10.1016/j.tre.2021.102265

Ghosh, K. (2020). Role of FinTech in green finance and sustainable development. *Gyan Management Journal*, *14*(2), 61–71. https://journals.acspublisher.com/index.php/gmj/article/view/187

Giera, J., & Kulińska, E. (2021). Risk factors in a logistics company using renewable energy sources. *Energies*, *14*(23), 8152. DOI: 10.3390/en14238152

Global Market Insights (GMI). (2024). Sustainable Aviation Fuel Market Size - By Fuel Type, Technology, Blending Capacity, End Use Analysis, Share, Growth Forecast, 2025 – 2034, December.

Goergen, M., Chahine, S., Wood, G., & Brewster, C. (2019). The relationship between public listing, context, multi-nationality and internal CSR. *Journal of Corporate Finance*, *57*, 122–141. DOI: 10.1016/j.jcorpfin.2017.11.008

Gomber, P., Kauffman, R. J., Parker, C., & Weber, B. W. (2018). On the FinTech revolution: Interpreting the forces of innovation, disruption, and transformation in financial services. *Journal of Management Information Systems*, *35*(1), 220–265. DOI: 10.1080/07421222.2018.1440766

Gond, J. P., & Mullenbach-Servayre, A. (2004). Les fondements théoriques de la responsabilité sociétale de l'entreprise. *La Revue des Sciences de Gestion: Direction et Gestion*, (205), 93.

Gonzalez-Garay, A., Heuberger-Austin, C., Fu, X., Klokkenburg, M., Zhang, D., van der Made, A., & Shah, N. (2022). Unravelling the potential of sustainable aviation fuels to decarbonise the aviation sector. *Energy & Environmental Science*, *15*(8), 15. DOI: 10.1039/D1EE03437E

Gorji, M. A., Akbarzadeh, M., & Shetab-Boushehri, S. N. (2022). Evaluation and improvement of the urban transportation networks resilience in short-term non-recurring traffic congestion: a novel graph connectivity-based criteria. *Transportation engineering, 10*, 100152.

Gourire, M., & Rouggani, K., & AMINE, N. B. (2025). Dévoiler les pratiques de la chaîne d'approvisionnement durable et leur impact sur la performance logistique des entreprises: An exploratory analysis of companies listed on the Moroccan stock exchange. *RMd• Economics, Management &. Social Sciences*, *2*(1), e202510–e202510.

Gowda, V. D., Surya, S. G., Kumar, N. M. G., Prasad, K., Prasad, V. S., & Kaur, M. (2024, May). Optimizing Renewable Energy Integration in Smart Grids through IoT-Driven Management Systems. In *2024 2nd International Conference on Advancement in Computation & Computer Technologies (InCACCT)* (pp. 783-788). IEEE.

Grandval, S., & Soparnot, R. (2005). Le développement durable comme stratégie de rupture: Une approche par la chaîne de valeur inter-sectorielle. *Management & Avenir*, *5*(3), 7–26. DOI: 10.3917/mav.005.0007

Grimme, W. (2023). The Introduction of Sustainable Aviation Fuels—A Discussion of Challenges, Options and Alternatives. *Aerospace (Basel, Switzerland)*, *10*(3), 218. DOI: 10.3390/aerospace10030218

Guan, R., Wang, H., Zheng, R., Shi, X., & Wang, K. (2025). Evaluating the impact of digital technologies on energy efficiency: Evidence from Chinese publicly listed companies. *Energy Policy*, *207*(1), 114847. DOI: 10.1016/j.enpol.2025.114847

Guan, Y., Sun, N., Wu, S. J., & Sun, Y. (2025). Supply Chain Finance, Fintech Development, and Financing Efficiency of SMEs in China. *Administrative Sciences*, *15*(3), 86. DOI: 10.3390/admsci15030086

Gumzej, R. (2023). Intelligent logistics systems in E-commerce and transportation. *Mathematical Biosciences and Engineering*, *20*(2), 2348–2363. DOI: 10.3934/mbe.2023110

Guo, J. (2022, November). Design and implementation of intelligent logistics system for photovoltaic enterprises. In *International Conference on Frontiers of Traffic and Transportation Engineering (FTTE 2022)* (Vol. 12340, pp. 266-271). SPIE. DOI: 10.1117/12.2652315

Guo, S., Li, M., & Chen, X. (2021). Influence mechanism of energy efficiency label on consumers' purchasing behavior of energy-saving household appliances. *Frontiers in Psychology*, *12*, 711854. DOI: 10.3389/fpsyg.2021.711854 PMID: 34733202

Gutiérrez-Diez, J. C., Castañón, A. M., & Bascompta, M. (2024). New method to study the effectiveness of mining equipment: A case study of surface drilling rigs. *Applied Sciences (Basel, Switzerland)*, *14*(5), 2185. DOI: 10.3390/app14052185

Gutu Ofgaa, S. U. (2020). The Integrated Exchange of Information and IoT for Smart Logistics Systems.

Hagiu, A., & Rothman, S. (2016, April). Network effects aren't enough. *Harvard Business Review*, *94*(4), 65–71.

Han, Y., Lam, J. C. K., & Li, V. O. K. "A Bayesian LSTM model to evaluate the effects of air pollution control regulations in China," in Proc. IEEE Big Data Workshop (Big Data), Dec. 2018, pp. 4465–4468. DOI: 10.1109/BigData.2018.8622417

Hanbury, O., & Vasquez, V. R. (2018). Life cycle analysis of geothermal energy for power and transportation: A stochastic approach. *Renewable Energy*, *115*, 371–381. DOI: 10.1016/j.renene.2017.08.053

Han, S., Ulhoi, J. P., & Song, H. (2024). Digital trust in supply chain finance: The role of innovative fintech service provision. *Journal of Enterprise Information Management*, *37*(3), 1596–1615. DOI: 10.1108/JEIM-07-2022-0238

Hansen, J., Kharecha, P., Sato, M., Masson-Delmotte, V., Ackerman, F., Beerling, D. J., Hearty, P. J., Hoegh-Guldberg, O., Hsu, S.-L., Parmesan, C., Rockstrom, J., Rohling, E. J., Sachs, J., Smith, P., Steffen, K., Van Susteren, L., von Schuckmann, K., & Zachos, J. C. (2013). Assessing "dangerous climate change": Required reduction of carbon emissions to protect young people, future generations and nature. *PLoS One*, *8*(12), e81648. DOI: 10.1371/journal.pone.0081648 PMID: 24312568

Han, Y., Shevchenko, T., Yannou, B., Ranjbari, M., Shams Esfandabadi, Z., Saidani, M., Bouillass, G., Bliumska-Danko, K., & Li, G. (2023). Exploring how digital technologies enable a circular economy of products. *Sustainability (Basel)*, *15*(3), 2067. DOI: 10.3390/su15032067

Hartley, P. R., & Medlock, K. B.III. (2017). The valley of death for new energy technologies. *The Energy Journal (Cambridge, Mass.)*, *38*(3), 33–61. DOI: 10.5547/01956574.38.3.phar

Hasan, M. (2024). FinTech and sustainable development / FinTech and sustainable finance. [Working paper or journal article]. (cronfa.swan.ac.uk)

Hayajneh, H. S., & Zhang, X. (2020). Logistics design for mobile battery energy storage systems. *Energies*, *13*(5), 1157. DOI: 10.3390/en13051157

Herbert, F. A., & Tchawa, Y. P. (2024). Do ICTs promote the renewable energy consumption? The moderating effects of economic growth and structural transformation in Africa. *Inter Economics*, *180*(C), 1–14. DOI: 10.1016/j.inteco.2024.100563

Hofer, C. W., & Schendel, D. (1978). *Strategy formulation: Analytical concepts*. West Publishing Co.

Holloway, S. (2024). Exploring the impact of real-time supply chain information on marketing decisions: Insights from service industries. In *Preprints*.

Hou, J., & Chen, C. (2022, April). Intelligent logistics supply chain management based on Internet of Things technology. In *2022 IEEE Asia-Pacific Conference on Image Processing, Electronics and Computers (IPEC)* (pp. 1266-1270). IEEE. DOI: 10.1109/IPEC54454.2022.9777588

Huang, H., McKinstry, S., & Weller, D. (2020). Quantum computing and operations management: A review and future research agenda. *Computers & Industrial Engineering*, *149*, 106802. DOI: 10.1016/j.cie.2020.106802

Huang, J., Xie, D., Qiu, Y., Wang, J., & Song, J. (2024). Green supply chain management: A renewable energy planning and dynamic inventory operations for perishable products. *International Journal of Production Research*, *62*(24), 8924–8951. DOI: 10.1080/00207543.2023.2220047

Huang, W., Chen, S., & Nguyen, L. T. (2020). Corporate social responsibility and organizational resilience to COVID-19 crisis: An empirical study of Chinese firms. *Sustainability (Basel)*, *12*(21), 8970. DOI: 10.3390/su12218970

Huang, W., & He, J. (2023). Impact of energy intensity, green economy, and natural resources development to achieve sustainable economic growth in Asian countries. *Resources Policy*, *84*(1), 1–12. DOI: 10.1016/j.resourpol.2023.103726

Hu, K., Raghutla, C., Chittedi, K. R., Zhang, R., & Koondhar, M. A. (2021). The effect of energy resources on economic growth and carbon emissions: A way forward to carbon neutrality in an emerging economy. *Journal of Environmental Management*, *298*(1), 1–15. DOI: 10.1016/j.jenvman.2021.113448 PMID: 34358940

Hurlimann, A., Moosavi, S., & Browne, G. R. (2021). Urban planning policy must do more to integrate climate change adaptation and mitigation actions. *Land Use Policy*, *101*, 105188. DOI: 10.1016/j.landusepol.2020.105188

Hussain, M., Al-Aomar, R., & Melhem, H. (2019). Assessment of lean-green practices on the sustainable performance of hotel supply chains. *International Journal of Contemporary Hospitality Management*, *31*(6), 2448–2467. DOI: 10.1108/IJCHM-05-2018-0380

Hussein, A. H. A., Jabbar, K. A., Mohammed, A., & Jasim, L. (2024). Harvesting the Future: AI and IoT in Agriculture. E3S Web Conf., 477, 90.

IATA (2023), *Air Passenger Forecast*

IATA (2023), *Industry Economic Performance Mid-Year Report*

IBM. (2020). *IBM Food Trust and Walmart: Revolutionizing food safety with blockchain*. https://www.ibm.com/blockchain/solutions/food-trust

IBM. (2022). *Case Study: Maersk and IBM Blockchain – TradeLens*. https://www.ibm.com/case-studies/tradelens-blockchain

ICAO (2023), *Long-Term Aspirational Goal Report*

ICAO. (2023). *Long-Term Aspirational Goal (LTAG) Report*. Montréal.

ICAO. (2023).. . *Environment Reporter*.

ICCT (International Council Clean Transportation). (2022). [*Aligning Aviation with Paris Agreement Goals*. International Energy Agency] [IEA]. *Vision (Basel)*, •••, 2050.

IEA (2020), World Energy Balances

IEA. (2023). *Tracking Clean Energy Progress Intergovernmental Panel on Climate Change. (2022). Climate Change 2022: Mitigation of Climate Change. Contribution of Working Group III to the Sixth Assessment Report of the Intergovernmental Panel on Climate Change*. Cambridge University Press.

Ilmarinen, P., & Akpinar, M. (2018). *Creating shared value in the banking industry: A case study from Finland. Finnish Business Review*. Published May 21, 2018.

Inês, A., Diniz, A., & Moreira, A. C. (2025). Driving eco-innovation in supply chains through multi-stakeholder collaboration: A review and research agenda. *Journal of Open Innovation, 11*(1), 100472. DOI: 10.1016/j.joitmc.2025.100472

Ingemarsdotter, E., Jamsin, E., & Balkenende, R. (2020). Opportunities and challenges in IoT-enabled circular business model implementation: A case study. *Resources, Conservation and Recycling, 162*, 105047. DOI: 10.1016/j.resconrec.2020.105047

Intergovernmental Panel on Climate Change Sixth Assessment Report. (2022). Impacts, Adaptation and Vulnerability.

International Energy Agency. (2020).. . *Energy Efficiency, 2020*, •••. https://www.iea.org/reports/energy-efficiency-2020

International Energy Agency. (2022). *Solar PV global supply chains*. International Energy Agency. https://www.iea.org/reports/solar-pv-global-supply-chains

International Monetary Fund. (2024). *Fintech applications for boosting climate finance*. IMF Staff Climate Note 2024/008. https://www.elibrary.imf.org/view/journals/066/2024/008/article-A001-en.xml

Issaoui, Y., Khiat, A., Bahnasse, A., & Ouajji, H. (2019). Smart logistics: Study of the application of blockchain technology. *Procedia Computer Science, 160*, 266–271. DOI: 10.1016/j.procs.2019.09.467

ITF. (2023). ITF Transport Outlook 2023. OECD. Available at: https://oecd.org/en/publications/itftransport-outlook-2023_b6cc9ad5-en.html

Ittmann, H. W. (2015). The impact of big data and business analytics on supply chain management. *Journal of Transport and Supply Chain Management, 9*(1). Advance online publication. DOI: 10.4102/jtscm.v9i1.165

Ivanov, D., & Dolgui, A. (2020). A digital supply chain twin for managing the disruption risks and resilience in the era of Industry 4.0. *Production Planning and Control, 32*(9), 775–788. DOI: 10.1080/09537287.2020.1768450

Ivanov, D., & Dolgui, A. (2020). Viability of intertwined supply networks: Extending the supply chain resilience angles toward survivability. *International Journal of Production Research, 58*(10), 2904–2915. DOI: 10.1080/00207543.2020.1750727

Ivanov, D., Dolgui, A., & Sokolov, B. (2019). The impact of digital technology and Industry 4.0 on the ripple effect and supply chain risk analytics. *International Journal of Production Research, 57*(3), 829–846. DOI: 10.1080/00207543.2018.1488086

Iyere, M., & Misopoulos, F. (2022). The degree of stakeholder influences and risks in sustainable supply chains: A systematic literature review. *International Journal of Contemporary Management, 58*(2), 9–26. DOI: 10.2478/ijcm-2022-0004

Jelti, F., Allouhi, A., Büker, M. S., Saadani, R., & Jamil, A. (2021). Renewable power generation: A supply chain perspective. *Sustainability (Basel), 13*(3), 1271. DOI: 10.3390/su13031271

Jensen, S. F., Kristensen, J. H., Christensen, A., & Wæhrens, B. V. (2024). An ecosystem orchestration framework for digital product passport design in a circular economy. *Business Strategy and the Environment, 33*(7), 7100–7117. DOI: 10.1002/bse.3868

Jia, F., Guo, J., Chen, L., & Islam, N. (2024). Does Fintech Adoption Improve Sustainable Supply Chain Management? An Innovation-Intensive Environment Perspective. *IEEE Transactions on Engineering Management*. DOI:10.1109/TEM.2024.3515471

Jia, M., Komeily, A., Wang, Y., & Srinivasan, R. S. (2019, May). Adopting Internet of Things for the development of smart buildings: A review of enabling technologies and applications. *Automation in Construction, 101*, 111–126. DOI: 10.1016/j.autcon.2019.01.023

Johnson, H. L. (1971). *Business in contemporary society: Framework and issues.* Wadsworth Publishing Company.

Kadhim Obaid, M., Abood, B. S. Z., Khalid Alazzai, W., & Jasim, L. (2024). From Field to Fork: The role of AI and IoT in Agriculture. E3S Web Conf., 491, 2006.

Kamble, S. S., Gunasekaran, A., & Gawankar, S. A. (2020). Sustainable Industry 4.0 framework: A systematic literature review identifying the current trends and future perspectives. *Process Safety and Environmental Protection, 133*, 262–279. DOI: 10.1016/j.psep.2019.11.016

Kamble, S. S., Gunasekaran, A., & Gawankar, S. A. (2021). Achieving sustainable performance through Industry 4.0 practices in Indian manufacturing: The role of digital supply chain maturity. *Journal of Cleaner Production, 279*, 123775. DOI: 10.1016/j.jclepro.2020.123775

Kanda, W., Geissdoerfer, M., & Hjelm, O. (2021). From circular business models to circular business ecosystems. *Business Strategy and the Environment, 30*(6), 2814–2829. DOI: 10.1002/bse.2895

Kao, Y.-C., Shen, K.-Y., Lee, S.-T., & Shieh, J. C. P. (2022). Selecting the fintech strategy for supply chain finance: A hybrid decision approach for banks. *Mathematics, 10*(14), 2393. DOI: 10.3390/math10142393

Karaeva, A., Magaril, E., & Al-Kayiem, H. H. (2023). *Review and comparative analysis of renewable energy policies in the European Union.*

Karthikeyan, K. S., & Nagaprakash, T. (2023a). Prioritizing IoT-driven sustainability initiatives in retail chains: Exploring case studies and industry insights. EAI Endorsed Trans. Internet Things, 10.

Karthikeyan, K. S., & Nagaprakash, T. (2023b). Prioritizing IoT-driven sustainability initiatives in retail chains: Exploring case studies and industry insights. EAI Endorsed Trans. Internet Things, 10.

Kashem, M. A., Shamsuddoha, M., & Nasir, T. (2024). Sustainable Transportation Solutions for Intelligent Mobility: A Focus on Renewable Energy and Technological Advancements for Electric Vehicles (EVs) and Flying Cars. *Future Transportation, 4*(3), 874–890. DOI: 10.3390/futuretransp4030042

Katana, K., Glaa, B., & Mirata, M. (2024). Facilitator roles for knowledge sharing in industrial symbiosis networks during emergence. *Business Strategy and the Environment, 33*(8), 8540–8558. DOI: 10.1002/bse.3923

Kemp, R., Rip, A., & Schot, J. W. (2001). Constructing transition paths through the management of niches. In Garud, R., & Karnoe, P. (Eds.), *Path Dependence.*

Kerényi, Á., & Molnár, J. (2018). Bank and FinTechs. *Economy and Finance, 5*(1), 86–97. https://www.bankszovetseg.hu/Public/gep/2018/ENG/086-097ig%20kerenyi-molnar-muller.pdf

Kezhu, W. (2024, November). A Novel Smart Logistics Platform based on Decentralized Sensor Tags and Cloud Scheduling. In *2024 5th International Conference on Data Intelligence and Cognitive Informatics (ICDICI)* (pp. 1014-1019). IEEE. DOI: 10.1109/ICDICI62993.2024.10810915

Khaleel, A. G., & Chakrabarti, M. (2018). Leap-Frogging Challenges and Possibilities for Renewable Energy Transition in Developing Countries.

Khan, N., Khan, A., Khan, D., Riaz, F., Farooq, M., Hadi, N. U., & Khan, M. I. (2025). The impact of digitalization and trade openness on energy efficiency: a comparative analysis across low, middle, and high-income countries. *Energy Conversion and Management: X*, 27(1), 1-17. DOI: 10.1016/j.ecmx.2025.101089

Khan, I., Hou, F., & Le, H. P. (2021). The impact of natural resources, energy consumption, and population growth on environmental quality: Fresh evidence from the United States of America. *The Science of the Total Environment, 754*(1), 1–12. DOI: 10.1016/j.scitotenv.2020.142222 PMID: 32920417

Khan, K., & Su, C. S. (2023). Does technology innovation complement the renewable energy transition? *Environmental Science and Pollution Research International*, *30*(1), 30144–30154. DOI: 10.1007/s11356-022-24336-3 PMID: 36427130

Khan, S. A. R., Godil, D. I., Yu, Z., Abbas, F., & Shamim, M. A. (2022). Adoption of renewable energy sources, low-carbon initiatives, and advanced logistical infrastructure—An step toward integrated global progress. *Sustainable Development (Bradford)*, *30*(1), 275–288. DOI: 10.1002/sd.2243

Klimczyk, V., Jasinski, R., Niklas, J., Siedlecki, M., & Ziolkowski, A. (2025). Sustainable Aviation Fuels: A Comprehensive Review of Production Pathways, Environmental Impacts, Lifecycle Assessment, and Certification Frameworks. *Energies*, *18*(14), 3705. DOI: 10.3390/en18143705

Kouhizadeh, M., Sarkis, J., & Zhu, Q. (2021). Blockchain technology and the sustainable supply chain: Theoretically exploring adoption barriers. *International Journal of Production Economics*, *231*, 107831. DOI: 10.1016/j.ijpe.2020.107831

Kuan-Jung, P. E. N. G. (2017). The FinTech revolution and financial regulation: The case of online supply-chain financing. *Asian Journal of Law and Society*, *4*(1), 109–132. https://doi.org/10.1017/als.2016.65

La Torre, M., Mango, F., Cafaro, A., & Leo, S. (2021). Does ESG performance affect market value? A study of FinTech companies. *Technological Forecasting and Social Change*, *162*, 120385. https://doi.org/10.1016/j.techfore.2020.120385

Labaran, M. J., & Masood, T. (2023). Industry 4.0 driven green supply chain management in renewable energy sector: A critical systematic literature review. *Energies*, *16*(19), 6977. DOI: 10.3390/en16196977

Lai, Y. (2024). Innovative strategies in logistics and supply chain management: Navigating modern challenges. SHS Web Conf., 183, 2020. DOI: 10.1051/shsconf/202418302020

Lee, D. S., Fahey, D. W., Skowron, A., Allen, M. R., Burkhardt, U., Chen, Q., Doherty, S. J., Freeman, S., Forster, P. M., Fuglestvedt, J., Gettelman, A., De León, R. R., Lim, L. L., Lund, M. T., Millar, R. J., Owen, B., Penner, J. E., Pitari, G., Prather, M. J., & Wilcox, L. J. (2021). The contribution of global aviation to anthropogenic climate forcing for 2000 to 2018. *Atmospheric Environment*, *244*, 244. DOI: 10.1016/j.atmosenv.2020.117834 PMID: 32895604

Lee, I., & Shin, Y. J. (2018). Fintech: Ecosystem, business models, investment decisions, and challenges. *Business Horizons*, *61*(1), 35–46. https://doi.org/10.1016/j.bushor.2017.09.003

Lei, W., & Tong, Z. (2022). Will the digital economy increase energy consumption? – an analysis based on the ICT application research. *Chinese Journal of Urban and Environmental Studies, 10*(1), 1–18. DOI: 10.1142/S2345748122500014

Li, J., He, Z., & Wang, S. (2022). A survey of supply chain operation and finance with FinTech: Research framework and managerial insights. *International Journal of Production Economics, 247*, 108431. DOI: 10.1016/j.ijpe.2022.108431

Li, L., Shi, Q., Wang, Z., Zhang, S., Jia, D., & Liu, W. (2025). Multi-time-scale scheduling of integrated port logistics-energy system considering piecewise linearization of high-order equations. *IET Generation, Transmission & Distribution, 19*(1), e70007. DOI: 10.1049/gtd2.70007

Liu, J., Chen, M., & Liu, H. (2020). The role of big data analytics in enabling green supply chain management: A literature review. *Journal of Data, Information and Management*, 2(2), 75–83. DOI: 10.1007/s42488-019-00020-z

Liu, L., Liu, J., Zhang, J., & Zhao, Y. (2025). Does digital economy promote renewable energy? – evidence from a cross national sample. *Asia & the Pacific Policy Studies, 12*(2), 1–12. DOI: 10.1002/app5.70015

Liu, Y., Tao, X., Li, X., Colombo, A. W., & Hu, S. (2023). Artificial intelligence in smart logistics cyber-physical systems: State-of-the-arts and potential applications. *IEEE Transactions on Industrial Cyber-Physical Systems*, *1*, 1–20. DOI: 10.1109/TICPS.2023.3283230

Liu, Z., Hansen, D. W., & Chen, Z. (2023). Leveraging Digital Twins to Support Industrial Symbiosis Networks: A Case Study in the Norwegian Wood Supply Chain Collaboration. *Sustainability (Basel)*, *15*(3), 2647. DOI: 10.3390/su15032647

Li, X., Jin, L., & Kan, H. (2019, June). Air pollution: A global problem needs local fixes. *Nature*, *570*(7762), 437–439. DOI: 10.1038/d41586-019-01960-7 PMID: 31239571

Li, Z., Liu, Y., & Zhang, W. (2022). The role of 5G technology in enhancing logistics and supply chain agility. *Journal of Logistics Research and Applications*, *25*(3), 245–262. DOI: 10.1080/13675567.2022.2040905

Lützkendorf, T., & Balouktsi, M. On net zero GHG emission targets for climate protection in cities: More questions than answers? In *Proceedings of IOP Conference Series: Earth and Environmental Science*; IOP Publishing: Bristol, UK, 2019; p. 012073. DOI: 10.1088/1755-1315/323/1/012073

Lv, W., Wei, Y., Li, X., & Lin, L. (2019). What dimension of CSR matters to organizational resilience? Evidence from China. *Sustainability (Basel)*, *11*(6), 1561. DOI: 10.3390/su11061561

Mackey, A., Mackey, T. B., & Barney, J. B. (2007). Corporate social responsibility and firm performance: Investor preferences and corporate strategies. *Academy of Management Review*, *32*(3), 817–835. DOI: 10.5465/amr.2007.25275676

Mahalik, M. K., Mallick, H., & Padhan, H. (2021). Do educational levels influence the environmental quality? The role of renewable and non-renewable energy demand in selected BRICS countries with a new policy perspective. *Renewable Energy*, *164*(1), 419–432. DOI: 10.1016/j.renene.2020.09.090

Makropoulos, C., Kritikos, N.-A., & Pantazis, C. (2024). Matchmaking for industrial symbiosis: A digital tool for the identification, quantification and optimization of symbiotic potential in industrial ecosystems. *Frontiers in Chemical Engineering*, *6*, 1363888. DOI: 10.3389/fceng.2024.1363888

Martinez-Valencia, L., Garcia-Perez, M., & Wolcott, M. P. (2021). Supply chain configuration of sustainable aviation fuel: Review, challenges, and pathways for including environmental and social benefits. *Renewable & Sustainable Energy Reviews*, *152*, 152. DOI: 10.1016/j.rser.2021.111680

Mashat, R. M., Abourokbah, S. H., & Salam, M. A. (2024). Impact of Internet of Things adoption on organizational performance: A mediating analysis of supply chain integration, performance, and competitive advantage. *Sustainability (Basel)*, *16*(6), 2250. DOI: 10.3390/su16062250

Master, I., Berliner, S., Shefy, N., Feldman, A., Bialik, B., Zivan, S., & Goldrat, M. (2011). *U.S. Patent Application No. 13/000,827.*

McKinsey and Company. (2020). A Fact-Based Study of Hydrogen Technology, Economics, and Climate Impact by 2050 Hydrogen-powered Aviation.

McKuin, B., Watson, J. T., Stohs, S., & Campbell, J. E. (2021). Rethinking sustainability in seafood. Elementa (Wash., DC), 9(1). DOI: 10.1525/elementa.2019.00081

Menaka, C., Awasthi, A., Yadav, D. C., & Jain, S. K. (2024). Designing a Renewable Energy System for Industrial IoT with Artificial Intelligence. In *E3S Web of Conferences* (Vol. 540, p. 13008). EDP Sciences. DOI: 10.1051/e3sconf/202454013008

Mexis, I., & Todeschini, G. (2020). Battery energy storage systems in the United Kingdom: A review of current state-of-the-art and future applications. *Energies*, *13*(14), 3616. DOI: 10.3390/en13143616

Meyer, J. W., & Rowan, B. (1977). Institutionalized organizations: Formal structure as myth and ceremony. *American Journal of Sociology*, *83*(2), 340–363. DOI: 10.1086/226550

Mia, S., Ahmed, F., Khan, I., Kabir, M. I., Roni, M. H., Cobra, K., ... & Mahmud, S. (2025). Integratıng Renewable Energy Wıth Internet of Thıngs (IoT): Pathways to A Smart Green Planet. *Kufa Journal of Engineering, 16*(1).

Miatto, A., Schandl, H., Fishman, T., & Tanikawa, H. (2025). Global Waste Generation and Embodied Material Streams: A detailed overview of global circularity and mitigation of greenhouse gas emissions. *Resources, Conservation and Recycling*, *208*, 108319. DOI: 10.1016/j.resconrec.2025.108319

Midhani, P. (2009). Resource Based View of Competitive Advantages: Importance, Issues and Implications. *Indian Management Research Journal*, *1*(2), 1–15.

Milian, E. Z., Spinola, M. M., & de Carvalho, M. M. (2019). FinTechs: A literature review and research agenda. *Electronic Commerce Research and Applications*, *34*, 100833. https://doi.org/10.1016/j.elerap.2019.100833

Min, H. (2010). Artificial intelligence in supply chain management: Theory and applications. *International Journal of Logistics*, *13*(1), 13–39. DOI: 10.1080/13675560902736537

Mintzberg, H. (1998). *Strategy safari: A guided tour through the wilds of strategic management*. Free Press.

Mire, R., Depraz, S., Collacott, B., & Collins, A. (2018, April). Sustainable Development Goals Atlas. In SPE International Conference and Exhibition on Health, Safety, Environment, and Sustainability? (p. D021S014R001). SPE. DOI: 10.2118/190550-MS

Mitchell, R. K., Agle, B. R., & Wood, D. J. (1997). Toward a theory of stakeholder identification and salience: Defining the principle of who and what really counts. *Academy of Management Review*, *22*(4), 853–886. DOI: 10.2307/259247

Mkhize, S., & Thopil, S. (2021). Evaluating the effectiveness of appliance energy labeling in South Africa: Challenges and outcomes. *Energy Research & Social Science*, *75*, 102052. DOI: 10.1016/j.erss.2021.102052

Moktadir, M. A., Rahman, T., Rahman, M. H., & Ali, S. M. (2020). Drivers to sustainable manufacturing practices and circular economy: A perspective of leather industries in Bangladesh. *Journal of Cleaner Production*, *273*, 123066. DOI: 10.1016/j.jclepro.2020.123066

Molla, A., & Licker, P. S. (2005). eCommerce adoption in developing countries: A model and instrument. *Information & Management, 42*(6), 877–899. https://doi.org/10.1016/j.im.2004.09.002

Mombekova, G., Nurgabylov, M., Baimbetova, A., Keneshbayev, B., & Izatullayeva, B. (2024). The relationship between energy consumption, population and economic growth in developing countries. *International Journal of Energy Economics and Policy, 14*(3), 368–374. https://ideas.repec.org/a/eco/journ2/2024-03-37.html. DOI: 10.32479/ijeep.15614

Moretto, A., & Caniato, F. (2021). Can supply chain finance help mitigate the financial disruption brought by COVID-19? *Journal of Purchasing and Supply Management, 27*(4), 100713. https://doi.org/10.1016/j.pursup.2021.100713

Morsing, M., & Schultz, M. (2006). Corporate social responsibility communication: Stakeholder information, response and involvement strategies. *Business Ethics (Oxford, England), 15*(4), 323–338. DOI: 10.1111/j.1467-8608.2006.00460.x

Mügge, J., Seegrün, A., Hoyer, T.-K., Riedelsheimer, T., & Lindow, K. (2024). Digital twins within the circular economy: Literature review and presentation of concepts. *Sustainability (Basel), 16*(7), 2748. DOI: 10.3390/su16072748

Muhammed, G., & Tekbiyik-Ersoy, N. (2020). Development of renewable energy in China, USA, and Brazil: A comparative study on renewable energy policies. *Sustainability (Basel), 12*(21), 9136. DOI: 10.3390/su12219136

Mullenbach-Servayre, A. (2007). L'apport de la théorie des parties prenantes à la modélisation de la responsabilité sociétale des entreprises. La revue des Sciences de Gestion, (1), 109-120.

Mumin, M. A., Yakubu, I. N., & Adam, I. O. (2025). Integrating Renewable Energy With Internet of Things (IoT) for Smart Logistics. In Developing Dynamic and Sustainable Supply Chains to Achieve Sustainable Development Goals (pp. 155-178). IGI Global Scientific Publishing.

Murshed, M. (2024). Can renewable energy transition drive green growth? The role of good governance in promoting carbon emission-adjusted economic growth in next eleven countries. *Innovation and Green Development, 3*(2), 1–13. DOI: 10.1016/j.igd.2023.100123

Murshed, M., Khan, U., Khan, A. M., & Ozturk, I. (2023). Can energy productivity gains harness the carbon dioxide-inhibiting agenda of the Next 11 countries? Implications for achieving sustainable development. *Sustainable Development (Bradford), 31*(1), 307–320. DOI: 10.1002/sd.2393

Musau, E. G. (2024). Optimizing transportation and distribution for environmental sustainability: Green logistics. In *Green Supply Chain Management Practice and Principles* (pp. 84–101). IGI Global. DOI: 10.4018/979-8-3693-3486-7.ch005

Muscio, A., Simonelli, F., & Vu, H. (2023, November). Bridging the valley of death in the EU renewable energy sector: Toward a new energy policy. *Business Strategy and the Environment*, *32*(7), 4620–4635. DOI: 10.1002/bse.3384

Nakano, Y., Sano, F., & Akimoto, K. (2022). Impacts of decarbonization technologies in air transport on the global energy system. *Transportation Research Part D, Transport and Environment*, *110*, 110. DOI: 10.1016/j.trd.2022.103417

Namahoro, J. P., Wu, Q., Zhou, N., & Xue, S. (2021). Impact of energy intensity, renewable energy, and economic growth on CO2 emissions: Evidence from African across regions and income levels. *Renewable & Sustainable Energy Reviews*, *147*(1), 1–13. DOI: 10.1016/j.rser.2021.111233

Nathaniel, S., Nwodo, O., Sharma, G., & Shah, M. (2020). Renewable energy, urbanization, and ecological footprint linkage in CIVETS. *Environmental Science and Pollution Research International*, *27*(16), 19616–19629. DOI: 10.1007/s11356-020-08466-0 PMID: 32219652

Naumann, S., Davis, M., Kaphengst, T., Pieterse, M., & Rayment, M. Design, implementation and cost elements of Green Infrastructure projects; Final report to the European Commission, DG Environment, Contract no. 070307/2010/577182/ETU/F.1, Ecologic institute and GHK Consulting; European Commission: Brussels, Belgium, 2011.

Navaretti, G. B., Calzolari, G., Mansilla-Fernandez, J. M., & Pozzolo, A. F. (2018). *FinTech and banking. Friends or foes?* European Economy: Banks, Regulation, and the Real Sector, 2, 9–30. https://doi.org/10.2139/ssrn.3099337

Naz, F., Ahmed, S., & Associates. (2022). FinTech growth during COVID-19 in the MENA region. *International Journal / PMC (open access).* 10.1007/s10660-022-09583-3 https://www.ncbi.nlm.nih.gov/pmc/articles/PMC9295879/

Nemmour, A., Inayat, A., Janajreh, I., & Ghenai, C. (2023, September). Green hydrogen-based E-fuels (E-methane, E-methanol, E-ammonia) to support clean energy transition: A literature review. *International Journal of Hydrogen Energy*, *48*(75), 29011–29033. DOI: 10.1016/j.ijhydene.2023.03.240

Neufville, R., Abdalla, H., & Abbas, A. (2022). Potential of connected fully autonomous vehicles in reducing congestion and associated carbon emissions. *Sustainability (Basel)*, *14*(11), 6910. DOI: 10.3390/su14116910

Nikitakos, N. (2012). Emissions Port. *FME Transactions*, *40*, 201–206.

Nishitani, K., Kokubu, K., Wu, Q., Kitada, H., Guenther, E., & Guenther, T. (2022). Material flow cost accounting for the circular economy: An empirical study of the triadic relationship between MFCA, environmental performance, and economic performance in Japanese firms. *Journal of Environmental Management*, *303*, 114219. DOI: 10.1016/j.jenvman.2021.114219

Noor, M. M., & Abidin, I. Z. (2022). An integrated renewable energy system for the supply of electricity and hydrogen energy for road transportation which minimizes greenhouse gas emissions. *International Journal of Sustainable Energy Planning and Management*, *35*, 66–79. DOI: 10.5278/ijsepm.2022.35.5

North, D. C. (1990). *Institutions, institutional change and economic performance*. Cambridge University Press.

Nujen, B. B., Kvadsheim, N. P., Mwesiumo, D., Reke, E., & Powell, D. (2023). Knowledge obstacles when transitioning towards circular economy: An industrial intra-organisational perspective. *International Journal of Production Research*, *61*(24), 8618–8633. DOI: 10.1080/00207543.2022.2158243

Nzeanorue, C. C., & Okpala, B. C. (2024). Smart Grids and Renewable Energy Integration: Challenges and Solutions. *Traektoriâ Nauki*, *10*(9), 3050–3060. DOI: 10.22178/pos.109-30

OECD. (2022). *Consumer behavior and energy efficiency: Insights from behavioral economics*. https://www.oecd.org/environment/

Okorie, O., Russell, J., Cherrington, R., Fisher, O., & Charnley, F. (2023). Digital transformation and the circular economy: Creating a competitive advantage from Net Zero Manufacturing. *Resources, Conservation and Recycling*, *189*, 106756. DOI: 10.1016/j.resconrec.2022.106756

Ordóñez-Borrallo, C., Gavilán, D., & García-Moreno, S. (2024). Green bonds and environmental performance: The effect of management's attention. *Corporate Social Responsibility and Environmental Management*, *31*(1), 413–430. DOI: 10.1002/csr.2858

Organisation for Economic Co-operation and Development. (2022). *Behavioural insights for energy labelling*. https://www.oecd.org/environment/

Oshilalu, A. Z., & Iboh, S. (2025). Integration of renewable energy technologies into transportation infrastructure to minimize carbon footprint and enhance long-term environmental resilience. *Global Journal of Engineering and Technology Advances*, *24*(03), 236–251. DOI: 10.30574/gjeta.2025.24.3.0281

Oueghlissi, R. (2013). La RSE et les PME. *Revue française de gestion, 236*(7), 163-180.

Ou, Y., Bao, Z., Ng, S. T., Song, W., & Chen, K. (2024). Land-use carbon emissions and built environment characteristics: A city-level quantitative analysis in emerging economies. *Land Use Policy*, *137*(1), 1–12. DOI: 10.1016/j.landusepol.2023.107019

Ozili, P. K. (2018). Impact of digital finance on financial inclusion and stability. *Borsa Istanbul Review*, *18*(4), 329–340. https://doi.org/10.1016/j.bir.2017.12.003

Pablo-Romero, M. D. P., Román, R., Sánchez-Braza, A., & Yñiguez, R. (2016). Renewable energy, emissions, and health. *Renewable Energy: Utilisation and System Integration, 173*.

Padder, S.A., Khan, R., Rather, R. A. (2024). Biofuel generations: New insights into challenges and opportunities in their microbe-derived industrial production. *Biomasse and Bioenergy*, Vol. 185, June.

Paine, L. S. (2001). *Value Shift: Why Companies Must Merge Social and Financial Imperatives to Achieve Superior Performance*. McGraw-Hill.

Pan, H., Bayanati, M., Vaseei, M., & Chobar, A. P. (2023). Empowering solar photovoltaic logistic operations through cloud-enabled blockchain technology: A sustainable approach. *Frontiers in Energy Research*, *11*, 1293449. DOI: 10.3389/fenrg.2023.1293449

Pan, Y., Liu, X., Zhou, Z., Guo, Y., Feng, Z., Wang, S., Wu, Q., Ouyang, D., Xie, Y., Wu, K., Xu, S., Guo, C., Welsch, E., Huang, C., & Li, J. (2024). Carbon neutrality and clean air acts can enable China to meet the Minamata Convention goals with substantial cost savings. *One Earth*, *7*(3), 483–496. DOI: 10.1016/j.oneear.2024.02.006

Paschalidou, A., Tsatiris, M., Kitikidou, K., Papadopoulou, C., Paschalidou, A., Tsatiris, M., ... & Papadopoulou, C. (2018). Methods (SWOT Analysis). Using Energy Crops for Biofuels or Food: The Choice, 39-44.

Peng, L., Li, Y., Raza, S. A., & Shahzadi, I. (2023). Natural resources and environmental sustainability: COP26 targets from resources-based perspective. *Resources Policy*, *83*(1), 1–13. DOI: 10.1016/j.resourpol.2023.103623

Petri, I., Amin, A., Ghoroghi, A., Hodorog, A., & Rezgui, Y. (2025). Digital twins for dynamic life cycle assessment in the built environment. *The Science of the Total Environment*, *993*, 179930. DOI: 10.1016/j.scitotenv.2025.179930

Pfitzer, M., Bockstette, V., & Stamp, M. (2013, September). Innovating for shared value. *Harvard Business Review*, *91*(9), 100–107.

Polat, O., Ozcan, B., Ertuğrul, H. M., Atılgan, E., & Özün, A. (2024). Fintech: A conduit for sustainability and renewable energy? Evidence from R^2 connectedness analysis. *Resources Policy*, *94*, 105098. Advance online publication. DOI: 10.1016/j.resourpol.2024.105098

Porter, M. E., Hills, G., Pfitzer, M., Patscheke, S., & Hawkins, E. (2012). Measuring shared value: How to unlock value by linking social and business results. FSG. https://ifcsia.org/wp-content/uploads/pdf/publications/Measuring-Shared-Value.pdf

Porter, M. E., & Kramer, M. R. (2011). Creating shared value: Redefining capitalism and the role of the corporation in society. *Harvard Business Review*, *89*(1), 2–17.

Poulsen, T., & Lema, R. (2017). Is the supply chain ready for the green transformation? The case of offshore wind logistics. *Renewable & Sustainable Energy Reviews*, *73*, 758–771. DOI: 10.1016/j.rser.2017.01.181

Powell, W. W., Koput, K. W., & Smith-Doerr, L. (1996). Interorganizational collaboration and the locus of innovation: Networks of learning in biotechnology. *Administrative Science Quarterly*, *41*(1), 116–145. DOI: 10.2307/2393988

Prajapati, H., Jauhar, S. K., Gunasekaran, A., Kamble, S., & Pratap, S. (2022). Blockchain and IoT-enabled virtual closed-loop supply chain for the circular economy: An integrated architecture. *Computers & Industrial Engineering*, *172*, 108530. DOI: 10.1016/j.cie.2022.108530

Pruess, J.-D. (2023). Unraveling the complexity of extended producer responsibility policy mix design, implementation, and transfer dynamics in the EU. *Journal of Industrial Ecology*, *27*(6), 1500–1520. DOI: 10.1111/jiec.13429

Psarommatis, F., & May, G. (2024). Digital Product Passport: A Path to Circularity and Sustainability in Modern Manufacturing. *Sustainability (Basel)*, *16*(1), 396. DOI: 10.3390/su16010396

Pu, Y., Liu, H., Wang, J., & Hou, Y. (2023). Collaborative scheduling of port integrated energy and container logistics considering electric and hydrogen-powered transport. *IEEE Transactions on Smart Grid*, *14*(6), 4345–4359. DOI: 10.1109/TSG.2023.3266601

Qin, Q., & Tan, L. (2025). Lighting a green future: The role of FinTech in the renewable energy transition. *Science Progress*, *108*(1), 00368504251323566. Advance online publication. DOI: 10.1177/00368504251323566 PMID: 40007333

Queiroz, M. M., Telles, R., & Bonilla, S. H. (2019). Blockchain and supply chain management integration: A systematic review of the literature. *Supply Chain Management*, *25*(2), 241–254. DOI: 10.1108/SCM-03-2018-0143

Quenum, A. C. (2025, June 9). EasyBank leverages AI to expand banking access in Tunisia. *WeAreTech Africa.* https://www.wearetech.africa/en/fils-uk/solutions/easybank-leverages-ai-to-expand-banking-access-in-tunisia

R. J. Hijmans and J. van Etten, "Raster: Geographic analysis and modeling with raster data," R Package Version 2.0-12, Jan. 12, 2012

Radanliev, P., De Roure, D., Page, K., Nurse, J. R. C., Mantilla Montalvo, R., Santos, O., Maddox, L., & Burnap, P. (2020). Cyber risk at the edge: Current and future trends on cyber risk analytics and artificial intelligence in the industrial internet of things and industry 4.0 supply chains. *Cybersecurity*, *3*(1), 13. DOI: 10.1186/s42400-020-00052-8

Rafiee, A., & Khalilpour, K. R. (2019). Renewable hybridization of oil and gas supply chains. In *Renewable Energy Integration with Conventional Energy Systems* (pp. 223–241). Elsevier., DOI: 10.1016/B978-0-12-813306-4.00011-2

Rajabzadeh, H., & Wiens, M. (2024). Resilient and sustainable energy supply chains: Insights on sourcing and pricing strategies in a non-collaborative and collaborative environment. *International Journal of Production Research, 62*(24), 9011–9042. Scholar+2KITopen+2IDEAS/RePEc+2DOI: 10.1080/00207543.2024.2331556Google

Ramasubramanian, B., & Ramakrishna, S. (2023). What's next for the Sustainable Development Goals? Synergy and trade-offs in affordable and clean energy (SDG 7). Sustainable earth reviews, 6(1), 17.

Raza, S. A., Shah, N., & Khan, K. A. (2020). Residential energy environmental Kuznets curve in emerging economies: The role of economic growth, renewable energy consumption, and financial development. *Environmental Science and Pollution Research International*, *27*(5), 5620–5629. DOI: 10.1007/s11356-019-06356-8 PMID: 31858417

Raza, S. F. J., Cullen, P., & Keeney, P. (2019). Implementation of low-cost, cloud-based air quality monitoring system using IoT sensors and machine learning techniques. *Procedia Computer Science*, *151*, 877–884.

Reddy, T. T., Devi, Y. R., & Kavita, G. (2023). Logistics, traceability in food supply chain management. E3S Web Conf., 391, 1075.

Rees, J. (1989). Natural resources, economy and society. In *Horizons in Human Geography* (pp. 364–394). Palgrave. DOI: 10.1007/978-1-349-19839-9_19

Rejeb, A., Appolloni, A., Rejeb, K., Treiblmaier, H., Iranmanesh, M., & Keogh, J. G. (2023). The role of blockchain technology in the transition toward the circular economy: Findings from a systematic literature review. *Resources, Conservation & Recycling Advances*, *17*, 200126. DOI: 10.1016/j.rcradv.2022.200126

Rendon-Benavides, R., Perez-Franco, R., Elphick-Darling, R., Plà-Aragonés, L. M., Gonzalez Aleu, F., Verduzco-Garza, T., & Rodriguez-Parral, A. V. (2023). In-transit interventions using real-time data in Australian berry supply chains. *The TQM Journal*, *35*(3), 759–777. DOI: 10.1108/TQM-11-2021-0319

Rönkkö, S., Hellström, K., Ritala, P., & Salminen, J. (2021). Remanufacturing in the heavy vehicle industry: A case study of a Finnish manufacturer. *Sustainability*, *13*(7), 4042. DOI: 10.3390/su13074042

Rout, C., Li, H., Dupont, V., & Wadud, Z. (2022). A comparative analysis of the total cost of ownership of on-highway and off-road heavy-duty vehicles powered by hydrogen, electricity, and diesel. *Heliyon*, *8*(12), e12417. DOI: 10.1016/j.heliyon.2022.e12417

Saba, C. S., & Ngepah, N. (2023). Empirics of convergence in industrialisation and their determinants: Global evidence. *Discover Sustainability*, *4*(1), 25. DOI: 10.1007/s43621-023-00136-8

Saberi, S., Kouhizadeh, M., Sarkis, J., & Shen, L. (2019). Blockchain technology and its relationships to sustainable supply chain management. *International Journal of Production Research*, *57*(7), 2117–2135. DOI: 10.1080/00207543.2018.1533261

Sadeghi, J., & Haapala, K. R. (2019). Optimizing a sustainable logistics problem in a renewable energy network using a genetic algorithm. *Opsearch*, *56*(1), 73–90. DOI: 10.1007/s12597-019-00356-5

Sadeghi, K., Ojha, D., Kaur, P., Mahto, R. V., & Dhir, A. (2025). Metaverse technology in sustainable supply chain management: Experimental findings. *Decision Support Systems*, *191*, 114423. DOI: 10.1016/j.dss.2025.114423

Safaei, M., Al Dawsari, S., & Yahya, K. (2024). Optimizing multi-channel green supply chain dynamics with renewable energy integration and emissions reduction. *Sustainability (Basel)*, *16*(22), 9710. DOI: 10.3390/su16229710

Salancik, G. R., & Pfeffer, J. (1978). A social information processing approach to job attitudes and task design. *Administrative Science Quarterly*, *23*(2), 224–253. DOI: 10.2307/2392563

Sambamurthy, V., Bharadwaj, A., & Grover, V. (2003). Shaping agility through digital options: Reconceptualizing the role of information technology in contemporary firms. *Management Information Systems Quarterly*, *27*(2), 237–263. https://doi.org/10.2307/30036530

Sánchez-García, E., Martínez-Falcó, J., Marco-Lajara, B., & Manresa-Marhuenda, E. (2024). Revolutionizing the circular economy through new digital technologies: A new era of sustainable progress. *Environmental Technology & Innovation*, *33*, 103509. DOI: 10.1016/j.eti.2023.103509

Sanyaolu, T. O. (2025). A conceptual model for inclusive lending through FinTech innovations: Expanding SME access to capital in the supply chain. *International Journal of Sustainable Finance*, *8*(2), 89–103.

Sarkis, J. (2001). Benchmarking for agility. *Benchmarking*, *8*(2), 88–107. https://doi.org/10.1108/14635770110389816

Sarkodie, S. A., Owusu, P. A., & Leirvik, T. (2020). Global effect of urban sprawl, industrialization, trade and economic development on carbon dioxide emissions. *Environmental Research Letters*, *15*(3), 034049. DOI: 10.1088/1748-9326/ab7640

Saulquin, J. Y., & Schier, G. (2007), Des perceptions managérials aux pratiques RSE: une étude exploratoire. *Gestion 2000*, vol. *24, n°*6.

Savage, G. T., Nix, T. W., Whitehead, C. J., & Blair, J. D. (1991). Strategies for assessing and managing organizational stakeholders. *The Academy of Management Perspectives*, *5*(2), 61–75. DOI: 10.5465/ame.1991.4274682

Sayeed, M. A., & Manikandan, K. (2024). Smart integration of renewable energy into transportation: Challenges, innovations, and future research directions. [Revue des Energies Renouvelables]. *Journal of Renewable Energies*, *27*(2), 319–337. DOI: 10.54966/jreen.v27i2.1300

Schlüter, L., Mortensen, L., Gjerding, A., & Kørnøv, L. (2023). Can we replicate eco-industrial parks? Recommendations based on a model of the IPE evolution process. *Revista de Producción Más Limpia*, *429*, 139499. DOI: 10.1016/j.jclepro.2023.139499

Schrotter, G., & Hürzeler, C. (2020). The digital twin of the city of Zurich for urban planning. *PFG–J. Photogramm. Remote Sens. Geoinf. Sci.*, *88*, 99–112.

Schultz, F. C., Valentinov, V., Reinhardt, R. J., & Pies, P. (2024). The rebound effect of the circular economy: Reconceptualization of rebound approaches and mitigation opportunities from an onomic perspective. *Journal of Industrial Ecology*, *28*, 374–385. DOI: 10.1111/jiec.13485

Scott, W. R. (2014). *Institutions and organizations: Ideas, interests, and identities* (4th ed.). Sage Publications.

Segara, S., Li, Q., Gallotta, A., Wang, Y., Gosling, J., & Rezgui, Y. (2024). A taxonomy of circularity indicators for the built environment: Integrating circularity through the RIBA plan of work. *Journal of Cleaner Production*, *446*, 141429. DOI: 10.1016/j.jclepro.2024.141429

Sellitto, M. A., de Lima, M. S., Ackermann, A. E. F., Kadel, N.Jr, & Butturi, M. A. (2025). Exploring industrial symbiotic networks: Challenges, opportunities and lessons for future implementations. *Sustainability (Basel)*, *17*(4), 1509. DOI: 10.3390/su17041509

Senyapar, H. N. D. (2024). Energy labels and consumer attitudes: A study among Gazi University staff. *Sustainability (Basel)*, *16*(5), 1754. DOI: 10.3390/su16051754

Seto, K. C., Churkina, G., Hsu, A., Keller, M., Newman, P. W., Qin, B., & Ramaswami, A. (2021). From low-to net-zero carbon cities: The next global agenda. *Annual Review of Environment and Resources*, *46*(1), 377–415. DOI: 10.1146/annurev-environ-050120-113117

Seuring, S., & Gold, S. (2013). Sustainability management beyond corporate boundaries: From stakeholders to performance. *Journal of Cleaner Production*, *56*, 1–6. DOI: 10.1016/j.jclepro.2012.11.033

Seuring, S., & Müller, M. (2008). From a literature review to a conceptual framework for sustainable supply chain management. *Journal of Cleaner Production*, *16*(15), 1699–1710. DOI: 10.1016/j.jclepro.2008.04.020

Shafique, M. N., Rashid, A., Bajwa, I. S., Kazmi, R., Khurshid, M. M., & Tahir, W. A. (2018). Effect of IoT capabilities and energy consumption behavior on green supply chain integration. *Applied Sciences (Basel, Switzerland)*, *8*(12), 2481. DOI: 10.3390/app8122481

Shahid, N., Shah, M. A., Khan, A., Maple, C., & Jeon, G. (2021). Towards greener smart cities and road traffic forecasting using air pollution data. *Sustainable Cities and Society*, *72*, 103062. DOI: 10.1016/j.scs.2021.103062

Sharath Kumar, C. R., & Praveena, K. B. (2023). SWOT analysis. *International Journal of Advanced Research*, *11*, 744–748. DOI: 10.21474/IJAR01/17584

Sharma, R., Sinha, A., & Kautish, P. (2021). Do economic endeavors complement sustainability goals in the emerging economies of South and Southeast Asia? *Management of Environmental Quality*, *32*(3), 524–542. DOI: 10.1108/MEQ-10-2020-0218

Sheffi, Y. (1985). *Urban transportation networks equilibrium analysis with mathematical programming methods*. Prentice-Hall.

Shen, Y., Han, M., & Zhang, X. (2024). Assessing the impact of digital technologies on energy efficiency: The role of ofdi and virtual agglomeration. *Journal of Resources and Ecology*, *15*(1), 117–129. DOI: 10.5814/j.issn.1674-764x.2024.01.010

Shimauchi, F., Nakagawa, R., & Watanabe, M. (2025). Bonds linked to sustainability, carbon emissions and credit risk. *International Journal of Financial Analysis*, *98*, 104127. DOI: 10.1016/j.irfa.2025.104127

Shi, Z., Fan, F., Tai, N., Qing, C., Meng, Y., & Guo, R. (2024). Coordinated Operation of the Multiple Types of Energy Storage Systems in the Green-seaport Energy-logistics Integrated System. *IEEE Transactions on Industry Applications*, *60*(3), 4482–4493. DOI: 10.1109/TIA.2024.3359585

Shuaibu, A. S., Mahmoud, A. S., & Sheltami, T. R. (2025). A review of last-mile delivery optimization: Strategies, technologies, drone integration, and future trends. *Drones (Basel)*, *9*(3), 158. DOI: 10.3390/drones9030158

Siems, E., Seuring, S., & Schilling, L. (2023). Stakeholder roles in sustainable supply chain management: A literature review. *Journal of Business Economics*, *93*(4), 747–775. DOI: 10.1007/s11573-022-01117-5

Singh, R., Dwivedi, A. D., & Srivastava, G. (2020). Internet of Things based blockchain for temperature monitoring and counterfeit pharmaceutical prevention. *Sensors (Basel)*, *20*(14), 3951. DOI: 10.3390/s20143951 PMID: 32708588

Singh, S., & Trivedi, A. (2021). Big data analytics and artificial intelligence applications in supply chain management: A review. *International Journal of Production Research*, *59*(7), 1993–2017. DOI: 10.1080/00207543.2020.1792000

Sirilertsuwan, P., Thomassey, S., & Zeng, X. (2020). A strategic location decision-making approach for multi-tier supply chain sustainability. *Sustainability (Basel)*, *12*(20), 8340. DOI: 10.3390/su12208340

Sloan, P. (2009). Redefining stakeholder engagement: From control to collaboration. *Journal of Corporate Citizenship*, *2009*(36), 25–40. DOI: 10.9774/GLEAF.4700.2009.wi.00005

Song, H., Han, S., Liu, W., & Ganguly, A. (2023). What role do FinTech companies play in supply chain finance? A signaling intermediary perspective. *Journal of Business and Industrial Marketing*, *38*(6), 1279–1294. DOI: 10.1108/JBIM-12-2021-0587

Soni, G., Kumar, S., Mahto, R. V., Mangla, S. K., Mittal, M. L., & Lim, W. M. (2022). A decision-making framework for Industry 4.0 technology implementation: The case of FinTech and sustainable supply chain finance for SMEs. *Technological Forecasting and Social Change*, *180*, 121686. DOI: 10.1016/j.techfore.2022.121686

Soulier, J. M. (2022). Introduction. La révolution supply chain, origines et définition. *Hors collection*, 9-30.

Souto, B. F. F. (2009). Crisis and corporate social responsibility: Threat or opportunity? *International Journal of Economic Sciences and Applied Research*, *2*(1), 36–50.

Srai, J. S., & Lorentz, H. (2019). Developing design principles for the digitalisation of purchasing and supply management. *Journal of Purchasing and Supply Management*, *25*(2), 100558. DOI: 10.1016/j.pursup.2018.07.001

Stam, E. (2015). Entrepreneurial ecosystems and regional policy: A sympathetic critique. *European Planning Studies*, *23*(9), 1759–1769. https://doi.org/10.1080/09654313.2015.1061484

Steinberg, D. C., Brown, M., Wiser, R., Donohoo-Vallett, P., Gagnon, P., Hamilton, A., & Prasanna, A. (2023). *Evaluating impacts of the inflation reduction act and bipartisan infrastructure law on the US power system (No. NREL/TP-6A20-85242). National Renewable Energy Laboratory*. NREL.

Stern, P. C. (2000). Toward a coherent theory of environmentally significant behavior. *The Journal of Social Issues*, *56*(3), 407–424. DOI: 10.1111/0022-4537.00175

Stevens, D., Dragicevic, S., & Rothley, K. (2007). iCity: A GIS–CA modelling tool for urban planning and decision making. *Environmental Modelling & Software*, *22*(6), 761–773. DOI: 10.1016/j.envsoft.2006.02.004

Strubell, E., Ganesh, A., & McCallum, A. (2021). Energy and policy considerations for deep learning in NLP. *Proceedings of the 57th Annual Meeting of the Association for Computational Linguistics*, 3645–3650. DOI: 10.18653/v1/P19-1355

Sudusinghe, J. I., & Seuring, S. (2022). Supply chain collaboration and sustainability performance in circular economy: A systematic literature review. *International Journal of Production Economics*, *245*, 108402. DOI: 10.1016/j.ijpe.2021.108402

Su, J. P., Wang, C. A., Mo, Y. C., Zeng, Y. X., Chang, W. J., Chen, L. B., & Chuang, C. H. (2017, May). i-Logistics: An intelligent Logistics system based on Internet of things. In *2017 International Conference on Applied System Innovation (ICASI)* (pp. 331-334). IEEE. DOI: 10.1109/ICASI.2017.7988419

Sultanova, G., Djuraeva, R., & Turaeva, S. (2022). The impact of the digital economy on renewable energy consumption and generation: evidence form European union countries. [Online]. DOI: 10.1145/3584202.3584218

Su, M., Wang, Q., Li, R., & Wang, L. (2022). Per capita renewable energy consumption in 116 countries: The effects of urbanization, industrialization, GDP, aging, and trade openness. *Energy*, *254*(1), 1–12. DOI: 10.1016/j.energy.2022.124289

Sun, X., Yu, H., & Solvang, W. D. (2022). Towards the smart and sustainable transformation of Reverse Logistics 4.0: A conceptualization and research agenda. *Environmental Science and Pollution Research International*, *29*(46), 69275–69293. DOI: 10.1007/s11356-022-22473-3

Sun, Y., Li, S., & Wang, R. (2023). FinTech: From budding to explosion—an overview of the current state of research. *Review of Managerial Science*, *17*, 715–755. https://doi.org/10.1007/s11846-021-00513-5

Sun, Y., Rameezdeen, R., Chow, C. W. K., & Gao, J. (2025). Information Sharing Barriers of Construction Projects Toward Circular Economy: Review and Framework Development. *Buildings (Basel, Switzerland)*, *15*(15), 2744. DOI: 10.3390/buildings15152744

Susan, W., Carsten, M., Daqiang, C., & Rosario, M. (2017). An IoT-enabled supply chain integration framework: Empirical case studies. In *Advances in Transdisciplinary Engineering*. IOS Press.

Szczepańska, A., Kaźmierczak, R., & Myszkowska, M. (2023). Smart City Solutions from a Societal Perspective—A Case Study. *International Journal of Environmental Research and Public Health*, *20*(6), 5136. DOI: 10.3390/ijerph20065136 PMID: 36982043

Takman, J., & Andersson-Sköld, Y. (2021). A framework for barriers, opportunities, and potential solutions for renewable energy diffusion: Exemplified by liquefied biogas for heavy trucks. *Transport Policy*, *110*, 150–160. DOI: 10.1016/j.tranpol.2021.05.021

Tang, Q., Li, H., & Zhang, Y. (2025). FinTech and corporate ESG performance: An empirical analysis based on the NEV industry. *Sustainability*, *17*(2), 434. https://doi.org/10.3390/su17020434

TechPression. (2025, January). Tunisian startup EasyBank secures $370,000. *TechPression*. https://techpression.com/tunisian-startup-easybank-secures-370000/?utm_source=chatgpt.com

Teece, D. J. (2007). Explicating dynamic capabilities: The nature and microfoundations of (sustainable) enterprise performance. *Strategic Management Journal*, *28*(13), 1319–1350. DOI: 10.1002/smj.640

Teece, D. J., Pisano, G., & Shuen, A. (1997). Dynamic capabilities and strategic management. *Strategic Management Journal*, *18*(7), 509–533. DOI: 10.1002/(SICI)1097-0266(199708)18:7<509::AID-SMJ882>3.0.CO;2-Z

Teng, Z., Xia, H., & He, Y. (2025). Rewiring sustainability: How digital transformation and FinTech innovation reshape environmental trajectories in the Industry 4.0 era. *Systems*, *13*(6), 400. https://doi.org/10.3390/systems13060400

Teramura, N., & Trakman, L. (2024). Confidentiality and privacy of arbitration in the digital era. *Arbitration International*, *40*(3), 277–297. DOI: 10.1093/arbint/aiae017

Thaler, R. H., & Sunstein, C. R. (2008). *Nudge: Improving decisions about health, wealth, and happiness*. Yale University Press.

Thanh, T. T., Ha, L. T., Dung, H. P., & Huong, T. T. L. (2023). Impact of digitalization on energy security: Evidence form European countries. *Environment, Development and Sustainability*, *25*(1), 11599–11644. DOI: 10.1007/s10668-022-02545-7 PMID: 35966336

The African Development Bank (AfDB) Group. (2012). Economic and Social Challenges Beyond the Revolution.

Thøgersen, J., & Schrader, U. (2012). From knowledge to action—New paths towards sustainable consumption. *Journal of Cleaner Production*, *29–30*, 1–3. DOI: 10.1016/j.jclepro.2012.02.023

Thompson, A. A. (2022). *Strategy: Core concepts and analytical approaches* (7th ed.). McGraw-Hill Education.

Thornbush, M. J., & Golubchikov, O. (2019). *Sustainable Urbanism in Digital Transitions: From Low Carbon to Smart Sustainable Cities*. Springer.

Tian, F. (2020). An agri-food supply chain traceability system for China based on RFID & blockchain technology. *IEEE Access : Practical Innovations, Open Solutions*, *8*, 92079–92091. DOI: 10.1109/ACCESS.2020.2992966

Tornatzky, L. G., & Fleischer, M. (1990). *The processes of technological innovation*. Lexington Books.

Torres-Rivera, A. D., Mc Namara Valdes, A. J., & Florencio Da Silva, R. (2023). The resilience of the renewable energy electromobility supply chain: Review and trends. *Sustainability (Basel)*, *15*(14), 10838. DOI: 10.3390/su151410838

Tretheway, M. W., & Markhvid, K. (2014). The aviation value chain: Economic returns and policy issues. *Journal of Air Transport Management*, *41*, 3–16. DOI: 10.1016/j.jairtraman.2014.06.011

Tseng, M. L., Lim, M. K., & Tan, K. H. (2018). Sustainable supply chain management: A review and research agenda. *International Journal of Production Economics*, *205*, 1–11. DOI: 10.1016/j.ijpe.2018.08.012

Tsolakis, N., Niedenzu, D., Simonetto, M., & Kumar, S. (2021). Supply network design for the circular economy: A review of current approaches and future directions. *Journal of Cleaner Production*, *292*, 126010. DOI: 10.1016/j.jclepro.2021.126010

Tuli, F. (2010). The basis of distinction between qualitative and quantitative research in social science: Reflection on ontological, epistemological and methodological perspectives. *Ethiopian Journal of Education and Sciences*, *6*(1). Advance online publication. DOI: 10.4314/ejesc.v6i1.65384

U.S. Department of Energy (DOE), U.S. Department of Transportation (DOT) & U.S. Department of Agriculture (USDA). (2022). SAF Grand Challenge Roadmap: Flight Plan for Sustainable Aviation Fuel. Bio*energy Technologies Office*. September.

U.S. Department of Energy, Federal Agencies Publish. (2025). SAF Grand Challenge Progress Report Highlighting Historic Efforts to Grow America's SAF Industry, *Bioenergy Technologies Office*. January.

Ubard, G. H., & Weiner, R. (1986). Regulation and long-term contracting in US natural gas markets. *The Journal of Industrial Economics*.

Uckelmann, D. (2008, September). A definition approach to smart logistics. In *International Conference on Next Generation Wired/Wireless Networking* (pp. 273-284). Berlin, Heidelberg: Springer Berlin Heidelberg.

Udeh, E. O., Amajuoyi, P., Adeusi, K. B., & Scott, A. O. (2024). The role of IoT in boosting supply chain transparency and efficiency. *Magna Sci. Adv. Res. Rev.*, *12*(1), 178–197. DOI: 10.30574/msarr.2024.11.1.0081

Ugochukwu, N. A., Goyal, S. B., & Arumugam, S. (2022). Blockchain-Based IoT-Enabled System for Secure and Efficient Logistics Management in the Era of IR 4.0. *Journal of Nanomaterials*, *2022*(1), 7295395. DOI: 10.1155/2022/7295395

Ummalla, M., & Samal, A. (2019). The impact of natural gas and renewable energy consumption on CO2 emissions and economic growth in two major emerging market economies. *Environmental Science and Pollution Research International*, *26*(20), 20893–20907. DOI: 10.1007/s11356-019-05388-4 PMID: 31115807

UNCTAD. (2024). *Digital Economy Report 2024*. [Online]. https://unctad.org/publication/digital-economy-report-2024

UNEP. (2025). Digitalization & Energy. [Online]. https://www.unep.org/topics/energy/digitalization/digitalization-energy

Unilever. (2023). *Sustainability performance 2023*. https://www.unilever.com/sustainable-living /

United Nations Environment Programme. (2020). *Monitoring, verification, and enforcement for appliance energy efficiency*. https://www.unep.org/

United Nations. (2015). *Transforming our world: The 2030 agenda for sustainable development*. https://sdgs.un.org/goals

Ursic, D., & Cestar, A. S. (2022). Crisis management and CSR in Slovenian companies: The impact of the COVID-19 pandemic. *Sustainability (Basel)*, *14*(5), 2690. DOI: 10.3390/su14052690

Uttam, N., Dutta, P., & Singh, A. (2024). Influence of stakeholders on supply chain social sustainability: New insights from small suppliers in the Indian manufacturing sector. *Journal of Cleaner Production*, *444*, 141015. DOI: 10.1016/j.jclepro.2024.141015

Van Dinter, R., Tekinerdogan, B., & Catal, C. (2022). Predictive maintenance using digital twins: A systematic literature review. *Information and Software Technology*, *151*, 107008. DOI: 10.1016/j.infsof.2022.107008

Venkatraja, B. (2020). Does renewable energy affect economic growth? Evidence from panel data estimation of BRIC countries. *International Journal of Sustainable Development and World Ecology*, *27*(2), 107–113. DOI: 10.1080/13504509.2019.1679274

Waechter, S., Klingenberg, H., & Meyer, R. (2022). Cross-national evidence on energy label effectiveness: A meta-analytic review. *Energy Policy*, *164*, 112891. DOI: 10.1016/j.enpol.2022.112891

Waller, M. A., & Fawcett, S. E. (2013). Data science, predictive analytics, and big data: A revolution that will transform supply chain design and management. *Journal of Business Logistics*, *34*(2), 77–84. DOI: 10.1111/jbl.12010

Wallner, T., Haslbeck, J. M. B., & Ruth Mugge, L. M. (2024). A network analysis of factors influencing the purchase of refurbished electronics. *Sustainable Production and Consumption*, *44*, 589–604. DOI: 10.1016/j.spc.2024.03.009

Wamba, S. F., & Akter, S. (2019). Big data analytics in supply chain: A review and bibliometric analysis. *Annals of Operations Research, 270*(1–2), 313–336. DOI: 10.1007/s10479-017-2703-3

Wamba, S. F., Gunasekaran, A., Akter, S., Ren, S. J. F., Dubey, R., & Childe, S. J. (2020). Big data analytics and firm performance: Effects of dynamic capabilities. *Journal of Business Research, 70*, 356–365. DOI: 10.1016/j.jbusres.2016.08.009

Wamda. (2025, January). Tunisian FinTech EasyBank secures $370,000 investment. *Wamda.* https://www.wamda.com/en/2025/01/tunisian-fintech-easybank-secures-370000-investment?utm_source=chatgpt.com

Wang, F., Harindintwali, J. D., Yuan, Z., Wang, M., Wang, F., Li, S., Yin, Z., Huang, L., Fu, Y., & Li, L.. (2021). Technologies and perspectives for achieving carbon neutrality. *Innovation, 2*, 100180. PMID: 34877561

Wang, G., Gunasekaran, A., Ngai, E. W. T., & Papadopoulos, T. (2016). Big data analytics in logistics and supply chain management: Certain investigations for research and applications. *International Journal of Production Economics, 176*, 98–110. DOI: 10.1016/j.ijpe.2016.03.014

WeAreTech Africa. (2025, February). EasyBank leverages AI to expand banking access in Tunisia. *WeAreTech Africa.* https://www.wearetech.africa/en/fils-uk/solutions/easybank-leverages-ai-to-expand-banking-access-in-tunisia?utm_source=chatgpt.com

Wee, H. M., Yang, W. H., Chou, C. W., & Padilan, M. V. (2012). Renewable energy supply chains, performance, application barriers, and strategies for further development. *Renewable & Sustainable Energy Reviews, 16*(8), 5451–5465. DOI: 10.1016/j.rser.2012.06.006

Wen, D., & Gao, W. (2023). Impact of renewable energy policies on solar photovoltaic energy: comparison of China, Germany, Japan, and the United States of America. In *Distributed energy resources: solutions for a low carbon society* (pp. 43–68). Springer International Publishing. DOI: 10.1007/978-3-031-21097-6_3

Wendel, L. M., Albers, S., & Dewulf, W. (2025). Towards a typology of airline vertical strategies. *Journal of Air Transport Management, 127*, 127. DOI: 10.1016/j.jairtraman.2025.102810

Williamson, O. (1989). Transaction cost economics, Chapter 3, *Handbook of Industrial Organization.* Elsevier. Vol.1.

Wirba, A. V. (2024). Corporate social responsibility (CSR): The role of government in promoting CSR. *Journal of the Knowledge Economy*, *15*(2), 7428–7454. DOI: 10.1007/s13132-023-01185-0

World Economic Forum. (2023). Sustainable Aviation Fuel, Offtake Manual, White Paper, June. Nakano, Y., Sano, F., & Akimoto, K. (2022). Impacts of decarbonization technologies in air transport on the global energy system. *Transportation Research Part D, Transport and Environment*, *110*, 103417.

World Intellectual Property Organization. (2025). *Global Innovation Index 2025: Tracking innovation through the COVID-19 crisis.* Cornell University, INSEAD, WIPO. https://www.globalinnovationindex.org/gii-2025-report

Wu, H., Han, M., & Shen, Y. (2023). Technology-driven energy revolution: The impact of digital technology on energy efficiency and its mechanism. *Frontiers in Energy Research*, *11*(1), 1–20. DOI: 10.3389/fenrg.2023.1242580

Wu, X., Zhang, Y., & Li, J. (2025). FinTech-driven green energy transition: Evidence from China. *Scientific Reports*, *15*(1), 18898. https://doi.org/10.1038/s41598-025-18898-8

Wu, Y., & Zhang, S. (2025). Research on the interaction between energy consumption pattern transformation and international trade in the digital economy era based on gig data analysis. *Sustainable Energy Research*, *12*(23), 1–17. DOI: 10.1186/s40807-025-00164-0

Xu, D., Abbasi, K. R., Hussain, K., Albaker, A., Almulhim, A. I., & Alvarado, R. (2023). Analyzing the factors contribute to achieving sustainable development goals in Pakistan: A novel policy framework. *Energy Strateg. Rev.*, *45*, 101050. DOI: 10.1016/j.esr.2022.101050

Xu, Q., Zhang, M., & Li, X. (2022). How does digitalization affect energy? International evidence. *Energy Economics*, *107*, 105879. DOI: 10.1016/j.eneco.2022.105879

Xu, S., Sun, C., & Liu, N. (2024). Road congestion and air pollution-analysis of spatial and temporal congestion effects. *The Science of the Total Environment*, *945*, 173896. DOI: 10.1016/j.scitotenv.2024.173896 PMID: 38880138

Yadav, G., & Singh, A. R. (2021). Role of IoT in enhancing supply chain visibility and traceability: A review. *Materials Today: Proceedings*, *46*(4), 11645–11650. DOI: 10.1016/j.matpr.2021.03.016

Yalcin, H. (2025). FinTech and the green transition: Exploring pathways to sustainable supply chains. *Technological Forecasting and Social Change*. Advance online publication. https://doi.org/10.1016/j.techfore.2025.121234

Yang, L., Wu, X., Huang, B., & Li, Z. (2025). Sustainable industrial energy supply systems with integrated renewable energy, CCUS, and energy storage: A comprehensive evaluation. *Sustainability (Basel)*, *17*(2), 712. DOI: 10.3390/su17020712

Yang, Q., Bolaji Samson, A., Muhammad Gummi, U., & Chen, D. (2023). Study on the use of industry 4.0 technology for the textile industry sustainable supply chain process. *Advances in Engineering Technology Research*, *4*(1), 140. DOI: 10.56028/aetr.4.1.140.2023

Yap, W. Y., & Notteboom, T. (2024). Renewable energy options for seaport cargo terminals with application to mega port Singapore. *Maritime Business Review*, *9*(4), 294–310. DOI: 10.1108/MABR-12-2023-0083

Yi, J., Dai, S., Li, L., & Cheng, J. (2024). How does digital economy development affect renewable energy innovation? *Renewable & Sustainable Energy Reviews*, *192*(1), 1–10. DOI: 10.1016/j.rser.2023.114221

Yin, R. K. (2018). *Case study research and applications: Design and methods* (6th ed.). Sage Publications.

Yin, S., Jia, F., Chen, L., & Wang, Q. (2023). Circular economy practices and sustainable performance: A meta-analysis. *Resources, Conservation and Recycling*, *190*, 106838. DOI: 10.1016/j.resconrec.2022.106838

Yu, H., & Sun, X. (2024). Uncertain remanufacturing reverse logistics network design in Industry 5.0: Opportunities and challenges of digitalization. *Engineering Applications of Artificial Intelligence*, *133*, 108578. DOI: 10.1016/j.engappai.2024.108578

Yüksel, S., & Dinçer, H. (2021). The role of technological development on renewable energy usage: an econometric analysis for G20 countries. In (eds.), *Research Anthology on Clean Energy Management and Solutions*, 1(4), 1-18. DOI: DOI: 10.4018/978-1-7998-9152-9.ch088

Yu, S., Liu, J., Hu, X., & Tian, P. (2022). Does development of renewable energy reduce energy intensity? Evidence from 82 countries. *Technological Forecasting and Social Change*, *174*(1), 1–13. DOI: 10.1016/j.techfore.2021.121254

Yu, Y., & Choi, Y. (2016). Stakeholder pressure and CSR adoption: The mediating role of organizational culture for Chinese companies. *The Social Science Journal*, *53*(2), 226–235. DOI: 10.1016/j.soscij.2014.07.006

Yu, Z., Khan, S. A. R., Ponce, P., Lopes de Sousa Jabbour, A. B., & Chiappetta Jabbour, C. J. (2022). Factors affecting carbon emissions in emerging economies in the context of a green recovery: Implications for sustainable development goals. *Technological Forecasting and Social Change*, *176*(1), 1–13. DOI: 10.1016/j.techfore.2021.121417

Zafar, M. W., Mirza, F. M., Zaidi, S. A. H., & Hou, F. (2019). The nexus of renewable and nonrenewable energy consumption, trade openness, and CO2 emissions in the framework of EKC: Evidence from emerging economies. *Environmental Science and Pollution Research International*, *26*(15), 15162–15173. DOI: 10.1007/s11356-019-04912-w PMID: 30927221

Zarrad, O., Mechichi, B. B., Bouattour, M., & Helali, K. (2025). The role of digitalization in the impact of renewable energy on sustainable development in the Eurozone. *Discover Sustainability*, *6*(1), 1–32. DOI: 10.1007/s43621-025-01639-2

Zedak, C., Belfqih, A., Boukherouaa, J., & El Mariami, F. (2023). An intelligent energy management system for optimum design and real-time operation. [IJPEDS]. *International Journal of Power Electronics and Drive Systems*, *14*(1), 480–495. DOI: 10.11591/ijpeds.v14.i1.pp480-495

Zein, M. E., & Gebresentbert, G. (2024). Digitalization in renewable energy sector. *Energies*, *17*(9), 1–20. DOI: 10.3390/en17091985

Zeng, L., Wong, W.-K., Fu, H., Mahmoud, H. A., Cong, P. T., Thuy, D. T. T., & Bach, P. X. (2024). FinTech and sustainable financing for low carbon energy transitions: A biodiversity and natural resource perspective in BRICS economies. *Resources Policy*, *88*, 104486. Advance online publication. https://doi.org/10.1016/j.resourpol.2023.104486

Zeng, L., Wong, W.-K., Fu, H., Mahmoud, H. A., Cong, P. T., Thuy, D. T. T., & Bach, P. X. (2024). How does FinTech promote a low-carbon energy transition? The role of green innovation and industrial development. *Journal of Cleaner Production*, *88*. Advance online publication. DOI: 10.1016/j.jclepro.2023.139860

Zhang, A. & Seuring, S. (2024). Towards a transparent digital product passport ecosystem for circular supply chains. *International Journal of Logistics Research and Applications*. DOI: 10.1080/13675567.2024.2374256

Zhang, Y. (2022, December). Smart logistics warehouse management system based on RFID and Internet of Things technology. In *2022 2nd International Conference on Networking, Communications and Information Technology (NetCIT)* (pp. 489-492). IEEE. DOI: 10.1109/NetCIT57419.2022.00121

Zhang, J., & Chen, J. (2013). Coordination of information sharing in a supply chain. *International Journal of Production Economics*, *143*(1), 178–187. DOI: 10.1016/j.ijpe.2013.01.005

Zhang, L., Wang, Y., & Li, H. (2022). Mathematical modeling of FinTech-enabled sustainable supply chain investments. *Mathematics*, *10*(14), 2439. https://doi.org/10.3390/math10142393

Zhao, K. (2021). *Comparative bank financial risk management models in traditional banks and challenger banks/fintechs* (Doctoral thesis, Sheffield Hallam University). Sheffield Hallam University.

Zhao, X., Taheripour, F., Malina, R., Staples, M. D., & Tyner, W. E. (2021). Estimating induced land use change emissions for sustainable aviation biofuel pathways. *The Science of the Total Environment*, *779*, 146238. DOI: 10.1016/j.scitotenv.2021.146238 PMID: 33744564

Zheng, J., Liu, B., & Huang, Y. (2024). Inhibiting or exacerbating? Digital financial inclusion and renewable energy efficiency. *Science, Technology & Society*. Advance online publication. DOI: 10.1177/0971721823115066

Zheng, M., & Wong, C. Y. (2024). The impact of digital economy on renewable energy development in China. *Innovation and Green Development*, *3*(1), 1–12. DOI: 10.1016/j.igd.2023.100094

Zhou, X., Chen, H., Mirza, N., & Safi, A. (2024). Greening mineral resources extraction: A pathway to carbon neutrality through energy productivity and institutional reform. *Resources Policy*, *91*(1), 1–12. DOI: 10.1016/j.resourpol.2024.104961

Zhou, Y., Liu, Z., Wang, M., Dong, R. K., & Yue, X. G. (2023). Evaluating the impacts of education and digitalization on renewable energy demand behaviour: New evidence from Japan. *Ekonomska Istrazivanja*, *36*(3), 1–16. DOI: 10.1080/1331677X.2022.2164033

Zhou, Y., & Zhang, L. (2023). Global consumer response to energy efficiency labels: Meta-analysis and policy implications. *Journal of Cleaner Production*, *389*, 136262. DOI: 10.1016/j.jclepro.2023.136262

Zhu, N., Bu, Y., Jin, M., & Mbroh, N. (2020). Green financial behavior and green development strategy of Chinese power companies in the context of carbon tax. *Journal of Cleaner Production*, *245*, 118908. DOI: 10.1016/j.jclepro.2019.118908

Zink, T., & Geyer, R. (2017). Rebound in the circular economy. *Journal of Industrial Ecology*, *21*(3), 593–602. DOI: 10.1111/jiec.12545

Zisopoulos, A., Steuer, B., Abussafy, C., Toboso-Chavero, S., Liu, C., Tong, X., & Schraven, D. (2023). Informal recyclers as stakeholders in the circular economy: A systematic and conceptual review. *Journal of Cleaner Production*, *415*, 137894. DOI: 10.1016/j.jclepro.2023.137894

Zsifkovits, H., & Woschank, M. (2019). Smart Logistics—Technology Concepts and Potentials. *BHM. Berg- und Hüttenmännische Monatshefte*, *164*(1), 42–45. DOI: 10.1007/s00501-018-0806-9

About the Contributors

Tharwa Najar (Ph.D., University of Gafsa, Tunisia), is an Associate Professor in Information Systems Management at the University of Gafsa, Tunisia. She holds a Ph.D. in Management from the University of Manouba, Tunisia. Her research interests include supply chain performance, interorganizational information systems impact, and inter-organizational relationships in the B2B environment within the automotive industry. Her contributions have been published in journals focusing on innovation, digital technologies, and supply chain management. Additionally, she has published works in the fields of knowledge management and open innovation in both large companies and SMEs. She is currently exploring the adoption of emerging technologies (e.g., blockchain and artificial intelligence) in the supply chain and within organizations.

Adel Aloui is Associate Professor of Management at ISTEC Business School. He holds a doctorate in Management Sciences and Industrial Engineering. His research contributes to the development of an understanding of the interaction between technical and organizational objects. His research focuses on the supply chain management, organization, CSR and innovation.

Moacir Godinho Filho is a full professor of Supply Chain Management, Operations Management and Decision Sciences. He joined EM Normandie in 2022. He has a PhD in Industrial Engineering from the Federal University of São Carlos – Brazil, awarded in 2004. He was also a visiting scholar at the University of Wisconsin in Madison (USA), North Carolina State University (USA), and Northwestern University (USA). His research interests are supply chain management, operations management, sustainable operations, industry 4.0, production planning and control, lean manufacturing, and Quick Response Manufacturing. Godinho Filho Moacir has published more than 130 papers in journals with highly selective editorial standards, the great majority with high Impact Factor in Operations Management field (JCR by Thomson Reuters) and indexed in ABS and FNEGE rankings.

***

Arakhita Behera, Assistant Professor in Economics, Department of Social Science and Humanities(DESSH), Regional Institute of Education (NCERT), Ajmer, Rajasthan -305004 (India).

Gézia Damergy is an Associate Professor of Economics with expertise in airline transportation, strategy, and industrial organization. She lectures at the Master's level and speaks regularly at international conferences. She holds Non-Executive Board positions with prominent airline operators and provides strategic counsel to national governments, regulatory bodies, and passenger advocacy groups. Her work focuses on analyzing competition and cooperation dynamics within the airline industry, offering insights that shape both corporate strategy and public policy.

Eya Frioui holds a Phd in management science from ISG Tunis. (Very honorable) "Strategist and Fintech's specialist with over 12 years of practical and academic experience. Assistant Professor of Management at MSB. Researcher at Tunis University, ARBRE Laboratory. Eya FRIOUI has worked in practice in management as a Director and have held many positions in leading media institutions since 2010. Besides the corporate experience and trainings, she has several teaching experiences within public and private universities, she has taught various courses. She has several communications at international conferences. She lectures on strategy and Fintech's innovation. Her recent work concentrates on three main topics, 1) Strategy and Fintech's innovation, 2) Fintech's Strategy design and implementation, 3) Fintech's strategists such as Strategy Directors. Eya FRIOUI is the moderator of several conferences, providing corporate trainings and advice, for institutions and Tunisian Fintech's. (IACE, WePay, BUSINESSMED...) She had several training courses at the International Monetary Fund IMF. Areas of Expertise Innovation management, Strategy, Innovative business models, Fintech's.

Suriya Ponnambalam specialized in Power Systems and is currently working as an Assistant Professor in department of EEE, Bharath Institute of higher Education and Research.

Ravikumar R. Natarajan is an experienced academician with over 12+ years of teaching and research expertise across institutions in India, the UAE, and the Maldives. He currently serves as an Assistant Professor at Marwadi University, Rajkot, specializing in Machine Learning, Recommender Systems, and AI-driven applications in healthcare and security. He holds a Ph.D. in Machine Learning and Recommender Systems from Amity University, Jaipur, a Master of Engineering in CSE from VMRF University, Salem, and a Bachelor of Engineering in CSE from

Anna University. He also has a MiniMBA from the Swiss eLearning Institute. Dr. Ravikumar has authored more than 8+ multiple SCIE/Scopus-indexed journal articles, 25+ IEEE/Springer conference papers, and co-authored a book on Artificial Intelligence. He has also published 9 patents and contributed 10+ book chapters in emerging domains such as Federated Learning, Blockchain, and Digital Twins for IGI Global, Wiley and Springer publisher. He actively mentors students through initiatives like the Research Activities Club (RAC) and Student Research Internship Training (SRIT), promoting innovation and research culture.

José De Jesús Reyes-Sánchez, Ph.D. in senior management, has six master's degrees, in education, in management, professional soccer technique, in physical preparation, in professional soccer, in physical rehabilitation sciences, in sports sciences, as well as in coaching and integral well-being, who is a research professor at the TECNM Instituto Tecnológico Superior De Fresnillo, with more than 20 years of teaching experience, PRODEP profile and coordinator of the academic body with vast experience in conferences and workshops, currently novelist and writer.

Aarthi S is a dedicated Assistant Professor in Computer Science with over six years of teaching experience. She is currently pursuing her Ph.D. at Marwadi University, Gujarat, and holds an M.Phil. and MCA from Bharathiar University, Tamil Nadu. Throughout her academic career, Aarthi has specialized in programming languages such as C++, Java, and C, and contributed to research in recommender systems and AI-related data privacy. She has published journal articles, contributed to book chapters, and actively participated in technical workshops. Known for her student mentorship and engagement in academic events, Aarthi remains committed to fostering innovation and excellence in computer science education.

Lalitha S D is working as an Associate Professor in CSE department at R.M.K. Engineering College, Kavaraipettai. She has completed her Ph.D. in Anna University, Chennai by November 2023. She has around 22 years of teaching experience. Her areas of interest are Web Programming & Artificial Intelligence.

Ajay K. Singh is working as an Associate Professor (Research) in the Department of Humanities and Social Science, Graphic Era (Deemed to be University) Dehradun. He has worked as Assistant Professor (Economics) in the School of Liberal Arts & Management, DIT University Dehradun for 6 years. He did Post-Doctorate Research with EDI of India, Ahmedabad, Gujarat (India). He received MPhil (Economics) from DAVV Indore (India), and PhD (Economics) from IIT Indore (India. He has published several research papers in the diversified area such as climate change, agricultural productivity, assessment of food security; estimation of GFSI, development of environmental sustainability index (ESI) and its association with socio-economic indicators; measurement and determinants of entrepreneurship ecosystem, and dimension of sustainable development and its interlinkages with economic development.

Srinivasan Vasan is an Assistant Professor in the Department of Humanities and Social Sciences at Graphic Era (Deemed to be University), with 14 years of teaching and field expertise. He has served in the Ministry of Health and Family Welfare, as a Guest Faculty, and as a Mobilization Training Specialist with the State Government. Recently, he contributed to a tuberculosis project funded by GFATM and USAID. He has published 60 papers, with 20 more under peer review in UGC-CARE and SCOPUS journals.

Index

A

B

C

D

M

P

Q

R

S

U